Real Analysis

A Long-Form Mathematics Textbook

Jay Cummings

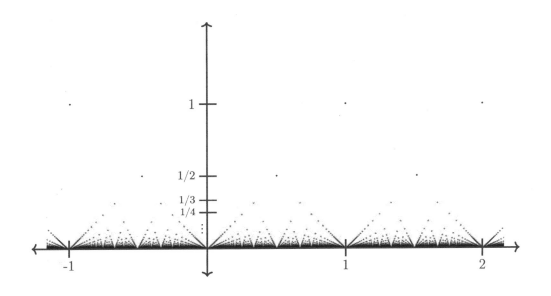

To my students, without whom I would have had to edit this whole damn book myself.

Editorial Board

Contents

List of Results

Chapter 1:   The Reals

Real analysis was developed in the 17th century as a tool to answer some truly fundamental questions from physics, and is now applied throughout the sciences. We begin, however, long before this point, in Greece.

1.1   Zeno's Paradoxes

Nearly 2,500 years ago, Zeno of Elea wrote down a set of paradoxes to support the philosophy that *change* is an illusion. Specifically, one type of change is *motion* — the change of an object's position. Zeno set out to show that objects do not actually move, and our senses are lying to us when we perceive motion.[1]

Zeno's First Paradox

How did he try to prove that motion is an illusion? He created a thought experiment. He said to imagine a race between Achilles (the Greek warrior of legend) and a tortoise (just a regular tortoise). Suppose the race is 1,000 meters long and the tortoise is given a 100 meter head start.

The gun is fired and both begin. Achilles is much faster than the tortoise, so of course you would expect him to overtake the tortoise and win the race, but Zeno said that he can prove that this is impossible. This was his reasoning: For Achilles to catch up to the tortoise, he would first have to reach the 100 meter mark, where the tortoise began the race.

---

[1] ...And to answer your first question, yes, Greek philosophers were believed to have experimented with psychedelic drugs.

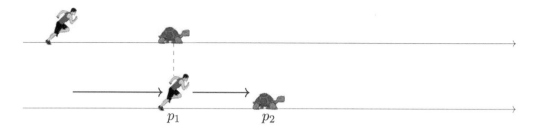

At this point we know that Achilles hasn't yet caught up to the tortoise because in the time it took Achilles to reach the 100 meter mark (call this point p_1), the tortoise has progressed forward some. Let's say the tortoise has made it to the point p_2. And now we repeat our above reasoning: Achilles is still trying to catch the tortoise, but before he can do so he will first have to reach the point p_2.

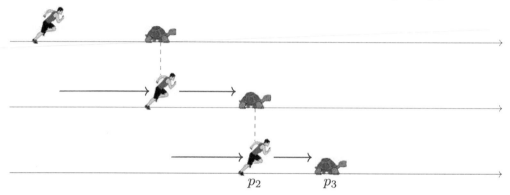

And you see, once again, while Achilles was reaching p_2 the tortoise progressed to a further point, p_3. So, Zeno says, this will go on forever. Achilles will always be behind the tortoise because he is always at a point that the tortoise has already been to, and the next point he must reach will always be a place that the tortoise is currently at. But by the time he gets to that point, the tortoise will have moved on at least a little bit farther, maintaining his lead. And repeat. By induction, the most recent stage of the picture always looks essentially like this.

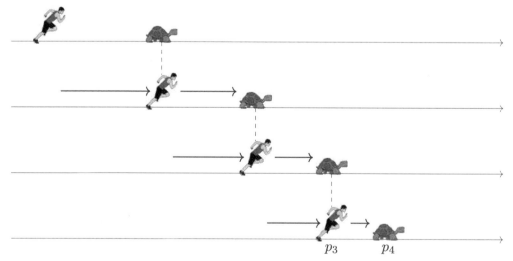

If he can never catch up, then he will always be behind, and therefore is guaranteed to lose the race.

So, what do you think? Does slow and steady win the race? It does seem paradoxical; our senses and life experiences tell us that this is obviously gibberish, but then again Zeno was a pretty smart guy, and his arguments do seem fairly logical and sound... So we have a challenge in front of us. When seemingly-good *reasoning* leads to a false *conclusion*, some part of our thinking must be wrong. In cases like this, it takes very careful thought to diagnose the issue.

Zeno's paradoxes were one of the initial motivations to study limits and develop the theory of real analysis. It will take over a hundred pages of this text before we fully understand why Zeno was mistaken, but we'll get there.

Zeno's Second Paradox

For the interested, Zeno had another similar paradox. Briefly, imagine an archer fires an arrow at a target. Zeno argued that the arrow will never reach the target. He said that for the arrow to reach it, it would first have to reach the halfway point, which takes some amount of time. Then, once there, there is still some distance to go and to get there it would first have to reach the point halfway between its current position and its goal, which again takes some amount of time. Then from its new position it is still not there and getting there again requires that it first gets halfway there, requiring some amount of time.

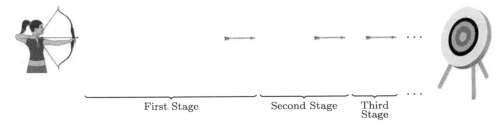

First Stage   Second Stage   Third
Stage

In general, the arrow will always be short of its goal, and to get to the target it will first have to get halfway there, which always requires some additional amount of time. If there is always some additional amount of time remaining, it will never actually make it... right?

Now, if you remember your calculus you might be able to figure out Zeno's mistake. He was correct that the arrow must first complete the first stage, and then must complete the second stage, and then must complete the third stage, and so on. And he was correct that each stage takes some finite amount of time. And it seems to make sense that if you have to do one thing at a time, and you have infinitely many things to do, that you will never complete all of those things... but is this correct?

We will come back to this problem in Chapter 4 (which is a chapter on series... hint hint), but the moral for now is that when things get smaller and smaller (like the time it takes Achilles to run from p_i to p_{i+1}, for increasingly large i), weird stuff can happen. And it takes careful, detailed thought to avoid getting stuck in paradoxical situations. Indeed, this is essentially our goal for the book.

Textbook Goal. By studying the infinite, develop a ground-up understanding of the real numbers and functions on the reals. Also, improve one's mathematical maturity; that is, understand mathematical statements and arguments, construct proofs and find counterexamples, and appreciate the intrinsic beauty in the mathematics.

So, that's what we are going to do. We are going to think carefully about the infinite, and as we progress through this book we will see how that careful reasoning can illustrate some pretty amazing properties of real numbers and functions. And unlike calculus where you spent most of your time studying super nice functions and intuitive situations, and very little time on weird paradoxes like the above, in analysis we will instead spend a significant amount of time on the weird situations, where our intuition may deceive us. We will build analysis from the ground-up, retraining our intuition on solid — and increasingly higher — ground.

So get ready, because I think you will enjoy it.

1.2   Basic Set Theory Definitions

We begin about as basic as possible — with sets. Here is a quick review of basic set theory definitions that you learned in your intro-to-proofs class.

> **Definition.**
>
> **Definition 1.1.**
>
> - A *set* is an unordered collection of distinct objects, which are called *elements*.[2]
>
> - If x is an element of a set S, we write $x \in S$. This is read as "x in S."
>
> - *Set-builder notation* looks like this:
>
> $$S = \{\text{elements} \;:\; \text{conditions used to generate the elements}\}.$$
>
>   For example,
> $$\{1, 4, 9, 16, 25, \dots\} = \{n^2 : n \in \mathbb{N}\}.$$
>
>   As a second example, the rationals can be built like so:
>
> $$\mathbb{Q} = \left\{ \frac{p}{q} : p, q \in \mathbb{Z}, q \neq 0 \right\}.$$

---

[2]Alternative definition: Everything. Everything is a set. Almost no definition in the world is as general as that of a set.

- The set containing no elements is called the *empty set* and is denoted $\emptyset$.

- If $x \in B$ for every $x \in A$, then A is a *subset* of B, which is denoted $A \subseteq B$.

- $A \cap B = \{x : x \in A \text{ and}^3 \ x \in B\}$.

- $A \cup B = \{x : x \in A \text{ or } x \in B\}$.

- $A \setminus B = \{x : x \in A \text{ and } x \notin B\}$.

- A and B are *disjoint* if $A \cap B = \emptyset$.

- If $A \subseteq U$ for a *universal set* U (typically $U = \mathbb{R}$), then the *complement* of A in U is $A^c = U \setminus A$.

- The *Cartesian product* $A \times B = \{(a, b) : a \in A \text{ and } b \in B\}$.

- The *power set* of a set A is $\mathcal{P}(A) = \{X : X \subseteq A\}$.

- If $A_1, A_2, A_3, \ldots, A_n$ are all sets, then the union of all of them is

$$\bigcup_{i=1}^{n} A_i = A_1 \cup A_2 \cup \cdots \cup A_n.$$

- If $A_1, A_2, A_3, \ldots, A_n$ are all sets, then the intersection of all of them is

$$\bigcap_{i=1}^{n} A_i = A_1 \cap A_2 \cap \cdots \cap A_n.$$

- The *cardinality* of a set A is the "number" of elements in the set, and is denoted $|A|$. (In Chapter 2 we will investigate what "number" means when A is infinite.)

---

[3]Example #628 of why English is a confusing language: The word 'and' can mean either a union or[4] an intersection. "Students and faculty are welcome to attend" means the *union* of the students and the faculty are invited. But, "If you are taking algebra and analysis this year, you'll enjoy this lecture" refers the *intersection* of those taking algebra and those taking analysis. Worse still, 'and' can refer to quantities as well, not just to collections! "These three pizzas and those five make eight." Mathematicians use symbols to avoid these ambiguities.

[4]The word 'or' is better.

Another way to think about a set is as a box, possibly with some things inside. When you look into a box, the things inside do not have any particular order; same thing with the elements of a set.

$$\boxed{\begin{array}{cc} 1 & 3 \\ & 2 \end{array}} = \{2,3,1\}$$

Of course, the above also corresponds to $\{1,2,3\}$, $\{3,2,1\}$, and $\{3,1,2\}$; with sets, order does not matter. Also, remember that the elements do not have to be numbers. Elements of a set can be *anything*.

$$\boxed{\begin{array}{cc} \text{apple} & \pi \\ & \text{Joe} \end{array}} = \{\text{apple, Joe, } \pi\}$$

Also, just as boxes can be empty, so can sets!

$$\boxed{} = \emptyset$$

Furthermore, it's certainly possible for one box to be inside another box. Likewise, it's certainly possible for one set to be a single element inside another set.

$$\boxed{\boxed{\begin{array}{cc} \text{apple} & \pi \\ & \text{Joe} \end{array}} \quad \begin{array}{c} \odot \\ 7 \end{array}} = \{\{\text{apple, Joe, } \pi\}, 7, \odot\}$$

Notice that the above set has three elements in it: (1) a set (containing three specific elements), (2) the number 7, and (3) a smiley face. Your box could also have just one thing in it: a smaller box with nothing inside it. This looks like this:

$$\boxed{\boxed{}} = \{\emptyset\}$$

Note the difference between the above example and three pictures back: Earlier we had the empty set; now we have a set whose one element is the empty set.

To test your understanding, think about what the union and intersection of two sets would look like, from this box interpretation.[5] Pushing this further, if A and B are two boxes (possibly with things inside), think if you can now describe the following in terms of boxes (Exercise 1.3):

- $A \setminus B$
- $\mathcal{P}(A)$
- $|A|$

Lastly, if $A_1, A_2, A_3, \ldots, A_n$ are all boxes, think if you can now describe the following in terms of boxes (Exercise 1.4):

- $\displaystyle\bigcup_{i=1}^{n} A_i = A_1 \cup A_2 \cup \cdots \cup A_n.$
- $\displaystyle\bigcap_{i=1}^{n} A_i = A_1 \cap A_2 \cap \cdots \cap A_n.$

Functions on sets

Now recall the following, which you also learned in your intro-to-proofs class.

Definition.

Definition 1.2.

- Given a pair of sets A and B, suppose that each element $x \in A$ is associated, in some way, to one element of B, which we denote $f(x)$. Then f is said to be a *function* from A to B. This is sometimes denoted "$f : A \to B$".[6]

  ○ Furthermore, A is called the *domain* of f, and B is called the *codomain* of f. The set $\{f(x) : x \in A\}$ is called the *range* of f.

- A function $f : A \to B$ is *injective* (or *one-to-one*) if $f(a) = f(b)$ implies that $a = b$.

- A function $f : A \to B$ is *surjective* (or *onto*) if, for every $b \in B$, there exists some $a \in A$ such that $f(a) = b$.[7]

- A function $f : A \to B$ is *bijective* if it is both injective and surjective.

---

[5]One answer: The union of two boxes A and B can be obtained by dumping everything in A and everything in B into a new box, and then removing any duplicate items. The intersection can be obtained by identifying everything in A that is also in B, and putting those items into a new box.

[6]To be completely precise, one can use relations to define a function $f : X \to Y$ as a subset of the cross product $X \times Y$. But that's enough of a pain that even in Baby Rudin that's avoided.

[7]Note that using the contrapositive you can come up with equivalent definitions for *injective* and *surjective*. For the former, the contrapositive gives the (equivalent) definition that f is injective if $a \neq b$ implies $f(a) \neq f(b)$. For the latter, an equivalent definition is that f is surjective if there does not exist a $b \in B$ for which $f(a) \neq b$ for all $a \in A$.

Example 1.3. Here are some examples of functions on finite sets—one which is surjective, and one which is not.

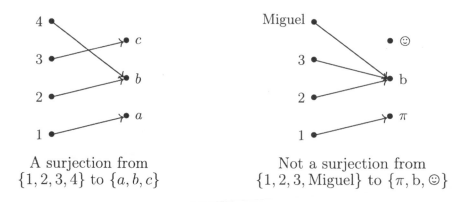

A surjection from
$\{1, 2, 3, 4\}$ to $\{a, b, c\}$

Not a surjection from
$\{1, 2, 3, \text{Miguel}\}$ to $\{\pi, \text{b}, \odot\}$

1.3   What is a Number?

How would you define a number? Some would say this is a philosophical question, while Appendix A demonstrates just how deep of a mathematical question this can be. The German philosopher, logician, and mathematician Gottlob Frege, on the other hand, believed these are one in the same.

> Every good mathematician is at least half a philosopher, and every good philosopher is at least half a mathematician.
>
> – Gottlob Frege

For now we won't get too philosophical, but we do want to discuss how our number system is built up, and what properties are essential at each stage.

Naturals

The natural numbers $\mathbb{N}$ is the set $\{1, 2, 3, \dots\}$. These are the counting numbers, and if anything is a number, they certainly are.[8] A sketch of how to formally, but painfully, construct all real numbers is included in Appendix A.

Integers

The integers $\mathbb{Z}$ is the set $\{\dots, -3, -2, -1, 0, 1, 2, 3, \dots\}$ where each number is 1 away from the next. The 19th century mathematician Leopold Kronecker said "God made the integers, all else is the work of man." What do you think he meant by this? And do you agree with him? I've found that a class of students will have many different opinions on this question, which I think is great.

---

[8]Note that 0 is *not* considered a natural number.[9] I will defend this to my grave.

[9]The set $\{0, 1, 2, 3, \dots\}$ is denoted $\mathbb{N}_0$. (Fun fact: '0' was first discovered by an ancient Babylonian who asked how many of his friends wanted to talk about numbers with him.)

Rationals

The next natural set of numbers to consider is the set of *rational numbers*:

$$\mathbb{Q} := \left\{ \frac{p}{q} : p, q \in \mathbb{Z}, q \neq 0 \right\}.$$

They have some nice properties:

- They are *nested*: Between any two rational numbers is another one.[10]

- They are closed under addition, subtraction, multiplication and division:
  - If $p, q \in \mathbb{Q}$, then $p + q \in \mathbb{Q}$.
  - If $p, q \in \mathbb{Q}$, then $p - q \in \mathbb{Q}$.
  - If $p, q \in \mathbb{Q}$, then $p \cdot q \in \mathbb{Q}$.
  - If $p, q \in \mathbb{Q}$, then $p \div q \in \mathbb{Q}$, provided $q \neq 0$.

But they also have some problems:

- It's possible that $p \in \mathbb{Q}$ but yet $\sqrt{p} \notin \mathbb{Q}$.

- It has lots of "holes": It's possible that $p_i \in \mathbb{Q}$ for each $i \in \mathbb{N}$, but yet the sequence $p_1, p_2, p_3, \dots$ "converges" to some $p \notin \mathbb{Q}$.

- They're not *algebraically closed*: There exist polynomials with rational coefficients whose roots are not rational.[11]
  - e.g. $f(x) = x^2 - 5$

- If we "fill in the holes," we in fact will see that "most" numbers are not rational!

Rational numbers were studied all around the world, but the Greeks' success is typically noted. One of the most influential of these mathematicians was Pythagoras,[12] who is often credited with being the first to prove the Pythagorean theorem, which, as you know, says that if a and b are the lengths of the legs of a right triangle, and c the length of the hypotenuse, then $a^2 + b^2 = c^2$. (Although, it might have been proven at the time by one of his students, and also perhaps earlier by the Babylonians.)

---

[10] And hence, between any two there are infinitely many others.

[11] This is the one problem that the reals will only partially solve. One needs the complex numbers to have algebraic closure.

[12] Pythagoras' story is cloaked in legend — but fortunately the legends are all highly amusing. Aristotle wrote that Pythagoras had a golden thigh, was born with a golden wreath upon his head, and that after a deadly snake bit him, he bit the snake back, which killed it; he was supposedly the son of Apollo, and it was said that a priest of Apollo gave him a magic arrow that allowed him to fly; the philosophers Porphyry and Iamblichus both reported that Pythagoras once persuaded a bull not to eat beans, and convinced a notoriously violent bear to swear that it would never harm a living thing again — and the bear was true to his word. What is odd is that none of his own writings have survived, and most of the credible writings about him were done long after his death. Some have even suggested that he was not a real person... but this is certainly a minority opinion among historians.

This theorem is significant not only for its own merits, but also because it is the key to proving that irrational numbers exist. Sadly, despite having the key, Pythagoras lived and died believing that all numbers were rational. After his death, his school of thought, called Pythagoreanism, lived on. About a century after his death, a Pythagorean named Hippasus proved what is now *the* classic proof of one of *the* classic theorems—that $\sqrt{2}$ is irrational. As the legend goes, the other Pythagoreans were so horrified by this theorem that they took Hippasus out to sea and threw him overboard, killing him. They then made a pact to never tell the world of his discovery. This has got to be one of the worst cover-ups in history, as today his proof is probably the second most known proof in the world, only behind Euclid's proof of the infinitude of primes.

In fact, even if only to stick it to the murderous, anti-intellectual Pythagoreans one last time, let's start with Hippasus' proof that $\sqrt{2}$ is irrational.[13]

Example 1.4. The number $\sqrt{2}$ is irrational. That is, the length of the hypotenuse of the following triangle is irrational.

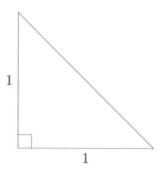

Proof. Assume for a contradiction that $\sqrt{2}$ is rational. Then there must be some non-zero integers p and q where

$$\sqrt{2} = \frac{p}{q}.$$

Moreover, we may assume that this fraction is written in lowest terms, meaning that p and q have no common divisors. Then,

$$\sqrt{2}q = p$$
$$2q^2 = p^2.$$

This implies[14] that $2 \mid p^2$, and hence $2 \mid p$. So we may write $p = 2k$ for some non-zero

---

[13]This is typically viewed as a classic *proof by contradiction*. But is it? What it means to be "irrational" is that you are "not rational." So is this in fact a *direct proof* that $\sqrt{2}$ is not rational (hence irrational)? Or is it a proof by contradiction? I'll let you mull that one over.

[14]By the fundamental theorem of arithmetic.

integer k. Plugging this in,

$$2q^2 = p^2$$
$$2q^2 = (2k)^2$$
$$2q^2 = 4k^2$$
$$q^2 = 2k^2.$$

Therefore $2 \mid q^2$, and hence $2 \mid q$. But this is a contradiction: We had assumed that p and q had no common factors, and yet we proved that 2 divides each. Therefore $\sqrt{2}$ can not be rational, meaning it is irrational.[15] $\square$

So the rationals are not quite enough. That said, they do have *almost* every other fundamental property we would want. To the point: They are what we call an *ordered field*. But first, what's a field? It's a set that satisfies the classic multiplicative and additive properties we know and love.

Definition.

Definition 1.5. A *field* is a nonempty set $\mathbb{F}$, along with two binary operations[16], addition ($+$) and multiplication ($\cdot$), satisfying the following axioms.

Axiom 1 (Commutative Law). If $a, b \in \mathbb{F}$, then $a + b = b + a$ and $a \cdot b = b \cdot a$.

Axiom 2 (Distributive Law). If $a, b, c \in \mathbb{F}$, then $a \cdot (b + c) = a \cdot b + a \cdot c$.

Axiom 3 (Associative Law). If $a, b, c \in \mathbb{F}$, then $(a + b) + c = a + (b + c)$ and $(a \cdot b) \cdot c = a \cdot (b \cdot c)$.

Axiom 4 (Identity Law). There are special elements $0, 1 \in \mathbb{F}$, where $a + 0 = a$ and $a \cdot 1 = a$ for all $a \in \mathbb{F}$.

Axiom 5 (Inverse Law). For each $a \in \mathbb{F}$, there is an element $-a \in \mathbb{F}$ such that $a + (-a) = 0$. If $a \neq 0$, then there is also an element $a^{-1} \in \mathbb{F}$ such that $a \cdot a^{-1} = 1$.

Example 1.6. Below are some examples and some non-examples of fields.

- The natural numbers $\mathbb{N}$ do not form a field; they fail the first half of Axiom 4 and both halves of Axiom 5.

- The integers $\mathbb{Z}$ *almost* form a field; they only fail the second half of Axiom 5: For example, given $a = 2$, there is no *integer* a^{-1} such that $2 \cdot a^{-1} = 1$. (The number $\frac{1}{2}$ is the multiplicative inverse of 2, but $\frac{1}{2}$ is not an integer.)

- One can check that the rationals $\mathbb{Q}$ form a field. (Run through the axioms on your own to verify this.)

---

[15]How does that taste, Pythagoreans? Bitter? Mmmhmm.

[16]Note: These operations are functions from $\mathbb{F} \times \mathbb{F} \to \mathbb{F}$, so implicit in this is that $\mathbb{F}$ is closed under these operations.

1.4   Ordered Fields

While I was a grad student, I once got an email from a student asking me about a homework question involving x, which was given to be a real number. Instead, he wrote: "Assume that x is really a number."

Now, without meaning to, his email did suggest a good question: what really is a number? As we demonstrated in Example 1.4, if we want to include in our number system all possible lengths of triangles, or we want our system to be closed under taking square roots, or we want our system to be closed under taking limits, then the rational numbers alone do not cut it. We need more...

Plug rationals into limits and roots
And you won't like what it reveals...
To apply such functions, and reap their fruits,
We need more...we need...The Reals!

But what essential properties are we missing? What does $\mathbb{R}$ have that a field does not? To get warmed up, think about $\mathbb{Q}$. With the definition of a field, we have *almost* reached the point where we have captured the essence of the rationals. Can you think of what else we need? There are three main properties that we are missing. . . .

First, there are infinitely many rationals[17] (and in some way they are "symmetric" about the 0 element). Second, the rationals have an ordering to them—not only are the elements different from each other, but given any two different ones, one is bigger than the other. So we need to add a "$<$" to the mix. Lastly, we would like *metric properties*; that is, we would like to talk about how big a number is (how far from zero), and the distance between two numbers.

Beautifully, there is a single elegant axiom that we can include to capture *all* of these properties, and we do it without defining any more functions or operations. And this is great: The simpler and more elegant your axiomatic system, the more general and appealing your theory is.

How can we do this? The beautiful idea is to define a subset of the field which we call the *positive* set P, and which acts just as the positive real numbers do.

Definition.

Definition 1.7. An *ordered field* is a field $\mathbb{F}$, along with the following additional axiom.

Axiom 6 (Order Axiom). There is a nonempty subset $P \subseteq \mathbb{F}$, called the *positive elements*, such that

(a) If $a, b \in P$, then $a + b \in P$ and $a \cdot b \in P$;

(b) If $a \in \mathbb{F}$ and $a \neq 0$, then either $a \in P$ or $-a \in P$, but not both.

---

[17] And for those who have studied abstract algebra, you know that the field axioms do not guarantee an infinite set. Indeed, there are many finite fields, e.g., $\mathbb{Z}/p\mathbb{Z}$ where p is a prime.

Why does this do the job? First, since 1 is positive[18], so are $1+1$, and $1+1+1$, and $1+1+1+1$, and so on, generating infinitely many different elements; likewise for $(-1)+(-1)+\cdots+(-1)$. Second, if you want to say that the number b is larger than the number a, it suffices to say that $b-a$ is positive. Lastly, positive elements are all you need to define the absolute value, and hence enough to give metric properties. We begin the formal treatment of this by defining the inequality.

> **Definition.**
>
> **Definition 1.8.** If $\mathbb{F}$ is an ordered field and $a, b \in \mathbb{F}$, then we say that "$a < b$" if $b - a \in P$. Likewise, $a \leq b$ means that either $a = b$ or $a < b$.

Define "$>$" similarly. Also, note that p is positive if $0 < p$. In many intro-to-proofs classes, students work out boring proofs of the basic properties of inequalities. Since a student should be tortured by those at most one time in their academic careers, we won't do them here.

As you are probably anticipating, $\mathbb{R}$ is going to be an ordered field. To make this clear when you look back at these pages, I will begin to indicate this fact parenthetically, like in the following note.

> **Note.**
>
> **Note 1.9.** You may use any of the standard properties of inequalities. For a, b, c in an ordered field $\mathbb{F}$ (like $\mathbb{R}$), these include:
>
> (a) If $a < b$, then $a + c < b + c$.
>
> (b) Transitivity: If $a < b$ and $b < c$, then $a < c$.
>
> (c) If $a < b$, then $ac < bc$ if $c > 0$, and $ac > bc$ if $c < 0$.
>
> (d) If $a \neq 0$, then $a^2 > 0$.

Finally, how does defining an ordered field (with positive subset P) give us a distance function? It all comes down to the absolute value function.

---

[18] *Proof.* If it weren't, then by part (b) of Axiom 6, -1 would be positive. But then, for any positive a, by part (a) we deduce that $a \cdot -1 = -a$ is also positive, contradicting part (b). $\square$

> **Definition.**
>
> **Definition 1.10.** If $\mathbb{F}$ is an ordered field (like $\mathbb{R}$), define the *absolute value* function $|\cdot| : \mathbb{F} \to \mathbb{F}$ to be
>
> $$|x| = \begin{cases} x, & x \geq 0 \\ -x, & x < 0 \end{cases}.$$

It might be a little odd to see "$-x$" when the absolute value function is supposed to make everything positive. But the point is that for those x values, $x < 0$. So x already has a negative sign! And one way to get rid of a negative sign is to add *another* negative sign, canceling out the one that x already has.[19] For example,

$|x| = x$ when $x \geq 0$	$|x| = -x$ when $x < 0$
• $\|7\| = 7$	• $\|-7\| = -(-7) = 7$
• $\|2\| = 2$	• $\|-2\| = -(-2) = 2$

> **Note.**
>
> **Note 1.11.** You may also use any of the standard properties of absolute values without proof, many of which were proven in your intro-to-proofs class. These include the following, where a and b are elements in an ordered field (like $\mathbb{R}$).
>
> (a) $|a| \geq 0$, with equality if and only if $a = 0$.
>
> (b) $|a| = |-a|$.
>
> (c) $-|a| \leq a \leq |a|$.
>
> (d) $|a \cdot b| = |a| \cdot |b|$.
>
> (e) $1/|a| = |1/a|$, if $a \neq 0$.
>
> (f) $|a/b| = |a|/|b|$, if $b \neq 0$.
>
> (g) $|a| \leq b$ if and only if $-b \leq a \leq b$.
>
> (h) The triangle inequality: $|a + b| \leq |a| + |b|$.
>
> (i) The reverse triangle inequality: $||a| - |b|| \leq |a - b|$.

We will prove the last three now, but as an exercise you may prove the rest.

---

[19]Sometimes two wrongs *do* make a right! Take that, Mom!

> **Proposition.**
>
> **Proposition 1.12** ($|a| \leq b$ *iff* $-b \leq a \leq b$). If $\mathbb{F}$ is an ordered field (like $\mathbb{R}$) and $a, b \in \mathbb{F}$, then
> $$|a| \leq b \qquad \text{if and only if} \qquad -b \leq a \leq b.$$

Proof. First assume that $|a| \leq b$. Multiplying both sides by -1, we also have that $-|a| \geq -b$, by Note 1.9 part (c). Now, Note 1.11 part (c) says that $-|a| \leq a \leq |a|$. Piecing these three facts together,

$$-b \leq -|a| \leq a \leq |a| \leq b,$$

which implies that $-b \leq a \leq b$, as desired.

To prove the backwards direction, assume that $-b \leq a \leq b$. In Case 1 below we will use the "$a \leq b$" part of this assumption, and in Case 2 we will use the "$-b \leq a$" part of this assumption.

Case 1: $a \geq 0$. If $a \geq 0$, then by the definition of the absolute value $a = |a|$, and so the assumption that $a \leq b$ implies that $|a| \leq b$, as desired.

Case 2: $a < 0$. If $a < 0$, then $-a = |a|$ and hence $a = -|a|$. So the assumption that $-b \leq a$ implies that $-b \leq -|a|$. Multiplying by -1 gives $|a| \leq b$, as desired. $\square$

> **Theorem.**
>
> **Theorem 1.13** (*The triangle inequality*). If $\mathbb{F}$ is an ordered field (like $\mathbb{R}$) and if $x, y \in \mathbb{F}$, then
> $$|x + y| \leq |x| + |y|.$$

Scratch Work.[20] Want to show:

$$|x + y| \leq |x| + |y|.$$

Which by the last proposition is true if and only if

$$-(|x| + |y|) \leq x + y \leq |x| + |y|.$$

---

[20]More so than in other math courses, in real analysis it is helpful to do some scratch work before your proof. This is the case for a few reasons, one of which is that in many of our proofs and in many of your homework problems, one discovers the proof by starting with what we are trying to prove, and then working backwards to reach things that we already know (from other theorems, known properties, or the problem's assumptions).

I.e.

$$-|x| - |y| \le x + y \le |x| + |y|.$$

And these sums look like they can be broken apart into

$$-|x| \le x \le |x|$$

and

$$-|y| \le y \le |y|,$$

both of which we know to be true by Note 1.11 part (c). So if we start at the bottom of our scratch work and move upwards, we should get our conclusion. . . Now here's the actual proof.

Proof. For $x, y \in \mathbb{F}$, by Note 1.11 part (c) we have

$$-|x| \le x \le |x| \qquad \text{and} \qquad -|y| \le y \le |y|.$$

Adding these two together (technically by applying Exercise 1.12 part (a)) gives

$$-|x| - |y| \le x + y \le |x| + |y|.$$

I.e.,

$$-(|x| + |y|) \le x + y \le |x| + |y|.$$

And so, by Proposition 1.12,

$$|x + y| \le |x| + |y|.$$

$\square$

Perhaps surprisingly, the triangle inequality is a very important tool in analysis, and will be used over and over again. Also, it's called the "triangle inequality" because of vectors forming a triangle. A vector $\vec{x}$ has length $|\vec{x}|$. So if you take a pair of vectors $\vec{x}$ and $\vec{y}$, and also consider the vector $\vec{x} + \vec{y}$, then the lengths of each of these can be pictured like this:

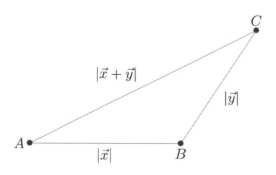

The shortest route between two points (on a plane) is a straight line. Therefore, the distance from A to C, which is given by $|\vec{x} + \vec{y}|$, is the shortest route to get from A to C. It's clearly a longer route if you go from A to B, and then from B to C, which is of distance $|\vec{x}| + |\vec{y}|$. Indeed, this is precisely what the triangle inequality says:

$$|\vec{x} + \vec{y}| \leq |\vec{x}| + |\vec{y}|.$$

Consequences of the Triangle Inequality

The triangle inequality has some nice *corollaries*; i.e., other theorems that follow quickly from it.

> ### Corollary.
>
> **Corollary 1.14** (*The reverse triangle inequality*). Assume that $\mathbb{F}$ is an ordered field (like $\mathbb{R}$) and $x, y \in \mathbb{F}$. Then,
>
> $$\Big||x| - |y|\Big| \leq |x - y|.$$

Proof. By Proposition 1.12, it suffices to show that

$$-|x - y| \leq |x| - |y| \leq |x - y|.$$

That is, we have to show two separate inequalities. We will first show the right-hand one (that $|x| - |y| \leq |x - y|$), and we will do so by applying a tricky application of the triangle inequality. Let $a = x - y$ and $b = y$. Then by the triangle inequality,

$$|a + b| \leq |a| + |b|.$$

That is,

$$|(x - y) + y| \leq |x - y| + |y|$$
$$|x| \leq |x - y| + |y|.$$

Rearranging,

$$|x| - |y| \leq |x - y|,$$

as desired.

We will use a similar approach to show that $-|x - y| \leq |x| - |y|$. Let $c = y - x$ and $d = x$. By the triangle inequality,

$$|c + d| \leq |c| + |d|.$$

That is,

$$|(y - x) + x| \leq |y - x| + |x|$$
$$|y| \leq |y - x| + |x|.$$

Rearranging,

$$-|y - x| \leq |x| - |y|,$$

which by Note 1.11 part (b) implies that

$$-|x - y| \leq |x| - |y|,$$

as desired. $\square$

Here are two final results that follow quickly from what we have already done.

Corollary.

Corollary 1.15 (*Triangle inequality corollaries*)**.** For both of the following, assume that $\mathbb{F}$ is an ordered field (like $\mathbb{R}$) and $x, y \in \mathbb{F}$.

- $|x - y| \leq |x| + |y|$.

    Proof: Replace y with $-y$ in the triangle inequality.

- $|x + y| \geq \big||x| - |y|\big|$.

    Proof: Replace y with $-y$ in the reverse triangle inequality.

And finally, with this, we can define a distance function just how you'd expect.

Definition.

Definition 1.16. Let $\mathbb{F}$ be an ordered field. Then define the *distance* function $d : \mathbb{F} \times \mathbb{F} \to \mathbb{F}$ by
$$d(x, y) = |x - y|.$$

And with that, we have described all of the axioms of the rationals. And in fact, to capture the reals we only need to add a single additional axiom, called the *completeness axiom*. This axiom fills in the "holes" in the rational numbers.

1.5 The Completeness Axiom

What's in a set? That which we call the reals,
By any other name would be as complete.

– Shakespeare in Dimension C-314[21]

We need just one final axiom to obtain the reals, and it is called *the completeness axiom*. Formally, it is an axiom about bounded sets, which requires a little motivation to understand. We begin there.

One of the first problems with the rationals that we mentioned was that if q is rational it need not be the case that $\sqrt{q}$ is rational—we worked out $q = 2$ in detail in Example 1.4. Said differently, if q is rational, there may not be a rational number x such that $x^2 = q$.

We want to include $\sqrt{2}$ in what will be the real numbers, but we also want all of the other irrational numbers. We could say "include all the numbers $\sqrt{q}$ where q is rational," but most irrational numbers are not of this form... We want an approach that gets all of them, identifies their ordering, etc. How do we use the rationals to identify *all* of the irrationals? There is a slick way to do this. Turning back to $\sqrt{2}$, consider the set

$$A := \{x \in \mathbb{Q} : x^2 < 2\}.$$

This set looks something like this:

$\approx$ -1.414 $\qquad\qquad$ $\approx$ 1.414

Here are the important ideas:

- This set is *bounded above*. For example, 2 and 6 and 1.5 are all bigger than everything in A, so these are all *upper bounds* on A.

- Among all these upper bounds, there is one that is special—the smallest of them all. This is the *least upper bound*, and as you might imagine, this bound will in fact be exactly $\sqrt{2}$ (which by Example 1.4 is not rational).

- In order to get all of $\mathbb{R}$, we will start with $\mathbb{Q}$ and then we will add in all the least upper bounds from all sets that are bounded above, like A. E.g., due to

---

[21]The Shakespeare in the parallel universe where all his plays were about math. And Jerry's happy.

the above set A, we include $\sqrt{2}$ into our set $\mathbb{R}$. Doing this for all such A makes $\mathbb{R}$ complete.[22]

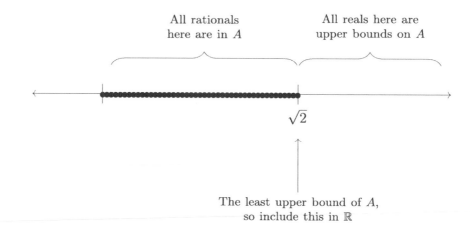

That is the blueprint for what we are going to do. Now here it is, formally.

Definition.

Definition 1.17. Let S be an ordered field (like $\mathbb{R}$) and $A \subseteq S$ be nonempty.

(i) The set A is *bounded above* if there exists some $b \in S$ such that $x \le b$ for all $x \in A$; in this case, b is called an *upper bound* of A.

(ii) The *least upper bound* of A — if it exists — is some $b_0 \in S$ such that (1) b_0 is an upper bound of A, and (2) if b is any other upper bound of A, then $b_0 \le b$. Such a b_0 is also called the *supremum* of A and is denoted $\sup(A)$.

(iii) Likewise, the set A is *bounded below* if there exists some $b \in S$ such that $x \ge b$ for all $x \in A$; in this case, b is called a *lower bound* of A.

(iv) Again, like above, the *greatest lower bound* of A — if it exists — is some $b_0 \in S$ such that (1) b_0 is a lower bound of A, and (2) if b is any other lower bound of A, then $b_0 \ge b$. Such a b_0 is also called the *infimum* of A and is denoted $\inf(A)$.

(v) If a set is both bounded above and bounded below, then it is simply called *bounded*.

Example 1.18. The below are left without proof for now. We will come back shortly and talk about how to prove these claims.

- The set $\mathbb{N} = \{1, 2, 3, \dots\}$ has no upper bounds. Lower bounds on $\mathbb{N}$ include $-17, 1, 0.123$ and $-\pi$. Note that $\sup(\mathbb{N})$ does not exist,[23] but $\inf(\mathbb{N}) = 1$.

- The set $\mathbb{Q}$ has no upper or lower bounds; consequently, $\sup(\mathbb{Q})$ and $\inf(\mathbb{Q})$ do not exist.

- $\sup\left(\left\{\frac{1}{n} : n \in \mathbb{N}\right\}\right) = 1$; $\inf\left(\left\{\frac{1}{n} : n \in \mathbb{N}\right\}\right) = 0$.

  $\rightarrow$ Note that the supremum here is in the set, while the infimum is not in the set.

- $\sup\left(\left\{\frac{n}{n+1} : n \in \mathbb{N}\right\}\right) = 1$; $\inf\left(\left\{\frac{n}{n+1} : n \in \mathbb{N}\right\}\right) = \frac{1}{2}$.

  $\rightarrow$ This is the set $\{\frac{1}{2}, \frac{2}{3}, \frac{3}{4}, \frac{4}{5}, \dots\}$. Note that the infimum here is in the set, while the supremum is not in the set. The proof of the above is asked of you in Exercise 1.29.

- As we have discussed, in $\mathbb{Q}$ the set $\{x \in \mathbb{Q} : x^2 < 2\}$ does not have a supremum. In $\mathbb{R}$ it will—in fact, $\sup\left(\{x \in \mathbb{Q} : x^2 < 2\}\right) = \sqrt{2}$.

Whew, ok, so those are the basic definitions plus examples thereof. We are now ready to state the crucial definition which will "fill in the holes" in the rationals and solve nearly all the problems we discussed.

Definition.

Definition 1.19. Let S be an ordered field (like $\mathbb{R}$). Then S has the *least upper bound property* if given any nonempty $A \subseteq S$ where A is bounded above, A has a least upper bound in S. In other words, $\sup(A) \in S$ for every such A.

Such a set S is also called *complete*.

This definition only explicitly guarantees the existence of least upper bounds, however as a corollary one can prove the existence of greatest lower bounds. Indeed, if S is complete, and $A \subseteq S$ is nonempty and bounded *below*, then $\inf(A) \in S$. This is a consequence of Exercise 1.33, in which you are asked to prove that $\inf(A)$ exists and moreover that $\inf(A) = -\sup(-A)$.

We have already argued that $\mathbb{Q}$ is not complete. To test your understanding, try to convince yourself that $\mathbb{N}$ is complete (this is asked for in Exercise 1.26).

No Sup For You!

− What ℝ we talkin' about? −

At last, we have reached the big, big idea.

Big idea. The reals are what you get when you start with $\mathbb{Q}$ and then add in just enough numbers so that it satisfies the least upper bound property. This is called *completing* $\mathbb{Q}$.

By doing this we will add in numbers like $\sqrt{2}$ and π, but we will not add in "numbers" like the complex number $2 + i$ or the ordinal number ω. Thus you see how important it is to carefully specify your axioms before developing a theory like that of real analysis.

All of these definitions have led up to the following huge theorem.

Theorem.

Theorem 1.20 (*Existence and uniqueness of* ℝ). There exists a unique[24] complete ordered field. We call this field *the real numbers*, $\mathbb{R}$.

A construction of $\mathbb{R}$ using so-called *Dedekind cuts* is sketched in Appendix A, and it is more nuanced and complicated than you might expect.

It is worthwhile now to sit back and reflect on what we have. If we add a completeness axiom to the six axioms we had for an ordered field, we end up with just seven axioms which collectively give us the real numbers. And everything else that we are going to do during this text will be derived from just these seven axioms. Math is often viewed as a pyramid consisting of a wide base of knowledge and results, followed by a smaller number of mid-level theorems, and topped off by a handful of super-specific peak theorems.

---

[24]That is, if you had any other complete ordered field, it really is the same as $\mathbb{R}$, you might just need to relabel the elements.

Sometimes this is phrased as "There exists a unique complete ordered field containing $\mathbb{Q}$" or "it is unique up to isomorphism." Conditions like "containing $\mathbb{Q}$" help clarify, but they are unnecessary.

The fact that there is *only one* complete ordered field in the entire universe highlights that there is something special about the reals. If we want the properties we have discussed, then the complex numbers $\mathbb{C}$, three-dimensional space $\mathbb{R}^3$, the extended reals $\mathbb{R} \cup \{\pm\infty\}$, the hyperreals $^*\mathbb{R}$ and all other options fail at least one axiom.

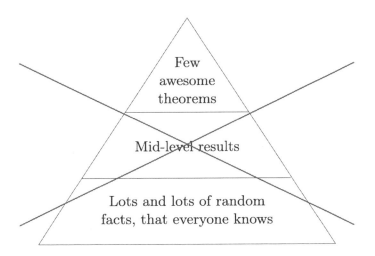

When you are proving something new, this pyramid may feel accurate, since in your mind you hold many small results and a handful of relevant mid-level results, all to prove a single new "peak" result. But for mathematics as a whole, the above pyramid is completely upside-down. Math begins with almost nothing — just a handful of structural or self-evident axioms — and from these, an entire universe is built.

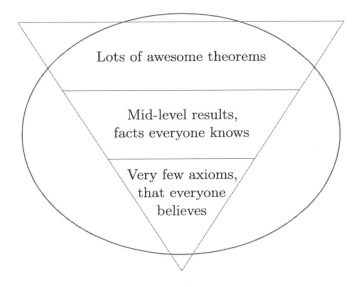

How is it that a handful of self-evident axioms can be combined to create startling theorems? It really is amazing. In my opinion, most of the best accomplishments of the (collective) human mind are in mathematics, and we have the rigors and axiomatic structure of mathematics to thank. So throughout this textbook, as we build derivatives, prove the existence of some remarkable function, or study series of functions, remember where it all started: with almost nothing.

Before moving on, below we collect the axioms which uniquely defined $\mathbb{R}$ — for your future reference.

Note.

Note 1.21. The set $\mathbb{R}$ has two binary operations, addition ($+$) and multiplication ($\cdot$), and is the unique set satisfying the following axioms.

Axiom 1 (Commutative Law). If $a, b \in \mathbb{R}$, then $a + b = b + a$ and $a \cdot b = b \cdot a$.

Axiom 2 (Distributive Law). If $a, b, c \in \mathbb{R}$, then $a \cdot (b + c) = a \cdot b + a \cdot c$.

Axiom 3 (Associative Law). If $a, b, c \in \mathbb{R}$, then $(a + b) + c = a + (b + c)$ and $(a \cdot b) \cdot c = a \cdot (b \cdot c)$.

Axiom 4 (Identity Law). There are special elements $0, 1 \in \mathbb{R}$, where $a + 0 = a$ and $a \cdot 1 = a$ for all $a \in \mathbb{R}$.

Axiom 5 (Inverse Law). For each $a \in \mathbb{R}$, there is an element $-a \in \mathbb{R}$ such that $a + (-a) = 0$. If $a \neq 0$, then there is also an element $a^{-1} \in \mathbb{R}$ such that $a \cdot a^{-1} = 1$.

Axiom 6 (Order Axiom). There is a nonempty subset $P \subseteq \mathbb{R}$, called the *positive elements*, such that

(a) If $a, b \in P$, then $a + b \in P$ and $a \cdot b \in P$;

(b) If $a \in \mathbb{R}$ and $a \neq 0$, then either $a \in P$ or $-a \in P$, but not both.

Axiom 7 (Completeness Axiom). Given any nonempty $A \subseteq \mathbb{R}$ where A is bounded above, A has a least upper bound. In other words, $\sup(A) \in \mathbb{R}$ for every such A.

1.6 Working with Sups and Infs

We now want to develop some properties of suprema and infima, to understand them better. Here is the first basic result.

Proposition.

Proposition 1.22 (*Suprema are unique*). If the supremum or infimum of $A \subseteq \mathbb{R}$ exists, then it is unique.

Recall that a supremum is the same thing as a least upper bound.

Proof. We will prove that suprema are unique, and leave the proof that infima are unique as an exercise; it's similar.

Assume for a contradiction that α and β are distinct least upper bounds of A. In particular, both are upper bounds of A, while $\alpha \neq \beta$. On one hand, since α is a least upper bound and β is an upper bound, we must have $\alpha \leq \beta$. On the other hand, since β is a least upper bound and α is an upper bound, we must have $\beta \leq \alpha$. In summary,

$$\alpha \leq \beta \quad \text{and} \quad \beta \leq \alpha.$$

This implies that $\alpha = \beta$, giving our contradiction. $\qquad\square$

Remember that there are always many ways to write up a proof. For example, the below is a second way to write up the same basic idea.

Second proof. Assume for a contradiction that α and β are distinct least upper bounds of A. In particular, both are upper bounds of A, while $\alpha \neq \beta$. Since $\mathbb{R}$ is an ordered field, either $\alpha < \beta$ or $\beta < \alpha$ (technically, this is a consequence of the "positive subset" that we defined); without loss of generality, assume that $\alpha < \beta$. But this contradicts β being the least upper bound: α is an upper bound but yet $\beta \not\leq \alpha$, as is required by the definition of the least upper bound. $\qquad\square$

I won't go through all the details, but I do want to include a final theorem to check off one of the first indications that we (and mathematicians over the millennia) should care about more than just $\mathbb{Q}$. It also ties nicely into where we began: with the square root of 2.

> **Theorem.**
>
> **Theorem 1.23** (*Square roots exist*). If $a \in \mathbb{R}$ and $a \geq 0$, then $\sqrt{a} \in \mathbb{R}$.

Proof idea. One can show that $\sqrt{a} = \sup\left(\{x \in \mathbb{R} : x^2 < a\}\right)$, which is in $\mathbb{R}$ by completeness. $\qquad\square$

Now consider the set $A = \{0.9, 0.99, 0.999, 0.9999, \ldots\}$. One upper bound of A is 2.7. This isn't the supremum, though, because it's not the *least* upper bound. One way to think about this is that we can subtract a bit off 2.7 and the result is still an upper bound. For instance, if we subtract 0.3 off then we get to 2.4, which is still an upper bound (but is still not the supremum because it's still not the *least* upper bound).

Now, 1 *does* happen to be the supremum. And like above, one way to think about it is this: First, it is an upper bound; second, if you subtract any bit off of 1, you no longer have an upper bound.

- If you subtract just 0.01 off, you get $1 - 0.01 = 0.99$. And this is not an upper bound because 0.999 is in A and is larger than $1 - 0.01$.

- If you subtract just 0.00012 off, you get $1 - 0.00012 = 0.99988$. And this is again not an upper bound because 0.99999999 is in A and is larger than $1 - 0.00012$.

As you can imagine, if you subtract off *any* positive number from 1, you no longer get an upper bound of A. For this reason, 1 is the *least* upper bound of A. In math: $\sup(A) = 1$.

This type of reasoning suggests the following theorem, which is an *analytic* way to think about suprema. Essentially, it says that "α is a least upper bound of a set A if, whenever you subtract a little bit off of α (written as "$\alpha - \varepsilon$"), it is no longer an upper bound (meaning that there is an element $x \in A$ that is now bigger than $\alpha - \varepsilon$). The reasoning is, if you could subtract off a little bit and still have an upper bound, then how is α the *least* upper bound? Clearly you could do better!

Theorem.

Theorem 1.24 (*Suprema analytically*). Let $A \subseteq \mathbb{R}$. Then $\sup(A) = \alpha$ if and only if

  (i) α is an upper bound of A, and

  (ii) Given any $\varepsilon > 0$, $\alpha - \varepsilon$ is *not* an upper bound of A. That is, there is some $x \in A$ for which $x > \alpha - \varepsilon$.

Likewise, $\inf(A) = \beta$ if and only if

  (i) β is a lower bound of A, and

  (ii) Given any $\varepsilon > 0$, $\beta + \varepsilon$ is *not* a lower bound of A. That is, there is some $x \in A$ for which $x < \beta + \varepsilon$.

Before we prove this, I will note that these ε-terms will be used lots and lots and lots and lots and lots in your real analysis course.[25] Always think about them as being some small, positive number. Also, in part (ii) note that the x depends on the ε. In the discussion before Theorem 1.24, for example, $\varepsilon = 0.01$ lead to $x = 0.999$; and $\varepsilon = 0.00012$ lead to $x = 0.99999999$. So different values of ε lead to different values of x.

Ok, let's prove the theorem now.

Proof. We will prove the supremum case and leave the infimum case as an exercise; it's similar.

First assume that $\sup(A) = \alpha$. We aim to prove (i) and (ii). The first of these is immediate: Since $\sup(A) = \alpha$, α is the least upper bound of A, which of course also implies that it is an upper bound of A.

Now we will show (ii). Let $\varepsilon > 0$. Since $\alpha - \varepsilon < \alpha$, we know that $\alpha - \varepsilon$ is not an upper bound of A, because if so that would contradict α being the *least* upper

---

[25] Based on a highly-scientific Ctrl-F search, this text refers to an 'ε' a total of 993 times, and includes "$\varepsilon > 0$" 156 times. So, yeah, it's pretty important in real analysis.

bound of A. And so, since $\alpha - \varepsilon$ is not an upper bound, there must be some $x \in A$ which is greater than $\alpha - \varepsilon$, as desired.

Now assume (i) and (ii). We aim to prove that $\sup(A) = \alpha$. That is, we wish to show that α is an upper bound of A (which is implied directly by (i)), and for any other upper bound β, we have $\alpha \leq \beta$. We have only the latter to prove. Assume that β is some other upper bound of A, and assume for a contradiction that $\beta < \alpha$. Note that $0 < \alpha - \beta$. We will use $(\alpha - \beta)$ as our ε, and then apply (ii) to contradict β being an upper bound. This will be the picture:

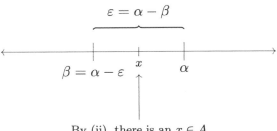

By (ii), there is an $x \in A$
between α and $\alpha - \varepsilon$.

Now we will work it out formally. Let

$$\varepsilon = \alpha - \beta.$$

Since $\varepsilon > 0$, by (ii) there exists some $x \in A$ such that $x > \alpha - \varepsilon = \alpha - (\alpha - \beta) = \beta$. But this is a contradiction, because we assumed that β was an upper bound of A, and yet we found an element $x \in A$ that is larger than β. This completes the proof. $\qquad\square$

Note: the forward direction of the proof also works well by contrapositive. The contrapositive of

$$\sup(A) = \alpha \quad \Longrightarrow \quad \begin{array}{c} \text{For all } \varepsilon > 0 \text{ there exists some} \\ x \in A \text{ such that } x > \alpha - \varepsilon \end{array}$$

is

$$\begin{array}{c} \text{There exists some } \varepsilon > 0 \text{ such that}^{28} \\ \text{for all } x \in A \text{ we have } x \leq \alpha - \varepsilon \end{array} \quad \Longrightarrow \quad \sup(A) \neq \alpha.$$

And to prove this, just observe that the left-hand side implies that $\alpha - \varepsilon$ is an upper bound of A, and so $\sup(A) \leq \alpha - \varepsilon$, which of course implies that $\sup(A) \neq \alpha$, as desired.

---

[28]Does it make sense why you don't have to change "$\varepsilon > 0$" to "$\varepsilon \leq 0$"? A negated statement, if true, falsifies the original statement. Now, the original statement only asserts that positive ε have some property, so proving anything about $\varepsilon \leq 0$ tells you nothing at all about the original statement. Make sense? It's like if you said "Every positive even integer is larger than 1," and I said "Nuh-uh, what about -2??" Clearly I have not falsified your statement, because you were only ever referring to the positive case.

Example 1.25. Let $A = \{x \in \mathbb{R} : x < 0\}$. Then $\sup(A) = 0$.

Proof. We will use the suprema analytically theorem (Theorem 1.24). Note that 0 is clearly an upper bound of A because A is defined as the reals less than 0. This proves (i) in the theorem.

To prove (ii), let $\varepsilon > 0$. We need to show that there is some $x \in A$ such that $x > 0 - \varepsilon$; that is, $x > -\varepsilon$. We claim that $x = -\varepsilon/2$ works. The following proves this.[29] We know that $\varepsilon > 0$, and hence

$$\varepsilon > 0$$
$$-\varepsilon > -2\varepsilon \qquad \text{(Subtracting } 2\varepsilon \text{ from both sides)}$$
$$-\varepsilon/2 > -\varepsilon. \qquad \text{(Dividing both sides by 2)}$$

Moreover, $\varepsilon > 0$ implies that $-\varepsilon < 0$, and hence $-\varepsilon/2 < 0$, which proves $-\varepsilon/2 \in A$. In summary, we have shown that $x = -\varepsilon/2$ has the property that $x \in A$ and $x > 0 - \varepsilon$. This proves condition (ii) of Theorem 1.24, and thus concludes the proof. $\qquad \square$

1.7   The Archimedean Principle

Earlier we asserted, without proof, that

$$\inf\left(\left\{\frac{1}{n} : n \in \mathbb{N}\right\}\right) = 0 \qquad \text{and} \qquad \sup\left(\left\{\frac{1}{n} : n \in \mathbb{N}\right\}\right) = 1.$$

We aim to prove these precisely. For the former, to show that $\inf\left(\left\{\frac{1}{n} : n \in \mathbb{N}\right\}\right) = 0$, we would need to show that 0 is a lower bound (this is easy), and also that $0 + \varepsilon$ is not a lower bound for any $\varepsilon > 0$. That is, for any $\varepsilon > 0$, there is some n such that $1/n < 0 + \varepsilon$. This may or may not seem obvious, but either way it is something we have to prove ourselves. How we do that is by using the so-called *Archimedean principle*.

Lemma.

Lemma 1.26 (*The Archimedean principle*). If a and b are real numbers with $a > 0$, then there exists a natural number n such that $na > b$.

In particular, for any $\varepsilon > 0$ there exists $n \in \mathbb{N}$ such that $\frac{1}{n} < \varepsilon$.

---

[29]The scratch work to find this proof: Start with what you want to show, and do algebra until you reach something we do know (like that $\varepsilon > 0$).

$$-\varepsilon/2 > -\varepsilon \qquad \text{(Want to show)}$$
$$-\varepsilon > -2\varepsilon \qquad \text{(Multiply both sides by 2)}$$
$$\varepsilon > 0 \qquad \text{(Add } 2\varepsilon \text{ to both sides)}$$

Now if we do this scratch work in the reverse direction, it works for our proof.

If a is already bigger than b, or not too much smaller, then this is easy — a small value of n accomplishes this. The challenge is when a is super small and b is super big. Then n has to be really, *really* big to make up for it. The way this is sometimes phrased is: Even a teaspoon can drain an ocean. What that means is this: Think about the number a as being the volume of water in a teaspoon, and b as the volume of water in the ocean, which is obviously much, much bigger. But according to the Archimedean principle there is some integer n where, if you take a teaspoon of water from the ocean n times in a row, then you will have completely emptied it.

Now let's prove it.

Proof. We aim to show that $na > b$ for some $n \in \mathbb{N}$; by dividing over the a, we aim to prove that there is some $n \in \mathbb{N}$ such that $n > b/a$. Now, the number b/a is just some real number that we know nothing about. In fact, let's just call it x. So, equivalently, we are trying to prove that given any real number x, there is some integer n such that $n > x$.

Assume for a contradiction that there is no integer larger than x. That is, assume that x is an upper bound on the set $\mathbb{N}$. Then $\mathbb{N}$ is a subset of $\mathbb{R}$ that is bounded above, and so by the completeness of $\mathbb{R}$ we deduce that $\sup(\mathbb{N})$ exists. Call this supremum α. Since α is the *least* upper bound of $\mathbb{N}$, we know that $\alpha - 1$ is not an upper bound. That is, there exists some integer $m > \alpha - 1$. Adding 1 to each side,

$$m + 1 > \alpha.$$

But this is a contradiction. If α is the supremum of $\mathbb{N}$, then it is an upper bound on $\mathbb{N}$. But we found $(m + 1) \in \mathbb{N}$ which is larger than α. This concludes the first statement in the principle.

The second part follows directly from the first by letting $a = \varepsilon$ and $b = 1$, and then dividing over the n. $\qquad\square$

Example 1.27. Show that $\inf\left(\left\{\dfrac{1}{n} : n \in \mathbb{N}\right\}\right) = 0$.

Solution. Let $A = \left\{\frac{1}{n} : n \in \mathbb{N}\right\}$. We will use the suprema analytically theorem (Theorem 1.24). We must then show that 0 is a lower bound of A and that, for all $\varepsilon > 0$, $0 + \varepsilon$ is not a lower bound of A.

The first of these is almost immediate: Since 1 and n are positive for each $n \in \mathbb{N}$, so is $1/n$. So $1/n > 0$, and thus 0 is indeed a lower bound for A.

Working toward the second, let $\varepsilon > 0$. Then by the Archimedean principle (Lemma 1.26), there exists some $n \in \mathbb{N}$ such that $\frac{1}{n} < \varepsilon$. This element, $\frac{1}{n}$, is in A and is less than $0 + \varepsilon$. So $0 + \varepsilon$ is not a lower bound of A. $\qquad\square$

Example 1.28. Show that $\sup\left(\left\{\dfrac{1}{n} : n \in \mathbb{N}\right\}\right) = 1$.

Solution. Let $A = \left\{\frac{1}{n} : n \in \mathbb{N}\right\}$. We will use the suprema analytically theorem (Theorem 1.24). We must then show that 1 is an upper bound of A and that, for all $\varepsilon > 0$, $1 - \varepsilon$ is not an upper bound of A.

For the first of these, note that since $n \geq 1$ for all $n \in \mathbb{N}$, and by dividing over the n we have that $1 \geq \frac{1}{n}$ for all $n \in \mathbb{N}$. So 1 is indeed an upper bound for A.

Working towards the second, let $\varepsilon > 0$. We need to show that there is some $x \in A$ such that $1 - \varepsilon < x$. But this is always accomplished by the number 1: Clearly $1 \in A$ and $1 - \varepsilon < 1$. □

Definition.

Definition 1.29. Suppose A and B are ordered field (like $\mathbb{R}$). Then A is *dense* in B if, for any $x, y \in B$, there exists $a \in A$ such that $x < a < y$.

For example, $\mathbb{Q}$ is dense in $\mathbb{Q}$ because given any $x, y \in \mathbb{Q}$, if we let $a = \frac{x+y}{2}$, then one can show that $a \in \mathbb{Q}$ and $x < a < y$. Likewise, $\mathbb{R} \setminus \mathbb{Q}$ is dense in $\mathbb{Q}$ since given $x, y \in \mathbb{Q}$, if we let $a = \frac{x+y}{\sqrt{2}}$, then one can show that $a \in \mathbb{R} \setminus \mathbb{Q}$ and $x < a < y$. Meanwhile, $\mathbb{Z}$ is *not* dense in $\mathbb{Q}$.

We are about to prove an important and far-reaching extension of these results, but first here is a quick lemma that we will use for the proof.

Lemma.

Lemma 1.30 ($y - x > 1 \Rightarrow \exists\, z \in \mathbb{Z}$ *where* $x < z < y$). Let $x, y \in \mathbb{R}$. If $y - x > 1$, then there exists $z \in \mathbb{Z}$ such that $x < z < y$.

For the below proof, recall that $\mathbb{N}_0 = \mathbb{N} \cup \{0\}$.

Proof. First assume that x and y are at least 0, and consider the set

$$A = \{n \in \mathbb{N}_0 : n \leq x\}.$$

Since $x \geq 0$, this set is non-empty, and since it is a set of nonnegative integers which is bounded above by x, this set is finite. So by Exercise 1.25, $\max(A)$ exists and is an element of A. Call this maximum M. We claim $z := M + 1$ works.

Note that since $M \in \mathbb{N}_0$, also $z \in \mathbb{N}_0$. Furthermore, since z is larger than the largest element of A, z is not in A, implying that $x < z$. Finally,

$$M \leq x \qquad \text{implies that} \qquad M + 1 \leq x + 1 < y.$$

So $z < y$. In summary, we have shown that $x < z < y$, as desired.

The cases where x and y are not at least 0 are similar. If both are negative, then by considering $-x$ and $-y$ the above argument gives an integer z where $-y < z < -x$, showing that $-z$ works, since $x < -z < y$. If one is positive and one is negative, then 0 works. □

The next theorem says that between any two real numbers is a rational number. The rationals — which are super nice to describe and work with — are everywhere.[30]

Theorem.

Theorem 1.31 ($\mathbb{Q}$ *is dense in* $\mathbb{R}$). The rational numbers are dense in the real numbers.

Proof Idea. The result follows from Lemma 1.30, and does so easily enough that it would be easy to miss what is what really going on. Assume that $x, y \in \mathbb{R}$ and assume both are positive with $x < y$. We aim to find a rational number $\frac{m}{n}$ between x and y. That is, with $x < \frac{m}{n} < y$.

Now, $y - x$ is some positive number, and so by Archimedean principle there is some $n \in \mathbb{N}$ with $\frac{1}{n} < y - x$.

Now think about all integer multiples of this $\frac{1}{n}$.

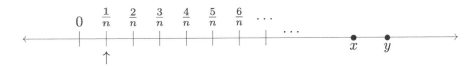

Note that since each one is $\frac{1}{n}$ away from the next one, but $y - x > \frac{1}{n}$, it is impossible for these dashes to completely hop over the interval between x and y. That is, at least one of these must fall between x and y.

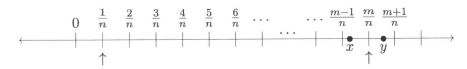

This is the idea, although by using Lemma 1.30 the proof hides some of these details. Now here's the proof of Theorem 1.31.

---

[30]In Chapter 2 when we study the cardinalities of these sets, this result will be even more impressive and important.

Proof. Pick any $x, y \in \mathbb{R}$ where $x < y$. We need to show that there exists some $\frac{m}{n} \in \mathbb{Q}$ (with $m, n \in \mathbb{Z}$) such that

$$x < \frac{m}{n} < y.$$

First note that if $x < 0 < y$ then we are done, since $0 \in \mathbb{Q}$. Furthermore, if we can show the theorem holds in the case that x and y are positive, then it implies that it holds when they are negative $\left(0 < x < \frac{m}{n} < y \text{ implies } -y < \frac{-m}{n} < -x < 0\right)$, so we may assume x and y are positive.

Since $y - x > 0$, by the Archimedean principle (Lemma 1.26) there exists some $n \in \mathbb{N}$ such that $n(y - x) > 1$; i.e., $ny - nx > 1$. And so, by Lemma 1.30, there is some integer m with

$$nx < m < ny.$$

That is,

$$x < \frac{m}{n} < y,$$

which concludes the proof. $\square$

Three final comments

Before concluding the first chapter, here a few miscellaneous comments. First, the proof of Lemma 1.30 also implies that, for any $x \in \mathbb{R}$, there exists an integer M such that $M \leq x \leq M + 1$. In particular, it implies that the floor and ceiling functions exist.

Definition.

Definition 1.32. Let $x \in \mathbb{R}$.

- The *ceiling* of x, denoted $\lceil x \rceil$, is the integer n such that $x \leq n < x + 1$

- The *floor* of x, denoted $\lfloor x \rfloor$, is the integer n such that $x - 1 < n \leq x$.

Second, we haven't formally defined the notion of an *interval*, but it's exactly as you would suspect.

Definition.

Definition 1.33. Define the *closed interval* $[a, b]$ to be $\{x \in \mathbb{R} : a \leq x \leq b\}$. Likewise the *open interval* (a, b) is defined to be $\{x \in \mathbb{R} : a < x < b\}$, and half-open intervals and intervals to $\pm\infty$ are again exactly as you would expect.

Lastly, the formal construction of $\mathbb{R}$ (see Appendix A) contains an assertion of the soundness of the *principle of mathematical induction*, and we will therefore use induction without hesitation. This principle is equivalent to the so-called *well-ordering principle*, which we record now before moving on to cardinality.

Principle.

Principle 1.34 (Well-ordering principle). Every non-empty subset of natural numbers contains a smallest element.

— Notable Exercises —

Exercises test your knowledge and develop your understanding. Many are written simply for this purpose, while others hold a greater significance for one reason or another. Before each chapter's exercises I will comment on a few of the exercises which are in this second camp[31]

- In Exercise 1.8 we prove *De Morgan's laws*, which are quite useful in set theory and beyond, as they describe how the compliment operation distributes over unions and intersections. They say this:

$$(A \cap B)^c = A^c \cup B^c \qquad \text{and} \qquad (A \cup B)^c = A^c \cap B^c.$$

  They will become useful again when we study topology in Chapter 5, where we will utilize an infinite form of these laws (see Fact 5.11). They are also particularly important in mathematical logic (under the guise of *laws of Boolean algebra*) — but that's a topic for another long-form textbook.

  De Morgan's laws highlight a certain symmetry between unions and intersections, and between a universal set X and the empty set $\emptyset$. This is an example of a *duality* property, which pop up all over mathematics.

- Exercise 1.13 is an early example of an *analytic* statement, in which a property is asserted to hold "for all $\varepsilon > 0$." Properties of this from will soon be filling up the pages of this text, and are ubiquitous in the field of real analysis. Recall that we already have one particular theorem of this form: The suprema analytically theorem (Theorem 1.24). Exercise 1.30 and others provide further work in this direction.

- Exercise 1.23 will be used several times in this book, most notably in Chapters 4 and 8.

- Exercise 1.34 is a special case of an important topological property we will study in Chapter 5. We will also look at the limitations of this property, which are alluded to in Exercise 1.35. You are asked to prove a generalization of it in Exercise 5.11 (b), although the best way to prove Exercise 1.34 at this point utilizes only things we discussed in this chapter.

---

[31]For instructors: I do not mean to single these out as suggestions as to what to assign for homework, and indeed the others are oftentimes just as good or better suited for assignments.

— Exercises —

- You must prove all of your answers (unless stated otherwise).

- Remember the proof techniques that you learned in your intro-to-proofs class: Direct, by contradiction, by using the contrapositive, by induction, and/or by cases. And you can disprove something by exhibiting a counterexample.

Note: Hints and solutions to select exercises can be found at `LongFormMath.com`.

Exercise 1.1. Explain the error in the following "proof" that $2 = 1$.
Let $x = y$. Then

$$x^2 = xy$$
$$x^2 - y^2 = xy - y^2$$
$$(x + y)(x - y) = y(x - y)$$
$$x + y = y$$
$$2y = y$$
$$2 = 1.$$

Exercise 1.2. Which of the following statements are true? Give a short explanation for each of your answers.

(a) For every $n \in \mathbb{N}$, there is an integer $m \in \mathbb{N}$ such that $m > n$.

(b) For every $m \in \mathbb{N}$, there is an integer $n \in \mathbb{N}$ such that $m > n$.

(c) There is an $m \in \mathbb{N}$ such that for every $n \in \mathbb{N}$, $m \geq n$.

(d) There is an $n \in \mathbb{N}$ such that for every $m \in \mathbb{N}$, $m \geq n$.

(e) There is an $n \in \mathbb{R}$ such that for every $m \in \mathbb{R}$, $m \geq n$.

(f) For every pair $x < y$ of integers, there is an integer z such that $x < z < y$.

(g) For every pair $x < y$ of real numbers, there is a real number z such that $x < z < y$.

Exercise 1.3. If A and B are two boxes (possibly with things inside), describe the following in terms of boxes:

(a) $A \setminus B$ (b) $\mathcal{P}(A)$ (c) $|A|$

Exercise 1.4. If $A_1, A_2, A_3, \ldots, A_n$ are all boxes (possibly with things inside), describe the following in terms of boxes:

(a) $\displaystyle\bigcup_{i=1}^{n} A_i = A_1 \cup A_2 \cup \cdots \cup A_n.$ (b) $\displaystyle\bigcap_{i=1}^{n} A_i = A_1 \cap A_2 \cap \cdots \cap A_n.$

Exercise 1.5. Prove that each of the following holds for any sets A and B.

(a) $A \cup B = A$ if and only if $B \subseteq A$.   (c) $A \setminus B = A$ if and only if $A \cap B = \emptyset$.

(b) $A \cap B = A$ if and only if $A \subseteq B$.   (d) $A \setminus B = \emptyset$ if and only if $A \subseteq B$.

Exercise 1.6. Suppose $f : X \to Y$ and $A \subseteq X$ and $B \subseteq Y$.

(a) Prove that $f(f^{-1}(B)) \subseteq B$.

(b) Give an example where $f(f^{-1}(B)) \neq B$.

(c) Prove that $A \subseteq f^{-1}(f(A))$.

(d) Give an example where $A \neq f^{-1}(f(A))$.

Exercise 1.7. Suppose that $f : X \to Y$ and $g : Y \to X$ are functions and that the composite $g \circ f$ is the identity function $\mathrm{id} : X \to X$. (The identity function sends every element to itself: $\mathrm{id}(x) = x$.) Show that f must be a one-to-one function and that g must be an onto function.

Exercise 1.8. The following are special cases of De Morgan's laws.

(a) Prove that $(A \cap B)^c = A^c \cup B^c$.   (b) Prove that $(A \cup B)^c = A^c \cap B^c$.

Exercise 1.9.

(a) Prove that $\sqrt{3}$ is irrational.

(b) What goes wrong when you try to adapt your argument from part (a) to show that $\sqrt{4}$ is irrational (which is absurd)?

(c) In part (a) you proved that $\sqrt{3}$ to be irrational, and essentially the same proof shows that $\sqrt{5}$ is irrational. By considering their product or otherwise, prove that $\sqrt{3} - \sqrt{5}$ and $\sqrt{3} + \sqrt{5}$ are either both rational or both irrational. Deduce that they must both be irrational.

Exercise 1.10. Prove that the multiplicative identity in a field is unique.

Exercise 1.11. Given an ordered field $\mathbb{F}$, recall that we defined the *positive elements* to be a nonempty subset $P \subseteq \mathbb{F}$ that satisfies both of the following conditions:

(i) If $a, b \in P$, then $a + b \in P$ and $a \cdot b \in P$;

(ii) If $a \in \mathbb{F}$ and $a \neq 0$, then either $a \in P$ or $-a \in P$, but not both.

(a) Give an example of some $P_1 \subseteq \mathbb{R}$ that satisfies (i) but not (ii).

(b) Give an example of some $P_2 \subseteq \mathbb{R}$ that satisfies (ii) but not (i).

Exercise 1.12. Assume that $\mathbb{F}$ is an ordered field and $a, b, c, d \in \mathbb{F}$ with $a < b$ and $c < d$.

(a) Show that $a + c < b + d$.

(b) Prove that it is not necessarily true that $ac < bd$.

Note whenever you use an axiom.

Exercise 1.13. Let a, b and ε be elements of an ordered field.

(a) Show that if $a < b + \varepsilon$ for every $\varepsilon > 0$, then $a \leq b$.

(b) Use part (a) to show that if $|a - b| < \varepsilon$ for all $\varepsilon > 0$, then $a = b$.

Note whenever you use an axiom.

Exercise 1.14. Prove that the equality $|ab| = |a| \cdot |b|$ holds for all real numbers a and b.

Exercise 1.15. For each of the following, find all numbers x which satisfy the expression.

(a) $|x - 4| = 7$

(b) $|x - 4| < 7$

(c) $|x + 2| < 1$

(d) $|x - 1| + |x - 2| > 1$

(e) $|x - 1| + |x + 1| > 1$

(f) $|x - 1| - |x + 1| > 1$

(g) $|x - 1| \cdot |x + 1| = 0$

(h) $|x - 1| \cdot |x + 2| = 3$

Exercise 1.16. Let $\max\{x, y\}$ denote the maximum of the real numbers x and y, and let $\min\{x, y\}$ denote the minimum. For example, $\min\{-1, 4\} = \min\{-1, -1\} = -1$. Prove that

$$\max\{x, y\} = \frac{x + y + |y - x|}{2} \qquad \text{and} \qquad \min\{x, y\} = \frac{x + y - |y - x|}{2}.$$

Then find a formula for $\max\{x, y, z\}$ and $\min\{x, y, z\}$.

Exercise 1.17. Prove that if $a, b \in \mathbb{R}$ and $0 < a < b$, then $a^n < b^n$ for any positive integer n.

Exercise 1.18. Prove that if $a_1, a_2, \ldots, a_n$ are real numbers, then

$$|a_1 + a_2 + \cdots + a_n| \leq |a_1| + |a_2| + \cdots + |a_n|.$$

Exercise 1.19. Prove that $\displaystyle\sum_{k=1}^{n} \frac{1}{k(k+1)} = \frac{n}{n+1}$ for every natural number n.

Exercise 1.20. Determine which natural numbers, n, have the property that $\sqrt{n}$ is irrational.

Exercise 1.21. Let $f : X \to Y$, and assume $A_1, A_2 \subseteq X$. Show that

$$f(A_1 \cap A_2) \subseteq f(A_1) \cap f(A_2).$$

Recall that if A is a set, then $f(A) = \{f(x) : x \in A\}$.

Exercise 1.22. Give an example of a function f, and a pair of sets A and B, for which

$$f(A \cap B) \neq f(A) \cap f(B).$$

Recall that if A is a set, then $f(A) = \{f(x) : x \in A\}$.

Exercise 1.23. Assume that $A \subseteq B$ and both are bounded above. Prove that $\sup(A) \leq \sup(B)$.

Exercise 1.24. Suppose $A \subseteq \mathbb{R}$ has a maximal element — that is, there is an element $M \in A$ such that $x \leq M$ for all $x \in A$. Likewise, assume $B \subseteq \mathbb{R}$ has a minimal element m.

(a) Prove that $\sup(A) = M$.   (b) Prove that $\inf(B) = m$.

Exercise 1.25. Suppose that A is a nonempty set containing finitely many elements. Prove by induction that A has a maximal element, and that $\max(A) \in A$.

Exercise 1.26. Prove that $\mathbb{N}$ is complete.

Exercise 1.27. For each item, compute the requested supremum or infimum or carefully explain why it does not exist. Either way, *prove* that your answer is correct.

(a) Determine $\sup A$ for $A = \left\{ \frac{(-1)^n}{n} : n \in \mathbb{N} \right\}$.

(b) Fix $\alpha \in (0, 1)$. Determine $\inf(B)$ for $B = \{\alpha^n : n \in \mathbb{N}\}$.

(c) Fix $\alpha \in (1, \infty)$. Determine $\sup(C)$ for $C = \{\alpha^n : n \in \mathbb{N}\}$.

Exercise 1.28. Prove the infimum case of Theorem 1.24.

Exercise 1.29. Prove that

$$\sup\left(\left\{\frac{n}{n+1} : n \in \mathbb{N}\right\}\right) = 1 \qquad \text{and} \qquad \inf\left(\left\{\frac{n}{n+1} : n \in \mathbb{N}\right\}\right) = \frac{1}{2}.$$

Exercise 1.30. Let $A, B \subseteq \mathbb{R}$, and assume that $\sup(A) < \sup(B)$.

(a) Show that there exists an element $b \in B$ that is an upper bound for A.

(b) Give an example to show that this is not necessarily the case if we instead only assume that $\sup(A) \leq \sup(B)$. You do not need to prove your answer.

Exercise 1.31. Suppose that $A, B \subseteq \mathbb{R}$ are nonempty and bounded above. Find a formula for $\sup(A \cup B)$ and prove that it is correct.

Exercise 1.32. Suppose $A \subseteq \mathbb{R}$ is bounded above and $c \in \mathbb{R}$. Define $c + A = \{c + a : a \in A\}$ and $cA = \{ca : a \in A\}$.

(a) Prove that $\sup(c + A) = c + \sup(A)$.

(b) Determine necessary and sufficient conditions on c and A for $\sup(cA) = c\sup(A)$. Give an example of a set A and number c where $\sup(cA) \neq c\sup(A)$.

Exercise 1.33. For $A \subseteq \mathbb{R}$, we denote $-A$ to be the set obtained by taking the opposite of everything in A. That is,

$$-A := \{-x : x \in A\}.$$

Suppose that $A \neq \emptyset$ and that A is bounded below. Prove that $-A \neq \emptyset$, $-A$ is bounded above, and $\sup(-A) = -\inf(A)$.

Exercise 1.34. For each $n \in \mathbb{N}$, assume we are given a closed interval $I_n = [a_n, b_n]$. Also, assume that each I_{n+1} is contained inside of I_n. This gives a sequence of increasingly smaller intervals,

$$I_1 \supseteq I_2 \supseteq I_3 \supseteq I_4 \supseteq \cdots.$$

Prove that $\bigcap_{n=1}^{\infty} I_n \neq \emptyset$. That is, prove that there is some real number x such that $x \in I_n$ for every $n \in \mathbb{N}$.

Exercise 1.35. Give an example showing that the conclusion of Exercise 1.34 need not hold if each I_n is allowed to be an open interval.

- For $A, B \subseteq \mathbb{R}$, we define
$$A + B = \{a + b : a \in A \text{ and } b \in B\}.$$

- For $A, B \subseteq \mathbb{R}$, we define
$$A \cdot B = \{a \cdot b : a \in A \text{ and } b \in B\}.$$

Exercise 1.36.

(a) Determine
$$\{1, 3, 5\} + \{-3, 0, 1\}$$

(b) Assume that $A, B \subseteq \mathbb{R}$ and $\sup(A)$ and $\sup(B)$ exist. Prove that
$$\sup(A + B) = \sup(A) + \sup(B).$$

Exercise 1.37.

(a) Determine
$$\{1, 3, 5\} \cdot \{-3, 0, 1\}$$

(b) Give an example of sets A and B where $\sup(A \cdot B) \neq \sup(A) \cdot \sup(B)$.

— Open Questions[32] —

Question 1. Consider the set

$$\left\{\frac{3}{2}, \left(\frac{3}{2}\right)^2, \left(\frac{3}{2}\right)^3, \left(\frac{3}{2}\right)^4, \ldots\right\},$$

but for each number remove the integer portion;[33] for example, $(3/2)^3 = 3.375$ would be reduced to 0.375. Is the resulting set dense in $[0, 1]$?

Question 2. Is $e + \pi$ rational?

Question 3. For points x and y in the plane, define $d(x, y)$ to be the distance between x and y. Does there exist a dense subset S of $\mathbb{R}^2$ where, for any $x, y \in S$, $d(x, y) \in \mathbb{Q}$?[34]

Question 4. Can every rational number x be represented as a quotient of *shifted primes*? That is, do there exist primes p and q such that

$$x = \frac{p+1}{q+1}.$$

---

[32] An "open question" is a question that is unsolved. No one knows the answer to any of these. If you are my student and you solve one of them, I will give you one point of extra credit.

[33] Another way to say this is "$(3/2)^n \pmod 1$." Or: $(3/2)^n - \lfloor (3/2)^n \rfloor$.

[34] That is, if $x = (x_1, x_2)$ and $y = (y_1, y_2)$ are in S, then $d(x, y) = \sqrt{(y_1 - x_1)^2 + (y_2 - x_2)^2} \in \mathbb{Q}$.

Chapter 2: Cardinality

2.1 Bijections and Cardinality

My research is in a field of math called *combinatorics*. One central problem in combinatorics is to count sets of things. The scientific philosopher Ernst Mach went as far as to say "Mathematics may be defined as the economy of counting. There is no problem in the whole of mathematics which cannot be solved by direct counting." It's a beautiful thought,[1] but even I would not go *quite* that far; nevertheless, I do think that the best solutions in math are those that use counting.

We have already defined the number of elements in a set A to be the *cardinality* of that set. For example,

$$|\{a, b, c\}| = 3,$$

and

$$|\{1, 4, 9, 16, 25, \ldots, 100\}| = 10,$$

and

$$|\mathbb{Z}| = \infty.$$

In general, we say that two sets have the same size if there is a way to pair up the elements between the two sets. Equivalently, if there is a bijection between them. This is known as *the bijection principle*.

Principle.

Principle 2.1 (*The bijection principle*). Two sets have the same size if and only if there is a bijection between them.

Example 2.2. One reason that the sets $\{1, 2, 3\}$ and $\{a, b, c\}$ have the same size is that the elements can be paired up like this:

$$1 \leftrightarrow b \qquad 2 \leftrightarrow a \qquad 3 \leftrightarrow c$$

---

[1] And it's at least semi-faintly-plausible. If he had said, "There is no problem in the whole of mathematics which cannot be solved by integration by parts," now *that* would have been bold.

And one reason that $\{x, y, z\}$ and $\{m, a, t, h\}$ do not have the same size is that the elements can not be paired up. Whenever you try, one element from the second set won't get a pair.   □

This next example is one I've given many times. Whenever someone asks what I do, I tell them I'm a math professor. Most then respond saying something like "oh, I was never good at math," or "I never liked math." But sometimes they follow-up and ask what sort of math I study, to which I tell them that I study combinatorics. After repeating that word slowly a few times, I usually tell them that combinatorics is the most fun math, has the most puzzle-like questions, and is the most insight- and problem-driven field. If they seem interested, I oftentimes offer to tell them a problem which I claim conveys what it *feels* like to do combinatorics. If they take the bait, I always give them the following problem.

Example 2.3. In the NCAA's March Madness Tournament there are 64 teams that play. The tournament is single-elimination (once you lose, you're out). How many games are played throughout the tournament?

> Sometimes the person will be able to realize that in the first round 32 games are played (since 64 teams pair up and play a round). I then note that this is indeed a way to solve the problem: If they apply the same reasoning they could find the number of games in the second round, the third round, and so forth, and then just add up the number of games from each round. But this, I tell them, is boring, it can't be easily generalized to other tournaments, and it doesn't give any insight into *why* the answer is what it is.
>
> Furthermore, I and many other combinatorists do not like computational solutions. Computations are boring. Plus, computers can do them better than I can, while also committing fewer errors. And computations often hide what is really going on. I want to know how many games are played not only because the answer might be of interest, but because I want to *understand* how many games are played.
>
> Finally, because people in the wild tend to like when math has applications, I sometimes also mention how the answer has practical value — the number of games played translates directly into the number of times the NCAA has to reserve TV time slots, how much revenue they should expect, the number of times they need to reserve a court, etc.
>
> But usually within a minute, and sometimes within 7 seconds, I give them the following neat solution. Here it is.

Solution. Note that in each game, precisely 1 team loses, and of course each loss happens in a game. So, number of games = number of losses. Now, the champion never loses, but every other team loses one game and then they're out. So with 64 teams, 63 of them lose once, so there must be 63 losses, so there must be 63 games.   □

That's it! Don't count games, count losses. There is a bijection between the games and the losses (every game has 1 loss, every loss occurs in 1 game), so counting losses is the same as counting games. And since 63 teams lose one time, it's easy to count those losses! Technically, we can view this solution as an application of the bijection principle, when we paired up each loss with the game it occurred in (and then counted the new set).

The really cool thing, though, is that this definition of the size of a set applies even to infinite sets. And that implies some truly fascinating things.

2.2   Counting Infinities

The ability to "pair up" elements between two sets is what it means for them to have the same size — this is perfectly intuitive for finite sets with nothing too counterintuitive resulting, but with infinite sets... well, some pretty neat stuff pops out. Indeed, the pluralization in this section's title was first your sign of the miracles to come.

One last thing before we do that: Here is a quick function-to-inequality fact that you probably learned in your intro-to-proofs class.

Fact.

Fact 2.4.   Let S and T be sets. Then,

- $|S| = |T|$ if and only if there is a bijection from S to T.

- $|S| \leq |T|$ if and only if there is an injection from S to T.

- $|S| \geq |T|$ if and only if there is a surjection from S to T.

Hilbert's Hotel

"No other question has ever moved so profoundly the spirit of man; no other idea has so fruitfully stimulated his intellect; yet no other concept stands in greater need of clarification, than that of the infinite."

– David Hilbert

We begin by talking about the set of problems related to the so-called Hilbert's Hotel. Assume that there is a hotel, called Hilbert's Hotel, which has infinitely many rooms in a row.

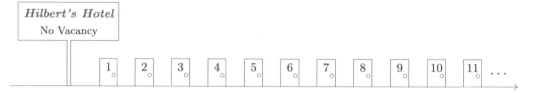

- Assume every room has someone in it, and so the "No Vacancy" sign has been turned on. With most hotels this would mean that if someone else arrives at the hotel, they will not be given a room. But this isn't the case with Hibert's Hotel. If, for $n \in \mathbb{N}$, the patron in room n moves to room $n + 1$, then nobody is left without a room and suddenly room 1 is completely open! So the new customer can go to room 1.[2]

- Now imagine 2 people arrived to the hotel. Can we accommodate them? Certainly! Now, just have everyone move from room n to room $n + 2$. This leaves rooms 1 and 2 open to the newcomers, and we are again good-to-go.

- What if, however, we have infinitely many people lined up wanting a room. Can we accommodate *all* of them? Yes! We still can! Just have the person in room n move to room $2n$. Then all of the odd-numbered rooms are vacant and the infinite line of people can take these rooms.[3]

The first point of this exercise is to simply realize that weird stuff can happen when dealing with the infinite. The second point, though, is to realize that each time the people switched rooms, those same exact people got new rooms. So in the first example when they each just moved one room down, that should mean that there are just as many rooms from 1 to ∞ as there are from 2 to ∞... And likewise for the others.

Indeed, with this in mind, let's talk about sizes of specific sets. But first, a ditty:

$$\infty \text{ bottles of beer on the wall,}$$
$$\infty \text{ bottles of beer.}$$
$$\text{Take one down, pass it around,}$$
$$\infty \text{ bottles of beer on the wall.}$$

(repeat)

Specific Sets

Example 2.5. There are the same number of natural numbers as there are natural numbers larger than 1 (that is, $|\mathbb{N}| = |\{2, 3, 4, \ldots\}|$). What's the bijection that shows this? Let

$$f : \mathbb{N} \longrightarrow \{2, 3, 4, \ldots\}$$
$$n \longmapsto n + 1.$$

---

[2]Make sure you take a moment to appreciate how remarkably, wonderfully weird this is.

[3]Make sure you take a moment to appreciate how remarkably, wonderfully weird this is.

REAL ANALYSIS, A LONG-FORM TEXTBOOK

(Note: This is notation for the function $f(x) = x + 1$ from the naturals to the set $\{2, 3, 4, \dots\}$.) In other (non-)words, this is the pairing

$$1 \leftrightarrow 2, \quad 2 \leftrightarrow 3, \quad 3 \leftrightarrow 4, \quad 4 \leftrightarrow 5, \quad \dots$$

$\square$

The Moral. Two sets can have the same size even though one is a *proper* subset of the other.[4]

Example 2.6. There are the same number of natural numbers as even natural numbers (that is, $|\mathbb{N}| = |2\mathbb{N}|$). What's the bijection that shows this? Let

$$f : \mathbb{N} \longrightarrow 2\mathbb{N}$$
$$n \longmapsto 2n.$$

(Note: This is notation for the function $f(x) = 2x$ from the naturals to the even naturals.) In other (non-)words, this is the pairing

$$1 \leftrightarrow 2, \quad 2 \leftrightarrow 4, \quad 3 \leftrightarrow 6, \quad 4 \leftrightarrow 8, \quad \dots$$

$\square$

The Moral. Two sets can have the same size even though one is a *proper* subset of the other and the larger one even has *infinitely* many more elements than the smaller one.[5]

Likewise:

Example 2.7. $|\mathbb{N}| = |\mathbb{Z}|$. $\square$

The proof of this is asked of you in Exercise 2.6.

> **Theorem.**
>
> **Theorem 2.8** ($|\mathbb{Z}| = |\mathbb{Q}|$)**.** There are the same number of integers as rational numbers.

Proof. What's the bijection that shows this? This one is a little complicated... It's difficult to write down the formula, but the formula can be described in a nice way. First note that if we can show that there are the same number of positive integers as positive real rational numbers, then by simply adding a minus sign to everything, we will get a full bijection. So we may focus on just the positive case.

---

[4]Make sure you take a moment to appreciate how remarkably, wonderfully weird this is.

[5]Make sure you take a moment to appreciate how remarkably, wonderfully weird this is.

Consider the rational numbers (with some duplication) written in this way:

	1	2	3	4	5	6	7	8	$\cdots$
1	1/1	2/1	3/1	4/1	5/1	6/1	7/1	8/1	$\cdots$
2	1/2	2/2	3/2	4/2	5/2	6/2	7/2	8/2	$\cdots$
3	1/3	2/3	3/3	4/3	5/3	6/3	7/3	8/3	$\cdots$
4	1/4	2/4	3/4	4/4	5/4	6/4	7/4	8/4	$\cdots$
5	1/5	2/5	3/5	4/5	5/5	6/5	7/5	8/5	$\cdots$
6	1/6	2/6	3/6	4/6	5/6	6/6	7/6	8/6	$\cdots$
7	1/7	2/7	3/7	4/7	5/7	6/7	7/7	8/7	$\cdots$
8	1/8	2/8	3/8	4/8	5/8	6/8	7/8	8/8	$\cdots$
$\vdots$	$\vdots$	$\vdots$	$\vdots$	$\vdots$	$\vdots$	$\vdots$	$\vdots$	$\vdots$	$\ddots$

Our bijection, which we call *the winding bijection,* can be pictured like this:

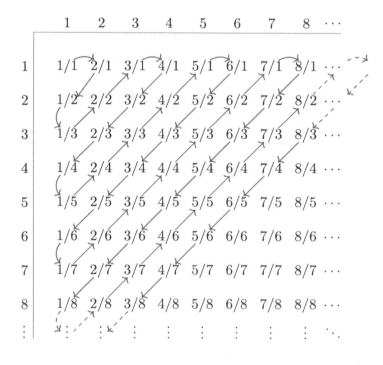

Weaving through this chart, you are guaranteed to hit every positive rational number. So if you pair up 1 with the first number you hit, 2 with the second number you hit, 3 with the third, and so on, then every positive rational number is in a pair. Now... there's just one small problem: each rational number is actually hit more than once. The number p/q will be written in positions $(p, q), (2p, 2q), (3p, 3q), \ldots$. But the fix is easy: When you come across a number that has already been hit, just skip it. Clearly you won't run out of rational numbers, so this does indeed pair up everything. So

$$f(n) = \text{ the } n^{\text{th}} \text{ new rational number you reach.}$$

And that's it![6] $\qquad\qquad\qquad\qquad\qquad\qquad\qquad\qquad\qquad\qquad\qquad\square$

The Moral. Even though there are infinitely many rational numbers between *every two consecutive integers*, the two sets *still* have the same size.[7]

Now, at this point you might be tempted to predict that the reason all these sets have the same size is that they all have infinitely many elements, and maybe all infinities are the same and that's all there is to it... But amazingly that's not actually the case. Theorem 2.9, which we will get to in a second, demonstrates this. But first, to get a hold on how the proof goes, here's a little game.

– A game of deduction –

Setup. Instructors: Here is a fun in-class game. Write down a set of five length-5 sequences of 0s and 1s. These are kept secret from the students.

Goal. The students' goal is to come up with a length-5 sequence of 0s and 1s which is not one of the instructor's five.

Game Rules. The students can ask the instructor any yes/no questions about the sequences, and receive a truthful reply. When the students make their final guess, they should be 100% confident that this guess is not one of the instructor's sequences. And they should try to figure it out while asking as few questions as possible.

One solution. One way to solve it is the following. Ask the instructor whether their first sequence has a 1 in the first position. Then ask the instructor whether their second sequence has a 1 in the second position. Then ask if the third sequence has a 1 in the third position, then whether the fourth sequence has a 1 in the fourth position, and finally whether the fifth sequence has a 1 in the fifth position.

---

[6]For the curious, one explicit bijection from $\mathbb{N}$ to $\mathbb{Q}^+$ is $h_+(n) = p_1^{(-1)^{a_1} \left\lfloor \frac{a_1+1}{2} \right\rfloor} \cdot p_2^{(-1)^{a_2} \left\lfloor \frac{a_2+1}{2} \right\rfloor} \cdots$, where p_i is the ith prime and $n = p_1^{a_1} p_2^{a_2} \cdots$ is the prime factorization of n. To get a bijection h from $\mathbb{Z}$ to $\mathbb{Q}$, just define $h(0) = 0$, $h(n) = h_+(n)$ if $n > 0$, and $h(n) = -h_+(-n)$ if $n < 0$.

[7]Make sure you take a moment to appreciate how remarkably, wonderfully weird this is.

Based on the instructor's yes/no answers, the students can deduce, for each $i \in \{1,2,3,4,5\}$, what the ith number of the ith sequence is.

Suppose they deduced that the first number has a 0 in the first position; the second number has a 1 in the second position; and the third, fourth and fifth numbers have a 1,0 and 0 in their third, fourth and fifth positions, respectively. Since those numbers, in order, were 0,1,1,0,0, a correct guess would be its opposite: 1,0,0,1,1. You just take everything the instructor has and flip it.

This is guaranteed to work. Consider the first sequence; you know that 1,0,0,1,1 is not the first sequence, because (by design) they differ in the first position! And you know that 1,0,0,1,1 is not the second sequence, because (by design) they differ in the second position! Indeed, each of the instructor's 5 sequences is different from 1,0,0,1,1 in one particular spot, and so you know that this sequence is different from each of the instructor's. Neat!

This clever idea is from Georg Cantor, and is known as "Cantor's diagonalization argument." It is the key idea to a beautiful proof that $|\mathbb{R}| > |\mathbb{N}|$,[8] which implies the amazing fact that there are different sizes of infinity.

Theorem.

Theorem 2.9 ($|\mathbb{R}| > |\mathbb{N}|$)**.** There are more real numbers than natural numbers.

This implies that some infinities are bigger than others.

Proof. Since $\mathbb{N} \subseteq \mathbb{R}$, clearly $|\mathbb{N}| \leq |\mathbb{R}|$. To show that they are not equal, we must prove that there is no bijection between $\mathbb{R}$ and $\mathbb{N}$. Let's again use the "pairing up" idea. We will prove it by contradiction. In fact, we will prove the stronger statement that there are more real numbers in $(0, 1)$ than there are natural numbers. (This of course would prove the larger statement since then we could say $|\mathbb{R}| \geq |(0,1)| > |\mathbb{N}|$.)

Assume for a contradiction that there does exist some way to pair up the naturals with the reals in $(0, 1)$. Writing the reals in decimal notation, assume the pairing is this:

$$1 \leftrightarrow 0 . a_{11}\, a_{12}\, a_{13}\, a_{14}\, a_{15}\, a_{16}\, a_{17}\, a_{18} \cdots$$
$$2 \leftrightarrow 0 . a_{21}\, a_{22}\, a_{23}\, a_{24}\, a_{25}\, a_{26}\, a_{27}\, a_{28} \cdots$$
$$3 \leftrightarrow 0 . a_{31}\, a_{32}\, a_{33}\, a_{34}\, a_{35}\, a_{36}\, a_{37}\, a_{38} \cdots$$
$$4 \leftrightarrow 0 . a_{41}\, a_{42}\, a_{43}\, a_{44}\, a_{45}\, a_{46}\, a_{47}\, a_{48} \cdots$$
$$5 \leftrightarrow 0 . a_{51}\, a_{52}\, a_{53}\, a_{54}\, a_{55}\, a_{56}\, a_{57}\, a_{58} \cdots$$
$$6 \leftrightarrow 0 . a_{61}\, a_{62}\, a_{63}\, a_{64}\, a_{65}\, a_{66}\, a_{67}\, a_{68} \cdots$$
$$7 \leftrightarrow 0 . a_{71}\, a_{72}\, a_{73}\, a_{74}\, a_{75}\, a_{76}\, a_{77}\, a_{78} \cdots$$
$$8 \leftrightarrow 0 . a_{81}\, a_{82}\, a_{83}\, a_{84}\, a_{85}\, a_{86}\, a_{87}\, a_{88} \cdots$$

$$\vdots$$

---

[8]By which we mean that the cardinality of $\mathbb{R}$ is strictly larger than the cardinality of $\mathbb{N}$.

So we are assuming that on the left of the arrows is every integer, and on the right of the arrows is every number in the interval $(0, 1)$, and they are just paired up in some way. (And note that each a_{ij} is some digit, from 0 to 9.) This proof is due to Georg Cantor and his next idea is quite brilliant. He said, focus now on the "diagonal" of the above. That is, focus on the numbers of the form a_{ii}.

$$1 \leftrightarrow 0.\boxed{a_{11}}\, a_{12}\, a_{13}\, a_{14}\, a_{15}\, a_{16}\, a_{17}\, a_{18}\, \cdots$$
$$2 \leftrightarrow 0.a_{21}\, \boxed{a_{22}}\, a_{23}\, a_{24}\, a_{25}\, a_{26}\, a_{27}\, a_{28}\, \cdots$$
$$3 \leftrightarrow 0.a_{31}\, a_{32}\, \boxed{a_{33}}\, a_{34}\, a_{35}\, a_{36}\, a_{37}\, a_{38}\, \cdots$$
$$4 \leftrightarrow 0.a_{41}\, a_{42}\, a_{43}\, \boxed{a_{44}}\, a_{45}\, a_{46}\, a_{47}\, a_{48}\, \cdots$$
$$5 \leftrightarrow 0.a_{51}\, a_{52}\, a_{53}\, a_{54}\, \boxed{a_{55}}\, a_{56}\, a_{57}\, a_{58}\, \cdots$$
$$6 \leftrightarrow 0.a_{61}\, a_{62}\, a_{63}\, a_{64}\, a_{65}\, \boxed{a_{66}}\, a_{67}\, a_{68}\, \cdots$$
$$7 \leftrightarrow 0.a_{71}\, a_{72}\, a_{73}\, a_{74}\, a_{75}\, a_{76}\, \boxed{a_{77}}\, a_{78}\, \cdots$$
$$8 \leftrightarrow 0.a_{81}\, a_{82}\, a_{83}\, a_{84}\, a_{85}\, a_{86}\, a_{87}\, \boxed{a_{88}}\, \cdots$$
$$\vdots$$

All real numbers were supposed to be paired up, but we are now going to create a real number that was not in that above list, and we will do so using the idea from the Game of Deduction. The new real number will be different than the first number in its 1st position, different than the second number in its 2nd position, and so on. The number will have decimal expansion

$$b \;=\; 0.b_1\, b_2\, b_3\, b_4\, b_5\, b_6\, b_7\, b_8\, \cdots$$

where $b_i \neq a_{ii}$ for all i. To keep it simple, let's just choose

$$b_i := \begin{cases} 1 & \text{if } a_{ii} \neq 1 \\ 2 & \text{if } a_{ii} = 1 \end{cases}.$$

Then notice that, although clearly $b \in (0, 1)$, b is nowhere in our list! We know b is not the number paired up with 1 because b and that number are different in the first position ($b_1 \neq a_{11}$). We know b is not paired up with 2 because b and that number are different in the second position ($b_2 \neq a_{22}$). In general, we know b is not paired up with k because b and that number are different in the kth position ($b_k \neq a_{kk}$). So this real number b is not anywhere to be found! Thus we have reached a contradiction; clearly we were unable to pair up all the reals, if b got left out.[9] $\square$

I want to quickly mention one other proof of Theorem 2.9 that I particularly enjoy, because it is visual and uses geometry. Imagine you were trying to map $\mathbb{N}$ on to the real number line (which clearly has $|\mathbb{R}|$ points).

---

[9]Don't feel too bad for b, though; it's in fact the case that "most" real numbers were left out. More on that soon.

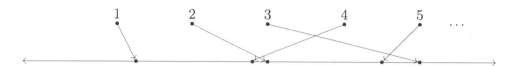

We aim to show that it is impossible for this map to be a bijection — there must be points on the real line that were missed. But instead of mapping just these points, let's make our job slightly harder. Around 1, let's put a little interval of length $\frac{1}{2}$. Around 2, let's put a little interval of length $\frac{1}{4}$. Around 3, let's place a little interval of length $\frac{1}{8}$. And so on. Now, when you map the points in $\mathbb{N}$ to the real line, send over the intervals too (possibly some intervals will overlap; this is ok). We'll now prove that not only are there points on the real line that weren't mapped to, but there are even points that these intervals don't cover!

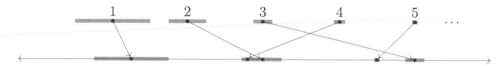

See what happened? Our intervals' lengths add up to $\frac{1}{2} + \frac{1}{4} + \frac{1}{8} + \cdots = 1$ (and with overlaps, their collective length when mapped to the real line may even be smaller than 1). But the whole real line has length ∞! So certainly there is no chance that all the points are covered; not only do the points of $\mathbb{N}$ not cover the line, but even if we fatten them up with these intervals, those intervals don't even cover the real line! And any point that's in the "$\infty - 1$" portion of the real line that is not covered by an interval was certainly not mapped to. So we have all sorts of points that were missed, and so the mapping is far from being a bijection.

(And we could instead pick intervals that add up to 0.0001, or any other tiny number. This proof therefore provides a visualization of how we aren't just missing the one point that Cantor's diagonalization argument finds, but "most" points are missed.)

Now here is the moral of Theorem 2.9, and its two proofs above.

The Moral. There are different sizes of infinity, and $|\mathbb{N}|$, $|\mathbb{Z}|$ and $|\mathbb{Q}|$ are all a smaller infinity than $|\mathbb{R}|$.[10]

The smallest infinity is $|\mathbb{N}|$ (Exercise 2.16), and since these are the counting numbers this infinity is called the *countable infinity*.

> ### Definition.
>
> **Definition 2.10.** If S is an infinite set, then S is *countable* if $|S| = |\mathbb{N}|$. Otherwise S is *uncountable*.

---

[10]Make sure you take a moment to appreciate how remarkably, wonderfully weird this is.

Notice that A being a countable set means there is a bijection $f : \mathbb{N} \to A$ where $f(1), f(2), f(3), f(4), \ldots$ is a listing of all elements in A. Therefore an infinite set A being countable is equivalent to being able to write $A = \{a_1, a_2, a_3, a_4, \ldots\}$.

Above we demonstrated a proof of the following.

> ### Theorem.
>
> **Theorem 2.11** (*Sizes of infinity*). There are different sizes of infinity, with countable infinity being the smallest. Moreover, $\mathbb{N}$, $\mathbb{Z}$ and $\mathbb{Q}$ are countable while $\mathbb{R}$ is uncountable.

Unprovable statements

We now know that $|\mathbb{N}| < |\mathbb{R}|$. Here's one natural question: Is there any infinity between these two? An astounding fact is that, based on the axioms of set theory (called ZFC), whether or not there exists such an infinity is *unprovable*. And what I don't mean is that mathematicians are not smart enough to find the answer; no, I mean that they *are* smart enough to have shown that *no proof can possibly exist*. That's right, there are statements in math which are impossible to prove and also impossible to disprove (but we *are* able to prove that they are unprovable, amazingly).

In fact, the question above is one of the most famous questions in mathematical history. It was asked by Georg Cantor and is called *the continuum hypothesis*. It is the very first of "Hilbert's problems," which are a collection of 23 problems that David Hilbert published in 1900 and presented in-part to the Paris conference of the International Congress of Mathematicians. They were selected as the most important problems for mathematicians to answer in the 20th century. It wasn't for another 40 years that Kurt Gödel proved his Earth-shaking Incompleteness Theorems — demonstrating that nearly every mathematical theory contains unprovable statements — that Cantor's question was "resolved."[11]

---

[11] Currently, 9 of the 23 problems have been decisively resolved (woo!), 2 have been resolved provided "it's unprovable" or "it's impossible" counts (see: Gödel's incompleteness theorem), 7 have been partially resolved (woo!), 3 remain completely unresolved (#Goals), and 2 are vaguely stated and hence hard to classify (¯_(ツ)_/¯). Many have driven research mathematics for over a century, and others (including Number 8 — better known as the Riemann Hypothesis) may drive mathematical research for centuries longer.

Here is the continuum hypothesis, formally stated.

Unprovable.

Unprovable 2.12 (*The continuum hypothesis*). There is no set whose cardinality is strictly between that of the naturals and the reals.

$$|\mathbb{N}| < |S| < |\mathbb{R}|.$$

But how does one prove that a statement is unprovable? First, we have to be careful with our language here. To say something is *unprovable* means that, within the axioms of our particular theory, the statement can neither be proven nor disproven. And so, since all proofs ultimately can be traced back and deduced from just the axioms, whether something is provable relies fundamentally on the chosen axioms.

That said, we are not picking axioms at random; the fundamental axioms that nearly all of mathematics is built on give unprovable statements. Before we discuss those, though, as a smaller, more concrete example let's consider the axioms for geometry from Euclid's *Elements*:

Axiom 1. Any two points can be joined by a straight line

Axiom 2. Any line segment can be extended to form a line.

Axiom 3. For any point P and choice of radius r we can form a circle centered at P of radius r.

Axiom 4. All right angles are equal to one another.

Axiom 5 (*The parallel postulate*). If ℓ is a straight line and P is a point not on the line ℓ, then there is at most one line that passes through P and is parallel to ℓ.

From these axioms, the first theorems of geometry were rigorously proven.[12] The first four axioms you probably agree are 100% obviously true and are fine to be assumed. However, if you are only 99% sure that the fifth axiom — called the parallel postulate — is obviously true, then you're not alone. The fact that the lines trail off to infinity have left countless mathematicians at least 1% uncomfortable, and over the last two millennia mathematicians have hoped that we could do away with the parallel postulate.

The problem is that it's an important property and many of our theorems rely on it. So we don't want to scrap it; ideally, we would like to deduce it from the other

---

[12]Well, almost. There are some subtle things that Euclid technically should have included as an axiom. For example, if there is a point inside a circle and one draws a line through that point, then that line must intersect the circle (twice). Things like this were eventually cleaned up a couple millennia later by David Hilbert.

four. If the first four axioms imply the parallel postulate, then the first theorem of geometry can simply be the parallel postulate. Then all of our other theorems would still hold just the same, that 1% doubt would be removed, and our axioms would be rock solid — as they should be.

As a theorem, though, it was shown that the parallel postulate is unprovable. But how was this done? The approach (discovered in the 19th century) is to show both of these:

- The five axioms above are consistent (they don't self-contradict).

- These five axioms are also consistent: Axioms 1-4 from above, along with an axiom that says "If ℓ is a straight line and P is a point not on the line ℓ, then there are <u>at least two</u> lines that pass through P and are parallel to ℓ."[13]

Basically, if you can show that the assumption that the parallel postulate is true leads to no contradictions (which means you can't *disprove* it, based on the first four axioms), and the assumption that the parallel postulate is false leads to no contradictions (which means you can't *prove* it, based on the first four axioms), then it must be the case that the parallel postulate is neither provable nor disprovable! Its truth value is completely independent of your axioms.

So, if your "theory" of geometry is based on just the first four axioms, then the above is an approach to show that the parallel postulate is an unprovable statement. In particular, the classic Euclidean geometry you learned in grade school is a consistent theory based on the first bullet above, and what is now called *hyperbolic geometry* is a consistent theory based on the second bullet point above (where we assume that for every line there are multiple other lines parallel to it).[14]

The continuum hypothesis was settled in this same way. As mentioned, ZFC set theory is the most fundamental collection of axioms in mathematics, which nearly all of math is based upon. And Kurt Gődel gave a consistent structure satisfying the axioms of ZFC set theory in which the continuum hypothesis was assumed to be true, while Paul Cohen gave a consistent structure satisfying the axioms of ZFC set theory in which the continuum hypothesis was assumed to be false. Combined, this proves that the continuum hypothesis is neither provable nor disprovable (in ZFC). That is, the continuum hypothesis — which asks what is presumably a basic question about the infinite — is unprovable.[15]

---

[13]Think about this as: Axioms 1-4 from before, as well as a fifth axiom that directly contradicts the previous Axiom 5.

[14]By the way, when someone asks you whether this crazy theoretical math is good for anything in the real world, remind them that although mathematicians developed their theories of hyperbolic geometry simply because they were curious and thought it was fun, decades later Albert Einstein later asserted in his theory of relativity that the the fabric of spacetime more closely resembles hyperbolic space than Euclidean space, and he was quite grateful that some mathematicians decades ago had gotten curious and thought it would be fun to develop this abstract geometric theory.

[15]A STRANGE THEOREM
THE ONLY WINNING MOVE IS
NOT TO PLAY.

2.3   How Many Infinities Are There?

We now know of two distinct infinities, but just how many infinities are there? The first theorem in this direction is the power set theorem. Recall that the power set $\mathcal{P}(A)$ of a set A is the set of all subsets of A.

Theorem.

Theorem 2.13 ($|A| < |\mathcal{P}(A)|$). If A is a set and $\mathcal{P}(A)$ is the power set of A, then
$$|A| < |\mathcal{P}(A)|.$$

Proof. Assume for a contradiction that $|A| \geq |\mathcal{P}(A)|$. That is, assume that there is a surjection f from A to $\mathcal{P}(A)$.[16] Since f is a surjection, for every $T \subseteq A$, there is some element $t \in A$ where $f(t) = T$. To reach our contradiction, we will construct a set $B \subset A$ which is not hit.

For each a there is one special property about the set $f(a)$ that we are going to care about: Is $a \in f(a)$ or is $a \notin f(a)$? (In the footnote, for instance, $3 \notin f(3)$ while $4 \in f(4)$.) In general, consider the set of all elements a such that $a \notin f(a)$, and call this set B:
$$B = \{a \in A : a \notin f(a)\}.$$

By the above, if we can show that that there is no b where $f(b) = B$, then we are done; we will have discovered an element of $\mathcal{P}(A)$ that was not hit by f, a contradiction.

<u>Claim.</u> There is no $b \in A$ such that $f(b) = B$.

<u>Proof of claim.</u> Assume for a contradiction that there does exist some $b \in A$ such that $f(b) = B$. Note by the definition of B that

$$b \in B \text{ if and only if } b \notin f(b).$$

But since we assumed that $f(b) = B$, this is equivalent to

$$b \in f(b) \text{ if and only if } b \notin f(b),$$

which is clearly a contradiction. $\square$

---

[16]For instance, if $A = \mathbb{N}$, then the domain of f is $\mathbb{N}$ and the the range of f is $\mathcal{P}(\mathbb{N})$, which is the collection of all subsets of $\mathbb{N}$. For example, maybe $f(3) = \{2, 5, 756\}$ and $f(4) = \{2, 4, 6, 8, 10, 12, \dots\}$. In general we only know that, for each $a \in A$, $f(a)$ is some *subset* of A.

Now, returning to the question about how many infinities are there, we can see from the above theorem that there are infinitely many infinities!

Super Cool Corollary.

Corollary 2.14 (*There exist infinitely many infinities*)**.** There exist infinitely many distinct infinite cardinalities.

Proof. By Theorem 2.13, the following is a chain of distinct infinite cardinalities

$$|\mathbb{N}| < |\mathcal{P}(\mathbb{N})| < |\mathcal{P}(\mathcal{P}(\mathbb{N}))| < |\mathcal{P}(\mathcal{P}(\mathcal{P}(\mathbb{N})))| < |\mathcal{P}(\mathcal{P}(\mathcal{P}(\mathcal{P}(\mathbb{N}))))| < \dots . \qquad \square$$

Ok, so we can identify infinitely many infinities. But then here's the next question: we have been talking all day[17] about how there are different infinities, so when we say there are *infinitely many* infinities, which infinity is it? By the above there are at least countably many infinities. But is there an uncountable number of infinities? More?

Similar to the continuum hypothesis, there is not even an answer to this question. In some sense, the "number" of infinities is more than *any* infinity. Why? Assume for a contradiction that A is a set which contains a set of each infinite size (a set of size $|\mathbb{N}|$, a set of size $|\mathbb{R}|$, and all the rest), and let B be the union of all of the sets in A. The contradiction comes when we consider $\mathcal{P}(B)$. By Theorem 2.13,

$$|B| < |\mathcal{P}(B)|. \qquad (\maltese)$$

Here's the contradiction: The cardinality $|\mathcal{P}(B)|$ can not possibly be in A. And here's why: If it were equal to $|b|$ for some $b \in A$, then since $b \subseteq B$ we would know that $|b| \leq |B|$. But combining this with $(\maltese)$,

$$|b| \leq |B| < |\mathcal{P}(B)| = |b|,$$

which is impossible. $\qquad \square$

So, in fact, there are more infinities than we can imagine. There are more than infinity of them, for any of the infinities! This seems to tear at the very definition of cardinality. Shouldn't every set have a cardinality? Well, yes. But here's the squeeze: it turns out that there was another contradiction in the above, when referring to A as "the set of all infinities." In fact, such a collection *is not even a set*.

 A day that will live in infinity!

17

If you find this material a little disquieting, you are not alone. When Cantor's theorems were first published a century and a half ago, many of the great mathematicians of the day responded with disgust. Henri Poincaré called it a "grave disease" infecting mathematics, Leopold Kronecker accused Cantor of being a "corrupter of youth," and many Christian theologians thought his work against the notion of a unique infinite was an affront to "God's exclusive claim to supreme infinity."[18] Cantor struggled with this for decades.

On the other hand, if this material interests you, then I applaud you and encourage you to read up on *Russell's paradox*, as it is the next step down a fascinating rabbit hole (which, quite literally, is a bottomless pit of mystery). And you should know that Cantor's legacy has been fully restored. The criticisms of the past have been replaced ten times over with praise and accolades. One early defender was the great David Hilbert. Towards the end of Cantor's career, Cantor was awarded the highly-prestigious Sylvester Medal by the Royal Society for his mathematical research. Some criticized this move, but Hilbert — characteristically ahead of his time — recognized the brilliance and importance of Cantor's work, saying:

"No one shall expel us from the paradise that Cantor has created."

---

[18]Or, less melodramatically:

There was a young fellow from Trinity,
Who took $\sqrt{\infty}$.
But the number of digits
Gave him the fidgets;
He dropped Math and took up Divinity.

—George Gamow

— Notable Exercises —

- Exercise 2.4 deals with a small and subtle point that is commonly assumed without argument. However, it's probably good to point out that such things do formally need justification. In this exercise you are proving that if, say, you know that $|A| = |B|$ and $|B| = |C|$, then you also know that $|A| = |C|$. Having these standard properties means that the equal sign here is an *equivalence relation*.

- The fact that the rationals are both dense and countable has enormous consequences throughout real analysis. Exercise 2.10 is an example of all this, and Exercise 2.13 emphasizes how few rationals there are in comparision to the numbers surrounding them. Related ideas will return importantly in the area of real analysis called *measure theory*. If you get a PhD in math there's a good chance that measure theory will be the first topic you'll cover in your graduate real analysis class.

- Exercise 2.12 highlights an important property of what are called *open sets*, which we will discuss in much greater detail in Chapter 5. See Theorem 5.5

- Exercise 2.16 is notable in that there are many mysteries about sizes infinites that remain either very difficult to solve, or impossible to solve. Indeed, work on the continuum hypothesis has told us that it is unprovable whether $|\mathbb{R}|$ is the second smallest infinity. Even the second smallest is unknowable! But as for the smallest infinity — *that* we do know. And this exercise determines the answer.

- Exercise 2.22 points out that most numbers we encounter in our lives — or which come to mind when try to think about a random number — are ridiculously well-behaved. As far as the real numbers go, conditions like rational and algebraic are the rarest of exceptions, rather than the rule.

— Exercises[19] —

Note: Hints and solutions to select exercises can be found at `LongFormMath.com`.

Exercise 2.1.

(a) List all the elements of $\mathcal{P}(\{a, b, c\})$.

(b) Determine a formula for the number of elements in the power set of an n-element set.

Exercise 2.2. Prove that $|\{e^n : n \in \mathbb{N}\}| = |\mathbb{N}|$.

Exercise 2.3. The following pairs of sets have the same size, and so there exists a bijection between them. Write down an explicit bijection in each case. You do not need to prove your answers.

(a) $(0, \infty)$ and $(1, \infty)$

(b) $(0, \infty)$ and $(-\infty, 3)$

(c) $(0, \infty)$ and $(0, 1)$

(d) $\mathbb{R}$ and $(0, \infty)$

(e) $\mathbb{R}$ and $(0, 1)$

(f) $\mathbb{Z}$ and $\{\ldots, \frac{1}{8}, \frac{1}{4}, \frac{1}{2}, 1, 2, 4, 8, \ldots\}$

(g) $\{0, 1\} \times \mathbb{N}$ and $\mathbb{N}$

(h) $[0, 1]$ and $(0, 1)$

Exercise 2.4. This problem shows that "equinumerosity is an equivalence relation." (This justifies the notation $|A| = |B|$.) Let A, B, and C be sets. For this problem only, we'll write $A \sim B$ to mean that A and B are equinumerous, meaning that there is a bijection $A \to B$.

(a) Show that $A \sim A$.

(b) Show that if $A \sim B$ then $B \sim A$.

(c) Show that if $A \sim B$ and $B \sim C$, then $A \sim C$.

Exercise 2.5.

(a) Prove that if A and B are countable sets, then $A \cup B$ is also a countable set.

(b) Prove that if A_n is a countable set for each $n \in \mathbb{N}$, then the set $\bigcup_{n=1}^{\infty} A_n$ is also countable.

---

[19] *Mo' chapters, mo' problems*

Exercise 2.6. Show that $|\mathbb{N}| = |\mathbb{Z}|$ by finding an explicit bijection from $\mathbb{N}$ to $\mathbb{Z}$. You do not need to prove your bijection works.

Exercise 2.7. Let $A, B \subseteq \mathbb{R}$. We define

$$A \cdot B = \{a \cdot b : a \in A \text{ and } b \in B\}.$$

(a) Give an example of sets A_1 and B_1 where $|A_1 \cdot B_1| < \max\{|A_1|, |B_1|\}$.

(b) Give an example of sets A_2 and B_2 where $|A_2 \cdot B_2| > \max\{|A_2|, |B_2|\}$.

(c) Give an example of sets A_3 and B_3 where $|A_3 \cdot B_3| = \max\{|A_3|, |B_3|\}$.

For which of the above does there exist an example where one or both of the sets are infinite?

Exercise 2.8.

(a) Describe a way to partition the set $\mathbb{N}$ into 6 subsets, each containing infinitely many elements

(b) Describe a way to partition the set $\mathbb{N}$ into infinitely many subsets, each containing infinitely many elements.

Exercise 2.9. Is $|\mathbb{Z} \times \mathbb{N}|$ countable or uncountable?

Exercise 2.10. Let S be the set of sequences (a_n) where, for each n, $a_n \in \{0, 1\}$. Is S countable or uncountable?

Exercise 2.11. Suppose that X is a nonempty set. Prove that the following three assertions are equivalent:

(a) X is finite or countably infinite.

(b) There is a one-to-one function $f : X \to \mathbb{N}$.

(c) There is an onto function $g : \mathbb{N} \to X$.

Exercise 2.12.

(a) Give an example of a collection of countably many disjoint open intervals, or prove that this does not exist.

(b) Give an example of a collection of uncountably many disjoint open intervals, or prove that this does not exist.

Exercise 2.13. Show that there are uncountably many irrational numbers.

Exercise 2.14. Prove that $\mathbb{N} \times \mathbb{N}$ is countably infinite by showing that the function $f : \mathbb{N} \times \mathbb{N} \to \mathbb{N}$ defined by $f(m, n) = 2^{n-1}(2m - 1)$ is a bijection.

Exercise 2.15. Let $\mathcal{F}$ be the collection of all functions $f : \mathbb{R} \to \mathbb{R}$. Prove that $\mathcal{F}$ is uncountable.

Exercise 2.16. Show that the smallest infinity is $|\mathbb{N}|$. That is, show that if $A \subseteq \mathbb{N}$, then either A is finite or $|A| = |\mathbb{N}|$.

Exercise 2.17. Prove that the set of all finite subsets of $\mathbb{N}$ is countable.

Exercise 2.18. Is the subset of rational numbers

$$\left\{ \frac{m}{n} : m, n \in \mathbb{Z} \text{ and } 1 \leq n \leq 10 \right\}$$

dense in $\mathbb{R}$?

Exercise 2.19. Let A be the set of polynomials with rational coefficients. Prove that $|A|$ is countable.

Exercise 2.20. Show that $|\mathcal{P}(\mathbb{N})| = |\mathbb{R}|$ by finding an explicit bijection from $\mathcal{P}(\mathbb{N})$ to $\mathbb{R}$. You do not need to prove your bijection works.

Exercise 2.21. Prove that the set of points on the unit circle in $\mathbb{R}^2$ (that is, $\{(x, y) : x^2 + y^2 = 1\}$) is uncountable.

Exercise 2.22. A real number x is said to be *algebraic* (over the rationals) if it satisfies some polynomial equation (of positive degree)

$$a_n x^n + a_{n-1} x^{n-1} + a_{n-2} x^{n-2} + \cdots + a_1 x + a_0 = 0$$

where each $a_i \in \mathbb{Q}$. If a real number is not algebraic, then it is *transcendental*.

(a) Prove that there are countably many algebraic numbers. (You may use the fundamental theorem of algebra which says that a polynomial with degree n has at most n real roots.)

(b) Prove that there are uncountably many transcendental numbers.

— Open Questions —

For subsets X and Y of $\mathbb{N}$, we say that X *splits* Y if both $Y \cap X$ and $Y \setminus X$ are infinite. A family F of subsets of $\mathbb{N}$ is *unsplittable* if no single subset of $\mathbb{N}$ splits every set in F; moreover, F is σ-unsplittable if even countably many sets don't suffice to split every member of F. (As an exercise, show that no countable family can be unsplittable.)

Question 1. Must the least size of an unsplittable family equal the least size of a σ-unsplittable family?

Notice that every unsplittable family is an infinite subset of the power set $\mathcal{P}(\mathbb{N})$, and under the continuum hypothesis all such subsets are countable or have size $|\mathbb{R}|$. Therefore, by applying the exercise, a positive answer to Question 1 would be directly provable, while a negative answer would have to be obtained in a universe where the continuum hypothesis fails.

Quesiton 1 is one of many questions about what you might call *almost-countable cardinals* but are typically called *cardinal characteristics of the continuum*.[20]

The next two open questions involve the *Axiom of Choice*, which is the innocuous assertion that for every family X of nonempty sets there is a function f assigning to each $x \in X$ a member of x (that is, $f(x) \in x$).[21] At the beginning of the 20th Century, the Axiom of Choice inspired some controversy. Today, the Axiom of Choice is mostly accepted by mathematicians, though it is still worthwhile in logic and set theory to understand when uses of it are necessary.

For sets X and Y, write $|X| \leq |Y|$ if there is a one-to-one function from X to Y. And write $|X| \leq^* |Y|$ if there is an onto function from Y to X (or if X is empty). In the presence of the Axiom of Choice, the orderings $\leq$ and $\leq^*$ are equivalent.

Question 2. Suppose that for all sets X and Y, the implication $|X| \leq^* |Y|$ implies $|Y| \leq |X|$ holds. Must the Axiom of Choice be true?

Question 3. Assuming the Axiom of Choice, there is no infinite sequence of sets strictly decreasing in cardinality:

$$|X_1| > |X_2| > |X_3| > \cdots .$$

(This is not obvious!) Now here's the open question: Does the non-existence of such a sequence imply the Axiom of Choice?

---

[20]For further reading, check out the delightfully-titled *Combinatorial cardinal characteristics of the continuum* by Andreas Blass.

[21] To make sure you understand this axiom, consider the difference between the cases (i) when every member of X is a set of integers and (ii) when every member of X is a set of real numbers. The Axiom of Choice for sets of integers is a theorem of set theory with the Axiom of Choice removed, but the Axiom of Choice for sets of real numbers is not.

Chapter 3: Sequences

3.1 Basic Sequence Definitions

The next major topic is that of sequences. You probably intuitively think of a sequence as an infinite list of numbers, like

$$1 \ , \ 4 \ , \ 9 \ , \ 16 \ , \ 25 \ , \ \ldots$$

This is a good way to think about them, but formally a sequence is defined to be a function.[1] If you think about mapping 1 to 1, 2 to 4, 3 to 9, and, in general, n to the nth number in the list, then we get the picture:

$$1 \ , \ 2 \ , \ 3 \ , \ 4 \ , \ 5 \ , \ \ldots$$

$$\downarrow \quad \downarrow \quad \downarrow \quad \downarrow \quad \downarrow$$

$$1 \ , \ 4 \ , \ 9 \ , \ 16 \ , \ 25 \ , \ \ldots$$

This gives us the formal definition of a sequence.

> **Definition.**
>
> **Definition 3.1.** A *sequence* of real numbers is a function $a : \mathbb{N} \to \mathbb{R}$.

So in the above example we would have

$$a(1) = 1, \quad a(2) = 4, \quad a(3) = 9, \quad a(4) = 16, \quad a(5) = 25, \ldots$$

and in general $a(n) = n^2$. As you have probably seen, it is more common to write this using subscript notation:

$$a_1 = 1, \quad a_2 = 4, \quad a_3 = 9, \quad a_4 = 16, \quad a_5 = 25, \quad \ldots.$$

---

[1] From a theoretical standpoint this is good thing, as it means we do not need to formally develop a new mathematical object in order to discuss sequences.

We will also often write sequences like this, just don't forget that a sequence is really a function from the $\mathbb{N}$ to $\mathbb{R}$ where $a(n)$ equals the nth number in the sequence.

> **Note.**
>
> **Note 3.2.**
>
> - As mentioned, oftentimes we write a_n instead of $a(n)$.
>
> - Sometimes we write $(a_n)_{n=1}^{\infty}$ or (a_n) or $(a_1, a_2, a_3, \dots)$ or $a_1, a_2, a_3, \dots$ to denote the sequence $a(n)$.
>
> - Sometimes we will give an explicit formula for a_n, like
>
> $$a_n = n^2$$
>
> while other times we may give it a recursive definition, like with the Fibonacci sequence:
>
> $$a_n = a_{n-1} + a_{n-2}$$
>
> with the initial conditions of $a_1 = 1$ and $a_2 = 1$. But of course, *any* function from $\mathbb{N}$ to $\mathbb{R}$ gives a sequence, not just ones with nice formulas. For example we could have
>
> $$\left(14.238, 7, \pi, -\sqrt{2}, e, 0, e^{\pi}, \dots\right).$$

Example 3.3.

(a) Write the sequence $a(n) = 2^{n-1}$ in the form $(a_1, a_2, a_3, \dots)$

(b) For the sequence

$$(3, 6, 11, 18, 27, \dots),$$

determine a formula for a_n.

(I stuck the solution in a footnote, in case you want to try it without accidentally glancing at the solution.[2])

3.2 Bounded Sequences

As is often the case in math, we want to describe how "nice" something is. In calculus, polynomials were quite nice because you knew how to differentiate and integrate them. Oftentimes you could even factor them. Exponential functions like $f(x) = 2^x$ were pretty nice, and logs weren't soooo bad. Likewise, we want to describe what it means for a sequence to be "nice."

We start with the notion of a sequence being *bounded*, which is one nice property.

---

[2] **Solution.** (a) $(1, 2, 4, 8, 16, \dots)$. (b) One answer: $a_n = n^2 + 2$. □

> **Definition.**
>
> **Definition 3.4.** A sequence (a_n) is *bounded* if the range $\{a_n : n \in \mathbb{N}\}$ is bounded. I.e., if there exists a lower bound $L \in \mathbb{R}$ and an upper bound $U \in \mathbb{R}$ where
>
> $$L \leq a_n \leq U$$
>
> for all n.

Here's a fact that follows quickly from definitions.

> **Proposition.**
>
> **Proposition 3.5** (*Bounded* $\Leftrightarrow |a_n| \leq C$). A sequence (a_n) is bounded if and only if there exists some $C \in \mathbb{R}$ for which $|a_n| \leq C$ for all n.

Proof. Recall again that boundedness means that there exists a lower bound $L \in \mathbb{R}$ and an upper bound $U \in \mathbb{R}$ where

$$L \leq a_n \leq U$$

for all n. Now let's prove each direction.

First we will prove the backwards implication. Assume that there exists a $C \in \mathbb{R}$ where

$$|a_n| \leq C.$$

Then

$$-C \leq a_n \leq C.$$

And so, by setting $L = -C$ and $U = C$ we have shown that

$$L \leq a_n \leq U,$$

which means that a_n is bounded. This concludes the backward direction.

Now we will prove the forward implication. If a_n is bounded, then there exists such an L and U. Let $C = \max\{|L|, U\}$.[3] Note that this implies that $C \geq U$ and (since $C \geq |L|$, that) $-C \leq -|L|$. Thus, for all n we have

$$-C \leq -|L| \leq L \leq a_n \leq U \leq C.$$

So we see that

$$-C \leq a_n \leq C,$$

which by Proposition 1.12 is the same as $|a_n| \leq C$. $\qquad\square$

---

[3]See pictures after this example for intuition on where this came from.

One way to think about the forward direction is to note that what "$|a_n| \leq C$ for all n" means is that you can draw a circle of radius C, center at 0, and catch all of the a_n points inside your circle. Below are examples of this.

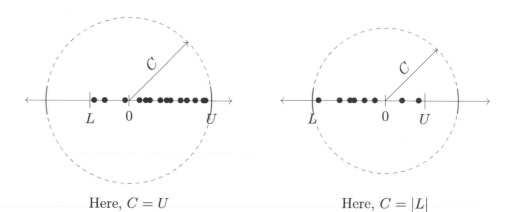

Here, $C = U$ Here, $C = |L|$

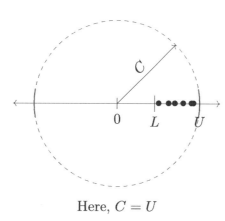
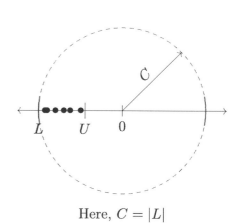

Here, $C = U$ Here, $C = |L|$

Example 3.6. Here are a couple examples.

- Let $a_n = (-1)^n \dfrac{1}{n}$. Then by Proposition 3.5, the sequence

$$(a_n) = (-1, \frac{1}{2}, -\frac{1}{3}, \frac{1}{4}, -\frac{1}{5}, \dots)$$

is bounded because $|a_n| \leq 1$ for all n.

- Let $a_n = 2 + \sin(n)$. Then the sequence

$$(a_n) = (2 + \sin(1), 2 + \sin(2), 2 + \sin(3), 2 + \sin(4), \dots)$$

is bounded. To see this, recall that $-1 \leq \sin(n) \leq 1$ for all n, and so $1 \leq 2 + \sin(n) \leq 3$ for all n. At this point we are done by the definition of boundedness. Or we could note that $|a_n| \leq 3$ for all n, and so we are done by Proposition 3.5. $\qquad\square$

3.3 Convergent Sequences

"The sequence $1, \frac{1}{2}, \frac{1}{3}, \frac{1}{4}, \frac{1}{5}, \ldots$ converges to 0" is a true statement, and after taking calculus you probably already have a pretty good understanding of what that statement means. But real analysis is an advanced math course, so we need to be precise about what *convergence* means. The following provides that precision.

Definition.

Definition 3.7. A sequence (a_n) *converges* to $a \in \mathbb{R}$ if for all $\varepsilon > 0$ there exists[4] some N such that $|a_n - a| < \varepsilon$ for all $n > N$.

When this happens, a is called the *limit* of (a_n).

This is an important definition. This is the first (of many) weird-looking definition(s) in this textbook, and at this stage it is *really* important that you *fully* understand it. So brew a pot of tea, find a comfortable Thinking Chair, and ponder it carefully; don't move on until that pot is emptied. The better you understand it now, the easier the rest of the course will be. Here are a few more comments to help:

<u>Comment 1.</u> The final condition that we are aiming to reach is that $|a_n - a| < \varepsilon$ for appropriate n. Note that this is the same as

$$-\varepsilon < a_n - a < \varepsilon,$$

which is equivalent to

$$a - \varepsilon < a_n < a + \varepsilon.$$

So our sequence has to live inside of this interval after some point N. And we think about ε as being really small. For example, if $a = 6$ and $\varepsilon = 0.01$, then after some point N, we know that *every* term of the sequence satisfies

$$6 - 0.01 < a_n < 6 + 0.01$$
$$5.99 < a_n < 6.01.$$

---

[4]Note: When more convenient, you may assume that $N \in \mathbb{N}$.

<u>Comment 2.</u> Below is the important picture to keep in your head. Note that after the N, all the points are inside the shaded region between $a - \varepsilon$ and $a + \varepsilon$.

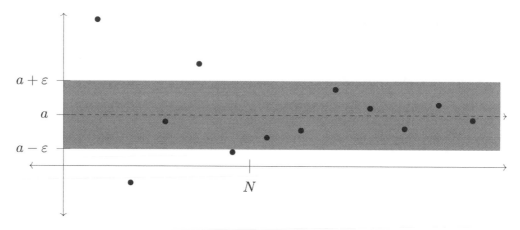

Now, if you picked a *smaller* value of ε, which we will call ε_2, you will need a *bigger* value of N (called N_2) to guarantee that all the points after N_2 are inside the shaded region.

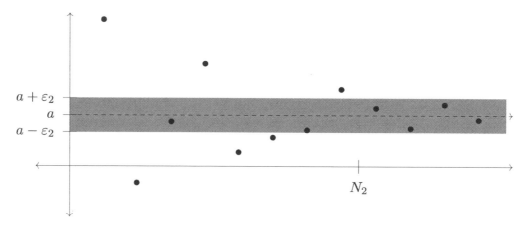

So you see, smaller values of ε typically require larger values of N.

<u>Comment 3.</u> You can think about the converging sequence as the tragic story of a rebellious youth. When young, he may be all over the place on any given day, living his life and getting into mischief. But eventually society settles him down. When $\varepsilon = 100$ (miles) he has a car, and with his gang of misfits he can roam wherever he pleases, causing mayhem. But when $\varepsilon = 10$ (miles), the value of N_1 represents the age at which he is placed on at-risk watch and is not allowed to leave the city (maybe $N_1 = 17$ years old). Then $\varepsilon = 0.1$ (miles) represents the age N_2 at which he really gets into trouble and is placed under house arrest. And $\varepsilon = 0.01$ represents the age N_3 at which he is sentenced to life in prison and will forever more be contained within a small wing of prison.[5]

---

[5]When I started this paragraph I did not anticipate it getting so dark. I even dropped the solitary confinement stage (or beyond) for that reason. Poor guy. Stay in school, kids.

<u>Comment 4.</u> What the definition means is that eventually the points get "arbitrarily close" to a. Maybe the first few points $(a_1, a_2, \ldots, a_{N-1})$ are not super close to a, but eventually they both get super close to a and stay super close to a.

And we really do mean *super* close, as it has to hold for *all* $\varepsilon > 0$. There is an N for which all the points from a_N onward are within 0.1 of a, but there's also some other N_2 for which all the points from a_{N_2} onward are within 0.0000000000000001 of a. *Any* positive value of ε has to work.

<u>Comment 5.</u> If you haven't spent considerable time thinking about this, or if you do not yet understand it *perfectly*, go brew another pot and settle in. You'll thank me later.

The final comment is about notation:

Notation.

Notation 3.8. The following are notationally equivalent.

- The sequence (a_n) *converges* to a

- $a_n \to a$ as $n \to \infty$

- $a_n \to a$

- $\lim_{n \to \infty} a_n = a$

Example 3.9. Show that the sequence

$$(a_n) = \left(1, \frac{1}{2}, \frac{1}{3}, \frac{1}{4}, \frac{1}{5}, \ldots\right)$$

converges to 0.

Solution 1. Fix any $\varepsilon > 0$. By the Archimedean property, there is an N for which $\frac{1}{N} < \varepsilon$. This N has the property that, for all $n > N$, we have $\frac{1}{n} < \frac{1}{N} < \varepsilon$. That is, for all $n > N$,

$$\left|\frac{1}{n} - 0\right| = \frac{1}{n} < \varepsilon.$$

And so, by the definition of sequence convergence (Definition 3.7), we may conclude that $a_n \to 0$. $\qquad\square$

Using the Archimedean property will not usually work, though. We need a more general approach. The fundamental idea behind Solution 2 (and most future solutions to this type of question) can be illustrated with the following.

Fingers-on-the-nose game. This is tough to type out, but here goes nothing.[6] Can you find a way to do this: Stick your left arm forward; stick your right arm forward, crossing your left so that your right wrist is in top of your left wrist, hands palm-to-palm; interlock your fingers; move your hands so that your pinkies are against your chest; stick just your two pointer fingers up so that your left pointer finger's fingernail is against the left side of your nose, and your right pointer finger's fingernail is against the right side of your nose. Now: Unwind your hands without your pointer fingers leaving your nose.

The game's lesson: If you don't know how to begin, then oftentimes by starting at the end one can unwind the whole problem and discover the key for how to begin.[7]

Now here is the outline.

> **Outline.**
>
> **Outline 3.10.** To show that $a_n \to a$, begin with preliminary work:
>
>   0. Scratch work: Start with $|a_n - a| < \varepsilon$ and unravel to solve for n. This tells you which N to pick for step 2 below.
>
> Now for your actual proof:
>
>   1. Let $\varepsilon > 0$.
>
>   2. Let N be the final value of n you got in your scratch work, and let $n > N$.
>
>   3. Redo scratch work (without ε's), but at the end use N to show that $|a_n - a| < \varepsilon$.

---

[6]Find me at a conference if you're confused and I'll show you it. It's fun.

[7] From *Alice in Wonderland*: "'Begin at the beginning,' the King said gravely, 'and go on till you come to the end: then stop.'" This is good advice, provided you can find your way from the beginning to the end. But if you find yourself going in loops without reaching the end, instead try to find your way from the end to the beginning. Sometimes that's much easier, just ask an 8 year old trying to solve one of those pencil maze puzzles.[8]

[8]Fun Fact: Essentially all[9] of those mazes can be solved by simply following the wall on your right wherever it goes.

[9]Technical condition is that it must be "simply connected." If it fails, though, then you may have to start over.[7]

Example 3.9, again. Show that the sequence

$$(a_n) = \left(1, \frac{1}{2}, \frac{1}{3}, \frac{1}{4}, \frac{1}{5}, \cdots\right)$$

converges to 0.

We follow Outline 3.10.

Scratch Work for Solution 2. This second solution is how most of our solutions will go, and it will use Outline 3.10. Given an arbitrary $\varepsilon > 0$, we will find what specific N guarantees that, for every $n > N$, we have $|a_n - 0| < \varepsilon$. For example, if $\varepsilon = \frac{1}{2}$, then $N = 2$ works. If $\varepsilon = \frac{1}{3}$, then $N = 3$ works. You see the pattern, but here is how we might come about it in general:

We want the following:

$$|a_n - a| < \varepsilon$$
$$\left|\frac{1}{n} - 0\right| < \varepsilon$$
$$\frac{1}{n} < \varepsilon$$
$$\frac{1}{\varepsilon} < n.$$

So as long as we choose $N = \dfrac{1}{\varepsilon}$, then for any $n > N$ we will have $n > \dfrac{1}{\varepsilon}$, which by the above will imply that $\frac{1}{n} \to 0$, as desired. The solution below is how we formally solve it.

(As you saw, in the above we essentially did the important steps of Outline 3.10 in reverse order. We started with Step 3, undoing a bunch of algebra to find an N that will work for Step 2.)

Solution 2. Fix any $\varepsilon > 0$. Set $N = \frac{1}{\varepsilon}$. Then for any $n > N$ (implying $\frac{1}{n} < \frac{1}{N}$),

$$|a_n - a| = \left|\frac{1}{n} - 0\right| = \frac{1}{n} < \frac{1}{N} = \frac{1}{1/\varepsilon} = \varepsilon.$$

That is, $|a_n - a| < \varepsilon$. So by Definition 3.7 we have shown that $\dfrac{1}{n} \to 0$. $\qquad\square$

Example 3.11. Let $a_n = 5 - \dfrac{1}{n^2}$. Show that $a_n \to 5$ as $n \to \infty$.

We follow Outline 3.10.

Scratch Work. Again, we first play around. We start with where we want to get to (that $|a_n - a| < \varepsilon$), and then do some algebra to figure out which values of N would give this.

We want the following:

$$|a_n - a| < \varepsilon$$
$$\left| \left(5 - \frac{1}{n^2} \right) - 5 \right| < \varepsilon$$
$$\left| -\frac{1}{n^2} \right| < \varepsilon$$
$$\frac{1}{n^2} < \varepsilon$$
$$\frac{1}{\varepsilon} < n^2$$
$$\frac{1}{\sqrt{\varepsilon}} < n.$$

So as long as we choose $N = \dfrac{1}{\sqrt{\varepsilon}}$, then for any $n > N$ we will have $n > \dfrac{1}{\sqrt{\varepsilon}}$, which by the above will imply that $5 - \frac{1}{n^2} \to 5$, as desired. But here's the formal solution:

Solution. Fix any $\varepsilon > 0$. Set $N = \frac{1}{\sqrt{\varepsilon}}$. Then for any $n > N$,

$$|a_n - a| = \left| \left(5 - \frac{1}{n^2} \right) - 5 \right| = \frac{1}{n^2} < \frac{1}{N^2} = \frac{1}{1/(\sqrt{\varepsilon})^2} = \frac{1}{1/\varepsilon} = \varepsilon.$$

That is, $|a_n - a| < \varepsilon$. So by Definition 3.7 we have shown that $5 - \dfrac{1}{n^2} \to 5$. $\square$

This next one looks a bit trickier, but the same procedure works.

Example 3.12. Let $a_n = \dfrac{3n+1}{n+2}$. Prove that $\lim\limits_{n\to\infty} a_n = 3$.

Scratch Work. Again, we first play around. We start with where we want to get to (that $|a_n - a| < \varepsilon$), and then do some algebra to figure out which values of n would give this.

We want the following:

$$|a_n - a| < \varepsilon$$

$$\left| \frac{3n+1}{n+2} - 3 \right| < \varepsilon$$

$$\left| \frac{3n+1}{n+2} - \frac{3(n+2)}{n+2} \right| < \varepsilon$$

$$\left| \frac{3n+1-3n-6}{n+2} \right| < \varepsilon$$

$$\left| \frac{-5}{n+2} \right| < \varepsilon$$

$$\frac{5}{n+2} < \varepsilon$$

$$\frac{5}{\varepsilon} < n+2$$

$$\frac{5}{\varepsilon} - 2 < n$$

So as long as we choose $N = \dfrac{5}{\varepsilon} - 2$, then for any $n > N$ we will have $n > \dfrac{5}{\varepsilon} - 2$, which by the above will imply that $\frac{3n-1}{n+2} \to 3$, as desired.

Solution. Fix any $\varepsilon > 0$. Set $N = \frac{5}{\varepsilon} - 2$. Then for any $n > N$,

$$|a_n - a| = \left| \frac{3n+1}{n+2} - 3 \right| = \left| \frac{3n+1}{n+2} - \frac{3n+6}{n+2} \right|$$

$$= \frac{5}{n+2} < \frac{5}{N+2} = \frac{5}{\left(\frac{5}{\varepsilon} - 2\right) + 2}$$

$$= \frac{5}{5/\varepsilon} = \varepsilon.$$

That is, $|a_n - a| < \varepsilon$. So by Definition 3.7 we have shown that $\dfrac{3n+1}{n+2} \to 3$. $\qquad\square$

Example 3.13. Let $a_n = 5$. Prove that $a_n \to 5$.

Scratch Work. Since a_n is the constant function 5, this means that $a_1 = 5$, $a_2 = 5$, $a_3 = 5$, etc. That is, this is the sequence $5, 5, 5, 5, 5, \ldots$. It makes perfect sense that $a_n \to 5$, but how do we show it? To use Definition 3.7, we need, for each $\varepsilon > 0$, to find an N for which, for every $n > N$, we have $|a_n - 5| < \varepsilon$. Sometimes in math the "easiest" problems are the hardest to work out, because some of the usual steps are missing... We want to figure out for which n we have $|a_n - a| < \varepsilon$. But:

$$|a_n - a| < \varepsilon$$
$$|5 - 5| < \varepsilon$$
$$0 < \varepsilon.$$

And this is always true, since we chose $\varepsilon > 0$! So if we choose any N at all, like $N = 1$ or $N = \pi$ or $N = 17$, we are guaranteed for it to work out.

Solution. Fix any $\varepsilon > 0$. Set $N = 17$. Then for any $n > N$,

$$|a_n - a| = |5 - 5| = 0 < \varepsilon.$$

That is, $|a_n - a| < \varepsilon$. So by Definition 3.7 we have shown that $5 \to 5$. $\square$

ε-neighborhoods

Related to all this is the notion of an ε-*neighborhood* of a point a.

Definition.

Definition 3.14. Let $\varepsilon > 0$. The ε-*neighborhood* of a point a is the interval

$$(a - \varepsilon, a + \varepsilon).$$

You can think of an ε-neighborhood as the interval centered at a of radius ε:

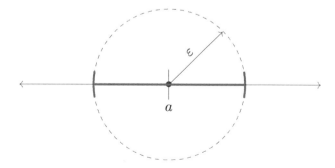

With this, one could rephrase the definition of convergence like so:

Definition.

Definition 3.7. A sequence (a_n) *converges* to $a \in \mathbb{R}$ if for all $\varepsilon > 0$ there exists[10] some N such that a_n is in the ε-neighborhood of a for all $n > N$.

On page 70, an ε-neighborhood (of y-values) is drawn as a shaded horizontal band. Following that figure is a second figure where an ε_2-neighborhood is shaded.

3.4 Divergent Sequences

Of course, not all sequences converge.

Definition.

Definition 3.15. If a sequence (a_n) does not converge, then it *diverges*.

Divergence can come in three forms.

- (a_n) *diverges to* ∞ $\left(\text{notation: } \lim_{n \to \infty} a_n = \infty\right)$ if, for all $M > 0$, there exists[11] some N such that $a_n > M$ for all $n > N$.

- (a_n) *diverges to* $-\infty$ $\left(\text{notation: } \lim_{n \to \infty} a_n = -\infty\right)$ if, for all $M < 0$, there exists[11] some N such that $a_n < M$ for all $n > N$.

- Otherwise, (a_n)'s limit *does not exist*.

Typically the first and second types are more straightforward to show, so we will start with an example of one of those.

Example 3.16. Let $a_n = n^2$. Show that $\lim_{n \to \infty} a_n = \infty$.

Scratch Work. We want

$$a_n > M$$
$$n^2 > M$$
$$n > \sqrt{M}.$$

So setting $N = \sqrt{M}$ should work.

---

[10]Note: When more convenient, you may assume that $N \in \mathbb{N}$.
[11]See Footnote 10.

Solution. Fix any $M > 0$. Set $N = \sqrt{M}$. Then for any $n > N$,

$$a_n = n^2 > N^2 = \left(\sqrt{M}\right)^2 = M.$$

So we have shown that if $n > N$, then $a_n > M$. Therefore $\lim_{n \to \infty} a_n = \infty$. □

What if the sequence's limit does not exist? Then how do we show a sequence diverges?

Note.

Note 3.17. One way to show that a_n diverges is to show that $a_n \not\to a$ for any a. Note first, by Definition 3.7, that "$a_n \to a$" means that

- For <u>every</u> $\varepsilon > 0$ there exists <u>some</u> N such that <u>for all</u> $n > N$ we have $|a_n - a| < \varepsilon$.

So to show that $a_n \not\to a$, we need to show the *negation* of that statement. That is, we must show that

- There exists <u>some</u> $\varepsilon > 0$ where <u>for all</u> N there exists <u>some</u> $n > N$ such that $|a_n - a| \geq \varepsilon$.

Note: In practice this is usually done with a proof by contradiction. You assume that $a_n \to a$ and then you demonstrate a specific ε where it fails, giving the contradiction.

Make sure this makes sense to you. How do you show that something does not happen for all $\varepsilon > 0$? Answer: You find one positive ε for which it fails. It's like saying "Every university in America has a student that can dunk a basketball." How would you show that that's not true? All you must do is find a single university where no student can dunk.[12] You don't have to check them all — as soon as you find one school that has no student that can dunk, you're done. Conversely, to disprove a "there exists" statement, you do have to demonstrate that all of the possibilities fail.

Now here's an example of showing a sequence diverges (of the "does not exist" variety).

Example 3.18. Let $a_n = (-1)^n$. Prove that (a_n) diverges.

Scratch Work. This is the sequence $-1, 1, -1, 1, -1, 1, \ldots$ It makes sense that there is no a for which $a_n \to a$. It certainly doesn't converge to 1, since half the time it is at -1 which is far away from 1. (If we let $\varepsilon = 1/2$, then there is no N for which, for every $n > N$, a_n is inside of the shaded band; see below.)

---

[12]But once you've found such a university, you do have to check all the students, since the negations of "<u>for all</u> universities <u>there exists</u> a student who can dunk" is "<u>there exists</u> a university such that <u>for all</u> of its students, they are unable to dunk."

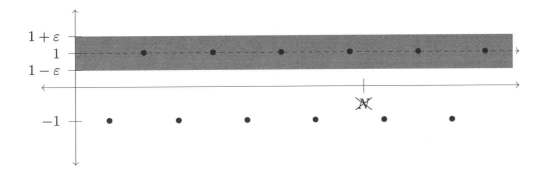

It likewise can't converge to -1. One might guess 0, since that is halfway between -1 and 1, but that also doesn't make sense since a_n is always of distance 1 from 0, so it's certainly not getting "closer and closer" to 0. (Or, $\varepsilon = 1/2$ works again.)

Ok, so we believe that it doesn't converge to anything, and we will use Note 3.17 to show it. We will use a proof by contradiction. We will assume there is some a for which $a_n \to a$, and then following Note 3.17's lead we will find a specific ε for which, for every N, there exists some larger n where a_n is *not* within ε of a. In particular, note that if we choose $\varepsilon = \frac{1}{2}$, then whatever we choose for a the sequence will be farther than $\frac{1}{2}$ from a either half of the time or all of the time.

Solution 1. Assume for a contradiction that there is some a for which $a_n \to a$. Let $\varepsilon = \frac{1}{2}$. Since we assumed that $a_n \to a$, there must be some N for which, for all $n > N$, we have $|a_n - a| < \frac{1}{2}$. That is, $|(-1)^n - a| < \frac{1}{2}$ for all $n > N$. We proceed by cases.

<u>Even n.</u> If n is even and $n > N$, then we have that

$$|1 - a| < \frac{1}{2}.$$

Unwinding this:

$$-\frac{1}{2} < 1 - a < \frac{1}{2}$$
$$-\frac{3}{2} < -a < -\frac{1}{2}$$
$$\frac{1}{2} < a < \frac{3}{2}.$$

<u>Odd n.</u> If n is odd and $n > N$, then we have that

$$|-1 - a| < \frac{1}{2}.$$

Unwinding this:

$$-\frac{1}{2} < -1 - a < \frac{1}{2}$$
$$\frac{1}{2} < -a < \frac{3}{2}$$
$$-\frac{3}{2} < a < -\frac{1}{2}.$$

But this is a contradiction; clearly no a can be inside of both $(\frac{1}{2}, \frac{3}{2})$ and $(-\frac{3}{2}, -\frac{1}{2})$. And so we must have $a_n \not\to a$. $\qquad\square$

Essentially, focusing on the even case created a ball around 1, of radius $\frac{1}{2}$, which a_n would have to live within for all $n > N$. The odd case created a ball around -1, of radius $\frac{1}{2}$, which a_n would have to live within for all $n > N$. But these two balls are disjoint, so a_n can't live in both, creating a contradiction.

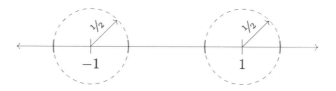

Now here's a second proof of, which is mostly black magic (AKA the triangle inequality).[13]

Solution 2. Again, let $\varepsilon = \frac{1}{2}$. The even case still gives $|1 - a| < \frac{1}{2}$, and the odd case still gives $|-1 - a| < \frac{1}{2}$; although in the odd case we will rewrite $|-1 - a|$ as $|1 + a|$. Then, by the triangle inequality

$$2 = |(1 - a) + (1 + a)| \leq |1 - a| + |1 + a|.$$

But each of the absolute values on the right are assumed to be less than $\frac{1}{2}$. So,

$$2 = |(1 - a) + (1 + a)| \leq |1 - a| + |1 + a| < \frac{1}{2} + \frac{1}{2} = 1.$$

So $2 < 1$? Preposterous! We have our contradiction. $\qquad\square$

---

[13]Ignore this if you like to see magic tricks without knowing how they're done: Intuitively, what the following proof is actually saying is that if $(-1)^n \to a$ (where a is some fixed number), then eventually all the points of the sequence must be within $\varepsilon = 0.5$ of whatever a is. However, the distance from -1 to 1 is equal to 2, and so the distance from -1 to a, plus the distance from a to 1, should be at least 2. But since $\varepsilon = 0.5$, those two latter distances are too small, giving a contradiction.

Next is something that you probably already assumed was true and wouldn't have thought we needed to prove, but when building a theory we shouldn't assume anything.

> **Proposition.**
>
> **Proposition 3.19** (*Limits are unique*). A sequence can not have more than one limit.

Scratch Work. The idea in the solution of Example 3.18 should be able to be tweaked to work here. The idea is to assume that you have a sequence a_n that converges to some a and also converges to some other number b. Since it converges to a, after some point all the sequence points are within ε of a, for some really tiny ε. But since it also converges to b, the same thing must happen there too. So if ε is small enough that those two regions are mutually exclusive, we will have a contradiction.

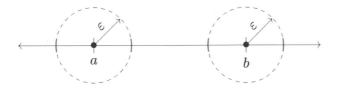

And this is how we will reach a contradiction: there is no way for a_n $(n > N)$ to be in both circles at the same time. Of course, if a and b are really close together we will need to choose a smaller ε to make sure the intervals remain disjoint, but that's the only difference.[14]

For instance, if the radius of those balls is a third of the distance between a and b, it should work out. With that intuition in mind, here's the proof.

---

[14]By now it has probably been ingrained in you that ε is always a small number; this will probably follow you throughout your mathematical career. Paul Erdős was one of the most interesting mathematicians in history. He published over 1,500 papers in his life — more than any mathematician in history — and yet never held a permanent job. He didn't want one. In an age before air travel was common, he would constantly be bouncing between universities and the homes of collaborators. He also had an interesting vocabulary. Alcohol was called "poison." Women were "bosses," and when a man marries a woman he has been "captured." People who retired from research mathematics had "died" (this one caused some terrible miscommunications), while those who actually did die had "left." And me telling you all that is just lead-up to what he called kids. He loved kids, but because they are small, he called them "epsilons."

Proof. Suppose for a contradiction that $a_n \to a$ and $a_n \to b$, where $a \neq b$; moreover, without loss of generality let's assume $a < b$. Let $\varepsilon = (b-a)/3 > 0$. Since $a_n \to a$ there exists some N_1 such that for $n > N_1$ we have $|a_n - a| < (b-a)/3$. Likewise, since $a_n \to b$ there exists some N_2 such that for $n > N_2$ we have $|a_n - b| < (b-a)/3$. Let $N = \max\{N_1, N_2\}$. Then for $n > N$ (implying $n > N_1$ and $n > N_2$), we have both $|a_n - a| < (b-a)/3$ and $|a_n - b| < (b-a)/3$. That is, for such n we have

$$-\frac{b-a}{3} < a_n - a < \frac{b-a}{3} \qquad \text{and} \qquad -\frac{b-a}{3} < a_n - b < \frac{b-a}{3}.$$

i.e.,

$$a - \frac{b-a}{3} < a_n < a + \frac{b-a}{3} \qquad \text{and} \qquad b - \frac{b-a}{3} < a_n < b + \frac{b-a}{3}.$$

In particular,

$$a_n < \frac{b+2a}{3} \qquad \text{and} \qquad \frac{2b+a}{3} < a_n.$$

But since $a < b$, we must have $b + 2a < 2b + a$.[15] But this then implies that

$$a_n < \frac{b+2a}{3} < \frac{2b+a}{3} < a_n,$$

which is clearly a contradiction. $\qquad\square$

Scratch work for second proof. There is a second proof that is even a little shorter. The idea is to show that $|a - b| < \varepsilon$ for all $\varepsilon > 0$, and therefore $|a - b| = 0$ (officially, this is because of Exercise 1.13), implying that $a = b$. Intuitively, the way to show this is to say that if a_n is getting really close to both a and b, then that forces a and b to be close to each other. If we demand that a_n is within $\varepsilon/2$ of both a and b, then the distance between a and b can't be more than ε:

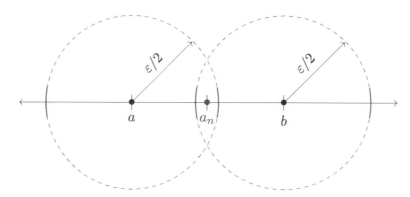

Using the triangle inequality we can then bound the distance from a to b by finding the distance from a to a_n, plus the distance from a_n to b. That's the idea, now here's the formal proof.

---

[15]Since b is larger than a, if you have 2 b's and 1 a, that's larger than if you have 2 a's and 1 b. Alternatively, just add "$a + b$" to both sides of "$a < b$."

Second proof. Let $\varepsilon > 0$. Since $\varepsilon/2 > 0$ and $a_n \to a$, there exists some N_1 such that for $n > N_1$ we have $|a_n - a| < \varepsilon/2$. Since $\varepsilon/2 > 0$ and $a_n \to b$, there exists some N_2 such that for $n > N_2$ we have $|a_n - b| < \varepsilon/2$. Let $N = \max\{N_1, N_2\}$ and pick any $n > N$. Then

$$
\begin{aligned}
|a - b| &= |a - a_n + a_n - b| \\
&\leq |a - a_n| + |a_n - b| && \text{(triangle inequality)} \\
&= |a_n - a| + |a_n - b| \\
&< \varepsilon/2 + \varepsilon/2 && (n > N \text{ implies } n > N_1 \text{ and } n > N_2) \\
&= \varepsilon.
\end{aligned}
$$

Since this holds for any $\varepsilon > 0$, we have shown that $|a - b| < \varepsilon$ for all $\varepsilon > 0$. Which (by Exercise 1.13) implies that $|a - b| = 0$, and hence $a = b$. So indeed, a_n can converge to only a single point. $\qquad\square$

We have been talking about bounded sequences and convergent sequences. The following is a nice proposition which combines these. It also emphasizes the idea that bounded sequences are nice, but convergent sequences are *really* nice. It demonstrates this by showing that the collection of convergent sequences are a (proper) subset of the collection of bounded sequences.

Proposition.

Proposition 3.20 (*Convergent $\Rightarrow$ bounded*)**.** If (a_n) is a convergent sequence, then (a_n) is bounded.

Scratch Work. We are told (a_n) converges to some a. By the definition of convergence (Definition 3.7), that means that, for any $\varepsilon > 0$, eventually all the terms of the sequences end up within ε of a. Suppose $\varepsilon = 1$ and N is the value for which $n > N$ implies $|a_n - a| < 1$. Perhaps the picture looks like this:

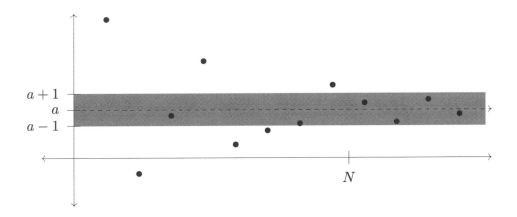

Now, if we only cared about the points after N, then we are certainly good-to-go, because all those points are bounded by $a-1$ and $a+1$. What about the other points? Well, the only others are $a_1, a_2, a_3, \ldots, a_N$. And since this is a finite collection, there must exist a largest one and the smallest one. The largest one, for instance, is $M := \max\{a_1, a_2, \ldots, a_n\}$. Combined, each a_i either is at most M (if $i \leq N$) or is at most $a+1$ (if $i > N$). Therefore the larger between M and $a+1$ is an upper bound on *all* a_i. A succinct way to write this largest value is to just include $a+1$ in the maximum. That is, $\max\{a_1, a_2, \ldots, a_n, a+1\}$ will be an upper bound on all a_i. Likewise for the lower bound.

That's the idea, now here's the proof.

***Proof*.** Since (a_n) is convergent, let a be the value it is converging to. By the definition of convergence (with $\varepsilon = 1$), there is some N where

$$|a_n - a| < 1$$

for all $n > N$. That is, $a - 1 < a_n < a + 1$ for all $n > N$. Let

$$U = \max\{a_1, a_2, a_3, \ldots, a_N, a+1\}$$

and

$$L = \min\{a_1, a_2, a_3, \ldots, a_N, a-1\}.$$

Note that if $n \leq N$, then $L \leq a_n \leq U$, since each such a_n is included in the sets which we are taking the minimum and maximum of. And if $n > N$, then we already noted that $a - 1 < a_n < a + 1$, which implies that

$$L \leq a - 1 < a_n < a + 1 \leq U,$$

and hence $L < a_n < U$. Combining these cases, we have that

$$L \leq a_n \leq U$$

for all n. And thus, by definition, (a_n) is bounded. $\square$

There is a second, related proof. It's a little shorter by using the absolute value version of boundedness (Proposition 3.5: (a_n) is bounded if and only if there is some C where $|a_n| \leq C$ for all n), which allows you to not worry about lower and upper bounds separately, and instead do them together.

***Second Proof*.** Since (a_n) is convergent, let a be the value it is converging to. By the definition of convergence (with $\varepsilon = 1$), there is some N where

$$|a_n - a| < 1$$

for all $n > N$. That is, $a - 1 < a_n < a + 1$ for all $n > N$. Consider

$$C := \max\left\{|a_1|, |a_2|, |a_3| \ldots, |a_N|, |a+1|, |a-1|\right\}.$$

Note that if $n \leq N$, then $a_n \leq |a_n| \leq C$. And if $n > N$, then $a - 1 < a_n < a + 1$, which implies that $|a_n| \leq \max\{|a-1|, |a+1|\} \leq C$. Combining these cases, we have that

$$|a_n| \leq C$$

for all n, which by Proposition 3.5 means that (a_n) is bounded.   □

Whenever you have deduced a result of the form "P implies Q," it is worthwhile to think about what the contrapositive says. The contrapositive "not-Q implies not-P" will look different, but it is logically equivalent. In the present case of Proposition 3.20, the contrapositive of

"If (a_n) is a convergent sequence, then (a_n) is bounded."

is

"If (a_n) is an unbounded sequence, then (a_n) is divergent."

3.5 Limit Laws

When you work with limits, it's good to know what the ground rules are for algebraically manipulating them. These five basic properties are usually called the "limit laws."

Theorem.

Theorem 3.21 (*Sequence limit laws*). Assume that (a_n) and (b_n) are convergent sequences of real numbers such that $a_n \to a$ and $b_n \to b$. Also assume that $c \in \mathbb{R}$. Then,

1. $(a_n + b_n) \to a + b$

2. $(a_n - b_n) \to a - b$

3. $(a_n \cdot b_n) \to a \cdot b$

4. $\left(\dfrac{a_n}{b_n}\right) \to \dfrac{a}{b}$, provided $b \neq 0$ and each $b_n \neq 0$.

5. $(c \cdot a_n) \to c \cdot a$

Laws 1 and 5 are Exercise 3.8, and we will prove Law 3 below. Once you've proven 1, then 2 is very similar (or follows from 1 and 5 by letting $c = -1$); likewise, once we have proven 3, then 4 is similar. Those two are Exercise 3.13. So we will prove 3 next and leave the rest as exercises.

Scratch Work. Here's the scratch work for Law 3. We want to find an N such that $n > N$ implies $|a_n b_n - ab| < \varepsilon$. Our two big assumptions are that $a_n \to a$ and $b_n \to b$. Here's what we get from that:

- Since $a_n \to a$, for any $\varepsilon_1 > 0$ there exists some N_1 such that $n > N_1$ implies that $|a_n - a| < \varepsilon_1$.

- Since $b_n \to b$, for any $\varepsilon_2 > 0$ there exists some N_2 such that $n > N_2$ implies that $|b_n - b| < \varepsilon_2$.

Now, going back to what we want to show, there is a clever trick that makes it all work. The idea is to first rewrite

$$|a_n b_n - ab| \quad \text{as} \quad |a_n b_n - ab_n + ab_n - ab|.$$

The middle terms cancel out, so those are the same. This is clever, though, because if you now use the triangle inequality, you get something we have control over:

$$|a_n b_n - ab_n + ab_n - ab| \le |a_n b_n - ab_n| + |ab_n - ab|$$
$$= |a_n - a| \cdot |b_n| + |a| \cdot |b_n - b|.$$

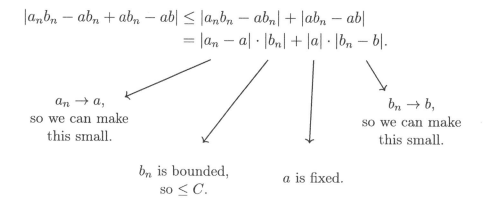

$a_n \to a$,
so we can make
this small.

$b_n \to b$,
so we can make
this small.

b_n is bounded,
so $\le C$.

a is fixed.

To get this all less than ε, we will make both of the products less than $\frac{\varepsilon}{2}$. To get $|a_n - a| \cdot |b_n|$ less than $\frac{\varepsilon}{2}$ we will have to ensure that $|a_n - a|$ is less than $\frac{\varepsilon}{2C}$. To get $|a| \cdot |b_n - b|$ less than $\frac{\varepsilon}{2}$ we will have to ensure $|b_n - b|$ is less than $\frac{\varepsilon}{2a}$. (And then be careful so we aren't dividing by 0.)

Proof. Let $\varepsilon > 0$. Since (b_n) converges, by Proposition 3.20 we know that (b_n) is bounded; that is, there exists some $C \in \mathbb{R}$ such that $|b_n| \le C$ for all n. Let $\varepsilon_1 = \frac{\varepsilon}{2C+1}$ and $\varepsilon_2 = \frac{\varepsilon}{2|a|+1}$. (Note: we need each "$+1$" to ensure we are not dividing by 0.)

Since $\varepsilon_1 > 0$ there exists some N_1 such that $|a_n - a| < \varepsilon_1$ for all $n > N_1$. And since $\varepsilon_2 > 0$ there exists some N_2 such that $|b_n - b| < \varepsilon_2$ for all $n > N_2$. Let

$N = \max\{N_1, N_2\}$. Then, for any $n > N$,

$$
\begin{aligned}
|a_n b_n - ab| &= |a_n b_n - a b_n + a b_n - ab| \\
&\leq |a_n b_n - a b_n| + |a b_n - ab| \\
&= |a_n - a| \cdot |b_n| + |a| \cdot |b_n - b| \\
&< \varepsilon_1 \cdot C + |a| \cdot \varepsilon_2 \\
&= \frac{\varepsilon}{2C + 1} \cdot C + |a| \cdot \frac{\varepsilon}{2|a| + 1} \\
&< \frac{\varepsilon}{2} + \frac{\varepsilon}{2} \\
&= \varepsilon.
\end{aligned}
$$

That is, for $n > N$ we have $|a_n b_n - ab| < \varepsilon$. Therefore $(a_n b_n) \to ab$, as desired. $\square$

Example 3.22. What is

$$
\lim_{n \to \infty} \frac{1}{2} \cdot \left(\frac{\frac{1}{n} + \frac{1}{n^2} + 4}{5 - \frac{1}{n^2}} \right) \cdot \left(\frac{3n + 1}{n + 2} + \frac{1}{\sqrt{n}} \right) ?
$$

Solution. Already in this chapter we have shown

- $\dfrac{1}{n} \to 0$
- $5 \to 5$
- $\dfrac{3n + 1}{n + 2} \to 3$
- $\dfrac{1}{n^2} \to 0$
- $5 - \dfrac{1}{n^2} \to 5$
- $\dfrac{1}{\sqrt{n}} \to 0$

So, by the limit laws,

$$
\lim_{n \to \infty} \frac{1}{2} \cdot \left(\frac{\frac{1}{n} + \frac{1}{n^2} + 4}{5 - \frac{1}{n^2}} \right) \cdot \left(\frac{3n + 1}{n + 2} + \frac{1}{\sqrt{n}} \right) = \frac{1}{2} \cdot \left(\frac{0 + 0 + 4}{5} \right) \cdot (3 + 0) = \frac{6}{5}.
$$

$\square$

Theorem.

Theorem 3.23 (*Sequence squeeze theorem*). Assume $a_n \leq x_n \leq b_n$ for all n. Furthermore, assume that

$$
a_n \to L \qquad \text{and} \qquad b_n \to L.
$$

Then,

$$
x_n \to L.
$$

Scratch Work. Just reading the theorem, it does jibe with our intuition of what it means for a sequence to converge.

We will prove it via the definition of convergence, and simply from that definition's vantage point there's good reason to believe it. We wish to show that eventually all the x_n is within ε of L (that is, $|x_n - L| < \varepsilon$ for all $n > N$, for some N). But since a_n and b_n are converging to L, there is some N where, for $n > N$, a_n and b_n are both within ε of L. Combining this with our assumption that $a_n \leq x_n \leq b_n$, it makes perfect sense that x_n would also have to be within ε of L. To formally prove it, it helps to consider two cases: whether x_n is smaller than L or larger than L. Suppose we are past the N at which both a_n and b_n are within ε of L, and suppose that for some $n > N$ we have $x_n \leq L$. Then the picture looks something like this:

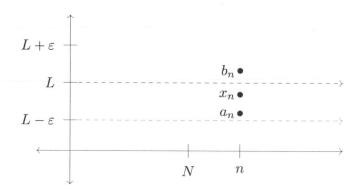

You can see that x_n is within ε of L, because a_n is! Similarly, if $L \leq x_n$, then the picture looks something like this:

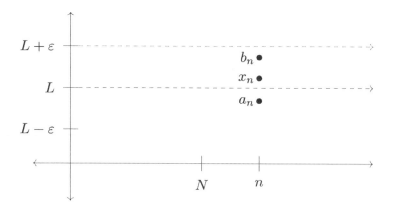

We again know that x_n is within ε of L, because b_n is!

Combined, the two pictures above show that $L - \varepsilon < x_n < L + \varepsilon$. To make this precise, though, we of course have to get our hands dirty with the definitions of the limits of (a_n) and (b_n).

Proof. Let $\varepsilon > 0$.

- Since $a_n \to L$, there exists some N_1 such that $n > N_1$ implies $|a_n - L| < \varepsilon$. That is, $-\varepsilon < a_n - L < \varepsilon$. Or,

$$L - \varepsilon < a_n < L + \varepsilon. \tag{$\maltese$}$$

- Since $b_n \to L$, there exists some N_2 such that $n > N_2$ implies $|b_n - L| < \varepsilon$. That is, $-\varepsilon < b_n - L < \varepsilon$. Or,

$$L - \varepsilon < b_n < L + \varepsilon. \tag{$\oslash$}$$

Let $N = \max\{N_1, N_2\}$, and let $n > N$. Combining the inequality $a_n \le x_n \le b_n$ with the left half of ($\maltese$) and the right half of ($\oslash$), we get

$$L - \varepsilon < a_n \le x_n \le b_n < L + \varepsilon$$
$$L - \varepsilon < x_n < L + \varepsilon$$
$$-\varepsilon < x_n - L < \varepsilon$$
$$|x_n - L| < \varepsilon.$$

$\square$

Here's one example using this theorem.

Example 3.24. Prove that

$$\lim_{n \to \infty} \frac{1}{n^{2.7} + \sqrt{n} + \pi} = 0.$$

Proof. Since

$$0 \le \frac{1}{n^{2.7} + \sqrt{n} + \pi} \le \frac{1}{n},$$

and since

$$\lim_{n \to \infty} 0 = 0 \qquad \text{and} \qquad \lim_{n \to \infty} \frac{1}{n} = 0,$$

by the sequence squeeze theorem (Theorem 3.23),

$$\lim_{n \to \infty} \frac{1}{n^{2.7} + \sqrt{n} + \pi} = 0.$$

$\square$

3.6   The Monotone Convergence Theorem

> **Definition.**
>
> **Definition 3.25.** A sequence (a_n) is *monotone increasing* if $a_n \leq a_{n+1}$ for all n. Likewise, a sequence (a_n) is *monotone decreasing* if $a_n \geq a_{n+1}$ for all n. If it is either monotone increasing or monotone decreasing, it is *monotone*.[16]

Example 3.26.

- (n) is monotone increasing.

- $\left(\dfrac{1}{n}\right)$ is monotone decreasing.

- (1) is both monotone increasing and monotone decreasing. □

Note that (n) is the sequence $(1, 2, 3, 4, \dots)$. Likewise, (1) is the sequence $(1, 1, 1, 1, \dots)$; it is the sequence you get when you take $a_n = 1$ for all n, so that $(a_n) = (a_1, a_2, a_3, \dots)$ is just $(1, 1, 1, \dots)$.

> **Theorem.**
>
> **Theorem 3.27** (*The monotone convergence theorem*). Suppose (a_n) is monotone. Then (a_n) converges if and only if it is bounded. Moreover,
>
> - If (a_n) is increasing, then either (a_n) diverges to ∞ or
>
> $$\lim_{n \to \infty} a_n = \sup\left(\{a_n : n \in \mathbb{N}\}\right).$$
>
> - If (a_n) is decreasing, then either (a_n) diverges to $-\infty$ or
>
> $$\lim_{n \to \infty} a_n = \inf\left(\{a_n : n \in \mathbb{N}\}\right).$$

---

[16]Note: "Monotone" means the same thing as "monotonic." But not "monotonous." *Never* "monotonous."

Proof Idea. Suppose (a_n) is monotonically increasing. If it's not bounded, then given any $M > 0$, this M is not an upper bound and so eventually (a_n) will get above it. And since it's monotonically increasing, once it gets above a number, it *stays* above that number.

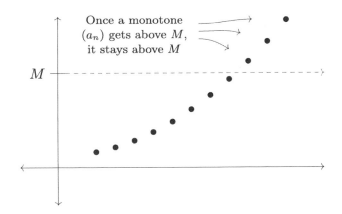

If, on the other hand, (a_n) is bounded, then the sequence must be leveling out (it's monotone, so it can't go up and down).

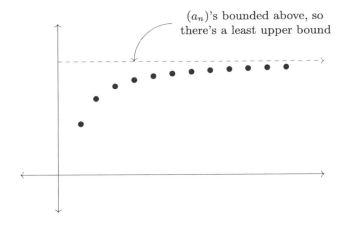

Proof. Assume that (a_n) is monotonically increasing, and let's suppose first that a_n is not bounded. Then for any $M > 0$ there exists some N such that $a_N > M$. But since (a_n) is monotonically increasing, for $n > N$ we have $a_n \geq a_N > M$. And so, by Definition 3.15, (a_n) diverges to ∞.

Next suppose that (a_n) is bounded. Then we have that $\{a_n : n \in \mathbb{N}\}$ is a subset of $\mathbb{R}$ which is bounded above, which by the completeness of $\mathbb{R}$ implies that $\sup(\{a_n : n \in \mathbb{N}\})$ exists.[17] Call this supremum α. We want to show that $\lim_{n\to\infty} a_n = \alpha$; that is, we want to show that for any $\varepsilon > 0$ there exists some N such that $n > N$ implies $|a_n - \alpha| < \varepsilon$.

---

[17]Remember that so far only sets have been defined to have suprema. Sequences don't. So we have to turn the sequence $(a_1, a_2, a_3, \ldots)$ into the set $\{a_n : n \in \mathbb{N}\}$ in order to talk about their supremum.

To show this, first let $\varepsilon > 0$. Since $\sup\left(\{a_n : n \in \mathbb{N}\}\right) = \alpha$, by the suprema analytically theorem (Theorem 1.24) there exists some $a_N > \alpha - \varepsilon$. And since (a_n) is monotonically increasing, we see that for any $n > N$ we have that $a_n \geq a_N > \alpha - \varepsilon$. And of course, $a_n \leq \alpha$ due to the fact that α is the supremum (and hence an upper bound) of $\{a_n : n \in \mathbb{N}\}$. And so, for $n > N$, we have $\alpha - \varepsilon < a_n < \alpha$. This implies $\alpha - \varepsilon < a_n < \alpha + \varepsilon$, and hence $-\varepsilon < a_n - \alpha < \varepsilon$, which, at last, gives

$$|a_n - \alpha| < \varepsilon,$$

completing the proof in the case when (a_n) is monotonically increasing.

The case where (a_n) is monotonically decreasing is *very* similar. It is asked of you in Exercise 3.25. $\qquad\square$

This theorem is nice for several reasons, but one of which is the fact that you do not have to know what the limit of (a_n) is in order to show that (a_n) is convergent. This is notable since in the definition of sequence convergence — which, until now, we have heavily relied on to show a sequence converges — requires that you already know (and can write down) what the limit is going to be. Below is an example where it would be quite a challenge to write down the limit of the sequence in a beneficial way. Yet by the monotone convergence theorem, we will be able to conclude that the limit exists.

Example 3.28. Let (a_n) be the sequence where $a_1 = 0.1$, $a_2 = 0.12$, $a_3 = 0.123$, $a_4 = 0.1234$, and so on. (And, to be clear, this pattern does not change when you reach double digits. For example, $a_{12} = 0.123456789101112$.) Prove that (a_n) converges.

Solution. Note that a_{n+1} and a_n match exactly until the last digits of a_{n+1}, which are the digits of $n + 1$. Therefore, $a_{n+1} - a_n$ is a number with a bunch of zeros followed the digits of $n + 1$. In particular,

$$a_{n+1} - a_n > 0$$

for all n. We have shown that $a_{n+1} > a_n$ for all n, proving that (a_n) is monotone increasing. Furthermore note that since all the terms begin with 0.1, we have that $a_n \leq 1$ for all n.

We have shown that (a_n) is monotone increasing and bounded above, therefore by the monotone convergence theorem (Theorem 3.27) the sequence converges, completing the proof. $\qquad\square$

> ## Proposition.
>
> **Proposition 3.29** (*Bounded S contains* $a_n \to \sup(S)$). Suppose $S \subseteq \mathbb{R}$ is bounded above. Then there exists a sequence (a_n) where $a_n \in S$ for each n and
>
> $$\lim_{n \to \infty} a_n = \sup(S).$$
>
> Likewise, if S is bounded below, then there exists a sequence (b_n) where $b_n \in S$ for each n and
>
> $$\lim_{n \to \infty} b_n = \inf(S).$$

Proof Idea. We need to find elements of S which are getting closer and closer to $\sup(S)$. What was our *super* useful theorem for doing just that? The suprema analytically theorem (Theorem 1.24)! That was the theorem that said for any $\varepsilon > 0$, there exists some $a \in S$ such that $a > \sup(S) - \varepsilon$. And since it works for any $\varepsilon > 0$, we can choose a sequence of ε values that are getting closer and closer to 0, which in turn produce a sequence of a values that are getting closer and closer to $\sup(S)$.

And once we have found a sequence of these elements which are getting closer and closer to $\sup(S)$, how do we formally show they converge to $\sup(S)$? The sequence squeeze theorem (Theorem 3.23) sure sounds like it's up to the job.

Proof. Suppose $S \subseteq \mathbb{R}$ is bounded above, implying that $\sup(S)$ exists. Let $\sup(S) = \alpha$ (implying $a_n \leq \alpha$ for all n). By the suprema analytically theorem (Theorem 1.24), for any $\varepsilon > 0$ there exists some $x \in S$ such that $x > \alpha - \varepsilon$. In particular, there exists some $a_1 \in S$ such that $a_1 > \alpha - 1$; and there exists some $a_2 \in S$ such that $a_2 > \alpha - \frac{1}{2}$; and there exists some $a_3 \in S$ such that $a_3 > \alpha - \frac{1}{3}$; and, in general, for each $n \in \mathbb{N}$ there exists some $a_n \in S$ such that $a_n > \alpha - \frac{1}{n}$.

Thus we obtain a sequence (a_n) from S. Moreover, this sequence does indeed converge to α: To see this, recall that in Example 3.9 we showed that $\frac{1}{n} \to 0$. Also, clearly (as in Example 3.13) the sequence $(\alpha, \alpha, \alpha, \dots)$ converges to α. Therefore, by Limit Law 2 we have that

$$\alpha - \frac{1}{n} \to \alpha - 0 = \alpha.$$

So we have an inequality

$$\alpha - \frac{1}{n} \leq a_n \leq \alpha,$$

where both the upper and lower bounds converge to α. By the sequence squeeze theorem (Theorem 3.23) we see that $\lim_{n \to \infty} a_n = \alpha$, as desired.

The infimum case is *very* similar. It's asked of you in Exercise 3.26. $\qquad\square$

3.7   Subsequences

Consider the sequence

$$(a_n) = (n^2).$$

Remember what this means: It means that when you plug in $n = 1$ you get the first term of the sequence, when you plug in $n = 2$ you get the second term, and so on. Therefore (a_n) is the sequence

$$(1, 4, 9, 16, 25, 36, 49, 64, 81, 100, \dots).$$

Now, a *subsequence* of (a_n), as you would expect, is something like

$$(4, 16, 36, 64, 100, \dots)$$
$$\uparrow \ \uparrow \ \uparrow \ \uparrow \ \uparrow$$
$$a_2, a_4, a_6, a_8, a_{10}, \dots \ = (a_{2n})$$

Or,

$$(1, 25, 81, 169, 289, \dots)$$
$$\uparrow \ \uparrow \ \uparrow \ \uparrow \ \uparrow$$
$$a_1, a_5, a_9, a_{13}, a_{17}, \dots \ = (a_{4n-3})$$

Of course, a subsequence does not have to be perfectly spaced like this. For example, $(1, 121, 144, 484, 10201, \dots)$ also is a subsequence. As long as you pick out one term after another in order, then you have a subsequence. This reasoning suggests the following definition.

> **Definition.**
>
> **Definition 3.30.** Let (a_n) be a sequence of real numbers and let
>
> $$n_1 < n_2 < n_3 < \dots$$
>
> be an increasing sequence of integers. Then
>
> $$a_{n_1}, a_{n_2}, a_{n_3}, \dots$$
>
> is called a *subsequence* of (a_n), and is denoted (a_{n_k}).

Here are some examples.

Example 3.31. Let $(a_n) = \left(1, \frac{1}{2}, \frac{1}{3}, \frac{1}{4}, \dots\right)$. Then

- $\left(1, \frac{1}{3}, \frac{1}{5}, \frac{1}{7}, \dots\right)$ is a subsequence where

$$n_1 = 1, \quad n_2 = 3, \quad n_3 = 5, \quad n_4 = 7, \quad \dots,$$

- $\left(\frac{1}{10}, \frac{1}{100}, \frac{1}{1,000}, \frac{1}{10,000}, \dots\right)$ is a subsequence where

$$n_1 = 10, \quad n_2 = 100, \quad n_3 = 1,000, \quad n_4 = 10,000, \quad \dots,$$

- And $\left(\frac{1}{2}, \frac{1}{25}, \frac{1}{16,129}, \frac{1}{100,007}, \dots\right)$ is a subsequence where

$$n_1 = 2, \quad n_2 = 25, \quad n_3 = 16,129, \quad n_4 = 100,007, \quad \dots.$$

- But $\left(1, \frac{1}{5}, \frac{1}{3}, \frac{1}{7}, \frac{1}{2}, \dots\right)$ is a *not* a subsequence (because we don't have $n_1 < n_2 < n_3 < \dots$),

- And $\left(\frac{1}{2}, \frac{1}{2}, \frac{1}{3}, \frac{1}{4}, \dots\right)$ is a *not* a subsequence (because $n_1 = n_2$).

$\square$

Notice that in this example, $a_n \to 0$. Also, each of the three actual subsequences had the property that $a_{n_k} \to 0$, as well. As you can imagine, this is true in general.

> ### Proposition.
>
> **Proposition 3.32** ($a_n \to a \Leftrightarrow$ *every* $a_{n_k} \to a$). A sequence (a_n) converges to a if and only if every subsequence of (a_n) also converges to a.

Proof. Assume that (a_n) converges to a and (a_{n_k}) is a subsequence of (a_n). Let $\varepsilon > 0$. Then there exists some N such that

$$|a_n - a| < \varepsilon \qquad\qquad (\circledast)$$

for all $n > N$.

We want to show that $a_{n_k} \to a$. That is, we want to show that there exists some N_1 such that $|a_{n_k} - a| < \varepsilon$ for all $k > N_1$.

Notice that since each $n_i \in \mathbb{N}$ and

$$n_1 < n_2 < n_3 < \dots,$$

that we have[18] $n_k \geq k$. Therefore by letting $N_1 = N$, for any $k > N_1$ we then know that $n_k > N$, so by $(\circledast)$ we have

$$|a_{n_k} - a| < \varepsilon.$$

This completes the forward direction. The backward direction is asked of you[19] in Exercise 3.32. $\square$

---

[18]For example, one subsequence is $(a_{n_\ell}) = (a_1, a_3, a_5, a_7, \dots)$, giving $n_1 = 1$, $n_2 = 3$, $n_3 = 5$, $n_4 = 7$, and so on. And so, $1 \leq n_1$, and $2 \leq n_2$, and $3 \leq n_3$, and $4 \leq n_4$, and so on.

[19]Hint: There is a 1-line proof for this exercise.

(It is also true that if (a_n) diverges to ∞, then every subsequence of (a_n) will also diverge to ∞. This is asked of you in Exercise 3.33.)

The contrapositive of Proposition 3.32 is that if there are two subsequences which don't converge to the same thing, then the original sequence diverges. Said differently, it is required that *all* subsequences converge to some a for (a_n) to converge as well. Can you see why it is not enough to know that just a single (a_{n_k}) converges? Try to think up an example on your own before reading the one below.

Example 3.33. Let $(a_n) = ((-1)^n)$. This sequence diverges despite many of its subsequences converging. For example, $(a_{2n}) = ((-1)^{2n}) = (1) = (1, 1, 1, 1, \dots)$ converges to 1.[20] Likewise, (a_{4n}) and (a_{28n}) converge to 1, while (a_{2n+1}) converges to -1. $\square$

The problem is that there are some subsequences which converge to one thing, and others which converge to something else. This suggests the following.

Corollary.

Corollary 3.34 (*Different subsequential limits $\Rightarrow$ diverges*). If (a_n) has a pair of subsequences converging to different limits, then (a_n) diverges.

This result is a special case of the contrapositive of the forward direction of Proposition 3.32. But it is perhaps easier to see via a proof by contradiction.

***Proof*.** Assume for a contradiction that (a_n) converges. Say, $a_n \to L$. By Proposition 3.32, this implies that every subsequence also converges to L. This contradicts the assumption that there are subsequences converging to different limits. $\square$

We showed in Example 3.33 that a subsequence of (a_n) can converge while (a_n) itself diverges. However, there is one important special class of sequences where having a single subsequence converge *does* imply that (a_n) also converges. Can you think of what sort of sequences have this property? Try to think it up on your own below looking at the footnote.[21]

---

[20] In the notation of Defintion 3.30, here we have $n_k = 2k$.

[21] If (a_n) is monotone, it has this property. We prove this in Proposition 3.35.

> ### Proposition.
>
> **Proposition 3.35** (*Monotone (a_n) has $a_{n_k} \to a \Rightarrow a_n \to a$*). If a monotone sequence (a_n) has a convergent subsequence, then (a_n) converges too, and has the same limit.

Proof Idea. First recall that by Proposition 3.32, a convergent sequence always has the same limit as any of its subsequences. So by this, our task reduces to showing that (a_n) converges.

A monotone sequence is convergent if and only if it is bounded (by the monotone convergence theorem). We are told that (a_n) is monotone and to show that it's bounded, we use the assumption that it has a convergent subsequence, (a_{n_k}). Here's the progression: (a_n) being monotone implies that its subsequence is too, and (a_{n_k}) being convergent (and now monotone) will mean that (a_n) is bounded too. Now here's the formal proof.

Proof. Assume that (a_n) is monotone increasing; the case where (a_n) is monotone decreasing is *very* similar.

Let (a_{n_k}) be a convergent subsequence of (a_n) and observe that since (a_n) is monotone increasing, being a subsequence means that (a_{n_k})'s terms comes from (a_n) and do so in the same order as they appear in (a_n), implying that (a_{n_k}) is also monotone increasing. That is, (a_{n_k}) was assumed to be convergent and was shown to be monotone — by the monotone convergence theorem (Theorem 3.27) this means (a_{n_k}) converges, and in particular converges to $\sup\left(\{a_{n_k} : k \in \mathbb{N}\}\right)$.

We now show that (a_n) is convergent, again by using the monotone convergence theorem. Since (a_n) is monotone increasing by assumption, we need only show that (a_n) is bounded above. To see this, observe that since the subsequence (a_{n_k}) is an infinite list of elements from (a_n), any particular element a_n from the sequence has an element a_{n_k} from this subsequence that comes after it.[22] And since both a_n and a_{n_k} are terms of (a_n), which is monotone increasing, we have $a_n \leq a_{n_k}$. So we have

$$a_n \leq a_{n_k} \leq \sup\left(\{a_{n_k} : k \in \mathbb{N}\}\right),$$

which proves that (a_n) is bounded above (by this supremum), and so by the monotone convergence theorem we have proven that (a_n) converges.

Finally, since we have shown that the sequence (a_n) is convergent, by Proposition 3.32 it has the same limit as any of its subsequences. This means that (a_n) must also be converging to $\sup\left(\{a_{n_k} : k \in \mathbb{N}\}\right)$, completing the proof. $\square$

---

[22] Probably not *immediately* after it, but certainly at some point after it. For instance, $k = n$ always works. To see this, recall that $\ell \leq n_\ell$ for any ℓ (See Footnote 18), which implies the headache-inducing expression $n \leq n_n$ (meaning the nth index of the subsequence is at least n), and hence $a_n \leq a_{n_n}$ (meaning the nth term of the subsequence is at least as large as the nth term of the original sequence, or equivalently that the $n_n{}^{\text{th}}$ term of the original sequence is at least as large as the nth term of the original sequence).

There is a second proof which, if you're a master of indices, is pretty snappy.

Second proof. Without loss of generality, assume that (a_n) is monotone increasing. There is a quick technical property to note first. Observe that since (a_n) is increasing, if $s < t$, then $a_s \le a_t$. Next, note that for any subsequence (a_{n_t}), we have $t \le n_t$ for all t.[23] Therefore, again since (a_n) is increasing, $a_t \le a_{n_t}$. Putting this all together,

$$a_s \le a_t \le a_{n_t} \qquad (\text{🚲})$$

whenever $s < t$.

Now we begin the main argument. Suppose that (a_{n_k}) is a convergent subsequence of (a_n). Let

$$a = \sup\left(\{a_{n_k} : k \in \mathbb{N}\}\right).$$

By the monotone convergence theorem and the fact that (a_{n_k}) converges, we know that $(a_{n_k}) \to a$. That is, there exists some $M \in \mathbb{N}$ such that

$$a - \varepsilon < a_{n_k} < a + \varepsilon$$

for all $k > M$. In particular, this holds when $k = M + 1$. Moreover, since a is in fact a supremum and hence an upper bound on the a_{n_k} terms, we can even say

$$a - \varepsilon < a_{n_{M+1}} \le a.$$

Combining this with (🚲) gives

$$a - \varepsilon < a_{n_{M+1}} \le a_\ell \le a_{n_\ell} \le a$$

for all $\ell > n_{M+1}$. So there does indeed exist an N (in particular, $N = n_{M+1}$) for which

$$a - \varepsilon < a_\ell \le a$$

for all $\ell > N$. Therefore $a_\ell \to a$, as desired. $\square$

---

[23]See Footnote 18.

3.8   The Bolzano-Weierstrass Theorem

We are now gearing up for a big theorem called the Bolzano-Weierstrass[24] theorem. The central question is this: Given an arbitrary sequence, how nice of a subsequence is that sequence guaranteed to have? And for which types of sequences are we guaranteed some *really* nice subsequence? One of the best properties a subsequence can have is being convergent. But after a moment's reflection we realize that not every sequence has a convergent subsequence. Try to think of an example on your own, and then you can check the footnote to see one answer.[25]

The Bolzano-Weierstrass theorem will tell us an important and general type of sequence which is guaranteed to contain a convergent subsequence, but we begin by asking whether there's a nice type of subsequence that *every* sequence contains. The answer is yes: every sequence contains a *monotone* subsequence. We prove this now.

> **Lemma.**
>
> **Lemma 3.36** ((a_n) *has monotone* a_{n_k}). Every sequence has a monotone subsequence.

Scratch Work. The proof of this is just straight-up snappy. Once you know precisely what to look for, it falls through perfectly. First, here's the idea behind it:

Usually we draw a sequence in the xy-plane by just plotting the points (e.g. if $a_3 = 6$, then we put a point at $(3, 6)$). This gives a picture like this:

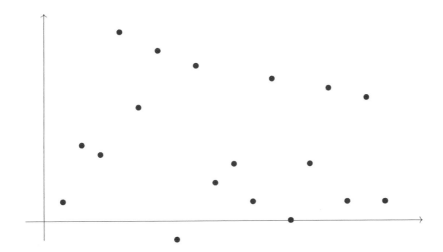

---

[24]Pro-tip: "Weierstrass" is pronounced with a hard 'v' at the start, like *vie-er-strahs*. It's German. It's also more fun to say this way.

[25]The sequence (n) works. This is the sequence $1, 2, 3, 4, 5, \ldots$, which diverges to ∞ and every subsequence of it also diverges to ∞.

This time, though, we will connect the dots to make a zig-zagged line:

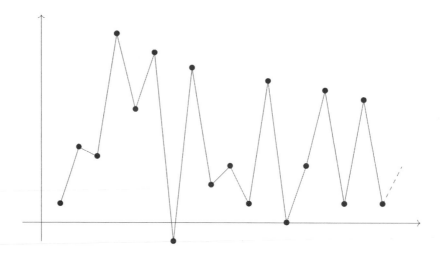

From this picture you can maybe spot a nice *decreasing* sequence, which is of course monotone:

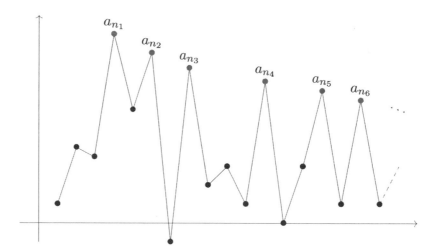

The magical definition that will solve everything is that of a *peak*; we want those labeled points to form *peaks*, and we want to be able to say that if we have a(n infinite) sequence of peaks, then we do indeed have a decreasing sequence. The definition that does this is this: define a *peak* to be a point a_n which is larger than every later point; that is, a_n is a peak if $a_n \geq a_m$ for all $m > n$.

So if we have infinitely many peaks, then we obtain a sequence like the one above which will be decreasing. So what if we don't? Then we only have finitely many peaks. In this case, we can find an *increasing* sequence, which is again monotone. To see how, just note that if you're past the last peak, then any point you pick is not a peak, which means there is some point after it which is larger. So one at a time you can pick larger and larger points, giving an increasing sequence.

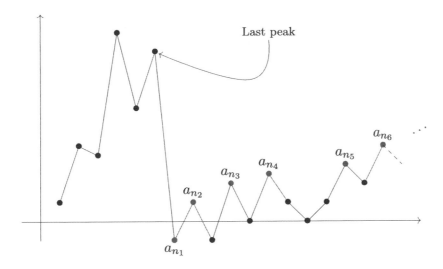

Let's now, at last, prove Lemma 3.36.

Proof. Call a_n a *peak* if a_n is larger than every later point. That is, if $a_n \geq a_m$ for all $m > n$.

Either there are infinitely many peaks or finitely many peaks. Assume first that (a_n) has infinitely many peaks. Then let a_{n_k} be the kth peak. Then, by the definition of a peak, $a_{n_k} \geq a_{n_{k+1}}$, implying that (a_{n_k}) is a decreasing subsequence.

Now assume that (a_n) has finitely many peaks, and let a_N be the last one. Then let $a_{n_1} = a_{N+1}$. Since a_N was the last peak, a_{N+1} is *not* a peak, implying that there is some later point a_{n_2} that is larger than it. Likewise, since a_{n_2} is after the last peak, it is also not a peak, and so there must be some later point a_{n_3} that is larger than it. Continuing in this way we construct an increasing subsequence a_{n_k}.

In either case we found a monotone subsequence, so we are done. $\square$

The lemma showed that any sequence has a monotone subsequence. It would be great if every sequence had a *convergent* subsequence, although unfortunately this is not the case (e.g., $(a_n) = (n)$). Indeed, any sequence which diverges to $\pm\infty$ will have no convergent subsequence (see Exercise 3.33). There are also examples of sequences whose limit does not exist, and none of whose subsequences converge — but every such example must be unbounded (See Exercise 3.35). Indeed, this is what the Bolzano-Weierstrass theorem says.

> ### Theorem.
>
> **Theorem 3.37** (*The Bolzano-Weierstrass theorem*). Every bounded sequence has a convergent subsequence.

Proof. Assume (a_n) is a bounded sequence. Then by Lemma 3.36 it has a monotone subsequence, (a_{n_k}). Also, since (a_n) is bounded, so is (a_{n_k}). Since (a_{n_k}) is both bounded and monotone, by the monotone convergence theorem it converges. $\square$

3.9   The Cauchy Criterion

Consider the following objective.

Objective: Show (a_n) converges without knowing its limit.

We have been talking a lot about what it means for a sequence (a_n) to "converge to a point a." The definition was:

Definition 3.7.   A sequence (a_n) converges to a point a if, for all $\varepsilon > 0$, there exists some N where

$$|a_n - a| < \varepsilon$$

for all $n > N$.

So what does it mean to just say that (a_n) "converges" (without knowing what a is)? It means that *there exists* some a for which $a_n \to a$. That is:

A sequence (a_n) converges if there exists some a such that, for all $\varepsilon > 0$, there exists some N where

$$|a_n - a| < \varepsilon$$

for all $n > N$.

The important thing to note here is that as of now, to show that a sequence (a_n) converges, we need to already know its limit, a, in order to do the calculation. But what if we don't know the limit? Can we somehow look at just the a_n terms of the sequence and figure out whether the sequence converges to *something* without knowing what that something is?

Proposed Solution 1: Maybe convergence means $|a_{n+1} - a_n|$ gets arbitrarily small.

One thing to note is that whenever a sequence converges, each term is deviating less and less from the previous term; the "jumps" are tending towards 0. So if consecutive terms of that sequence are getting closer and closer together, does that mean the sequence converges? Maybe. But with this intuition, one possible condition would be this:

Possibly it's the case that a sequence (a_n) converges if, for all $\varepsilon > 0$, there exists some N where

$$|a_{n+1} - a_n| < \varepsilon$$

for all $n > N$.

Is this condition necessary for convergence? The answer seems like it should be 'yes.' If a sequence is convergent, even if we don't know what the limit is there still must be some a which the terms are getting closer and closer to, and if the terms are getting close to a, then they must also be getting closer to each other. So yes, saying "the terms are getting closer and closer to each other" does seem necessary.

Is it sufficient for convergence? Is it possible that the terms are getting closer and closer but yet the sequence diverges anyways? If your intuition says 'no,' then that's quite reasonable. However, counterexamples do exist.

Example 3.38. The sequence $(a_n) = (\sqrt{n})$ diverges, and yet we can show that the terms of this sequence are getting closer and closer together.[26] Indeed, this sequence has the property that, for all $\varepsilon > 0$, there exists some N such that

$$|a_{n+1} - a_n| < \varepsilon$$

for all $n > N$.

Solution. Note that

$$\begin{aligned}
|a_{n+1} - a_n| &= |\sqrt{n+1} - \sqrt{n}| \\
&= \sqrt{n+1} - \sqrt{n} \\
&= (\sqrt{n+1} - \sqrt{n}) \cdot \frac{\sqrt{n+1} + \sqrt{n}}{\sqrt{n+1} + \sqrt{n}} \\
&= \frac{(n+1) - n}{\sqrt{n+1} + \sqrt{n}} \\
&= \frac{1}{\sqrt{n+1} + \sqrt{n}} \\
&< \frac{1}{\sqrt{n}}.
\end{aligned}$$

Let $\varepsilon > 0$ and let $N = 1/\varepsilon^2$. Then by the above, for any $n > N$ we have

$$|a_{n+1} - a_n| < \frac{1}{\sqrt{n}} < \frac{1}{\sqrt{N}} = \frac{1}{\sqrt{1/\varepsilon^2}} = \varepsilon.$$

$\square$

So simply having the terms get close to each other is not enough. What can go wrong? Well, even if $|a_{n+1} - a_n|$ is small, and $|a_{n+2} - a_{n+1}|$ is small, and $|a_{n+3} - a_{n+2}|$ is small, and so on, when you add up enough small things the result could still be big. For instance, possibly $|a_{n+10000} - a_n|$ is big. How do we fix this?

---

[26] This makes some sense. I mean, how far apart are the numbers $\sqrt{1,000,000}$ and $\sqrt{1,000,001}$?

<u>Proposed Solution 2: Maybe convergence means $|a_m - a_n|$ gets arbitrarily small.</u>

We fix the above shortcoming by making not just consecutive terms (past some N) close to each other, but by making *all* terms (past some N) close to each other. This is so important that it has a name; it is known as *the Cauchy criterion*.

This criterion is named after 18th century French mathematician Augustin-Louis Cauchy, a brilliant man who just happens to have looked remarkably like Vladimir Putin. Seriously, you be the judge:

    Cauchy, age 49   Putin, age 46

Definition.

Definition 3.39. A sequence (a_n) is *Cauchy* if for all $\varepsilon > 0$ there exists[27] some N such that
$$|a_m - a_n| < \varepsilon$$
for all $m, n > N$.

Example 3.40. Show that $\left(7 + \frac{1}{n}\right)$ is Cauchy.

Scratch Work. Looking back at the definition, we need to think about $a_n = 7 + \frac{1}{n}$ and $a_m = 7 + \frac{1}{m}$. For an $\varepsilon > 0$ we want to show that there exists some N such that
$$|a_n - a_m| < \varepsilon \qquad n, m > N.$$

We now work backwards. We start with this conclusion and unwind it, trying to get

---

[27]Note: When more convenient, you may assume that $N \in \mathbb{N}$.

to something that we can control.

$$|a_n - a_m| < \varepsilon$$

$$\left|\left(7 + \frac{1}{n}\right) - \left(7 + \frac{1}{m}\right)\right| < \varepsilon$$

$$\left|\frac{1}{n} + \frac{-1}{m}\right| < \varepsilon$$

Moving to the next line, we are applying the triangle inequality. The left-hand side below at least as big as the left-hand side above. So, if the below inequality holds, then the above will as well. This is what we want, since for the actual solution we will be doing this process in reverse.

$$\left|\frac{1}{n}\right| + \left|\frac{-1}{m}\right| < \varepsilon$$

$$\left|\frac{1}{n}\right| + \left|\frac{1}{m}\right| < \varepsilon$$

So if we make each of these two terms less than $\varepsilon/2$, we will be good to go. The Archimedean principle should allow us to do that.

We now do this in reverse for the actual solution.

Solution. Let $\varepsilon > 0$. Since $\varepsilon/2 > 0$, by the Archimedean principle there exists some $N > 0$ such that $\frac{1}{N} < \varepsilon/2$. Then, for any $m, n > N$,

$$
\begin{aligned}
|a_n - a_m| &= \left|\left(7 + \frac{1}{n}\right) - \left(7 + \frac{1}{m}\right)\right| \\
&= \left|\frac{1}{n} + \frac{-1}{m}\right| \\
&\leq \left|\frac{1}{n}\right| + \left|\frac{-1}{m}\right| \qquad \text{(Triangle inequality)} \\
&= \left|\frac{1}{n}\right| + \left|\frac{1}{m}\right| \\
&< \frac{1}{N} + \frac{1}{N} \\
&< \frac{\varepsilon}{2} + \frac{\varepsilon}{2} \\
&= \varepsilon.
\end{aligned}
$$

$\square$

In words: A sequence is Cauchy if, after some point N, *any* two terms of the sequences differ by at most ε. That is, after some point, all the terms of the sequences are close to each other. Notice how this is different than the definition of convergence:

(a_n) converges to a point a if after some point all the terms are getting close to a; for a Cauchy sequence, though, you do not need to know what the limit is, the definition depends only on the terms of the sequence. That said, as we had hoped and as the next theorem shows, these two concepts are actually the same. To prove it, though, we will need the following lemma.

Lemma.

Lemma 3.41 (*Cauchy $\Rightarrow$ bounded*). If (a_n) is Cauchy, then (a_n) is bounded.

Proof Idea. This proof is similar to that of Proposition 3.20, where we showed that convergent sequences are bounded. The reasoning there was: if $a_n \to \alpha$, then after some N, all the terms are bounded between $\alpha - 1$ and $\alpha + 1$. And before N there are only finitely many points, so those are bounded too. A very similar idea works here.

Proof. Assume (a_n) is Cauchy. Then (for $\varepsilon = 1$) there exists some $N \in \mathbb{N}$ such that

$$|a_m - a_n| < 1$$

for all $m, n > N$. In particular, for all $n > N$,

$$|a_n - a_{N+1}| < 1.$$

Therefore for $n > N$, we know that the terms are bounded:

$$a_{N+1} - 1 < a_n < a_{N+1} + 1.$$

And there are only finitely many points before a_N, so these are certainly bounded. Consequently, we can find a general bound on all a_n. Indeed, if we let

$$L = \min\left(\{a_1, a_2, a_3, \dots, a_N, a_{N+1} - 1\}\right)$$

and

$$U = \max\left(\{a_1, a_2, a_3, \dots, a_N, a_{N+1} + 1\}\right),$$

then

$$L \leq a_n \leq U$$

for all $n \in \mathbb{N}$. So (a_n) is bounded. $\square$

> **Theorem.**
>
> **Theorem 3.42** (*Cauchy criterion for convergence*). A sequence converges if and only if it is Cauchy.

Proof Sketch. For the forward direction, suppose a_n converges to a. If a_n and a_m are both within $\varepsilon/2$ of a, then they must be within ε of each other. We can formally show this using the triangle inequality.

For the backwards direction, the idea is this: It will be helpful to identify what the Cauchy sequence is converging to. And by Bolzano-Weierstrass we can find a subsequence that is converging to some a. Our goal will then be to prove that a is in fact the sequence's limit. How? Well, that subsequence gets super close to a. What about the terms in (a_n) which are not in the subsequence? By the Cauchy criterion, they get super close to the elements of the subsequence! And if a term in the sequence is super close to a term of the subsequence, which is in turn super close to the limit... then the sequence must be close to the limit, too.

Proof. First we will prove the forwards direction. Assume that (a_n) converges to some $a \in \mathbb{R}$. Let $\varepsilon > 0$. Since $\varepsilon/2 > 0$, there exists some $N \in \mathbb{N}$ such that for any $n > N$ we have

$$|a_n - a| < \frac{\varepsilon}{2}.$$

Then, for any $n, m > N$,

$$\begin{aligned} |a_n - a_m| &= |a_n - a + a - a_m| \\ &\leq |a_n - a| + |a - a_m| \\ &< \frac{\varepsilon}{2} + \frac{\varepsilon}{2} \\ &= \varepsilon. \end{aligned}$$

Next, we will prove the backwards direction. Assume that (a_n) is Cauchy, and note that by Lemma 3.42, (a_n) is bounded. Combining these two facts, and applying the Bolzano-Weierstrass theorem (Theorem 3.37), we conclude that some subsequence of (a_n) converges. Say, (a_{n_j}) converges to a. Our goal is to show that (a_n) converges; we will in fact prove that $a_n \to a$.

Let $\varepsilon > 0$. Since $\varepsilon/2 > 0$ and (a_n) is Cauchy, there exists some N_1 such that

$$|a_n - a_m| < \varepsilon/2 \qquad (\star)$$

for all $n, m > N_1$. Since $\varepsilon/2 > 0$ and (a_{n_j}) converges to a, there exists some N_2 such that $j > N_2$ implies

$$|a_{n_j} - a| < \varepsilon/2. \qquad (\bigstar)$$

We want to choose some J so that, for subscripts past this point, both $(\star)$ and $(\bigstar)$ hold. Choose $J = \max\{N_1, N_2\}$. Notice, from the definition of a subsequence, that $n_j \geq j$.[28] In particular, $n_{J+1} > J$. And so, for any $j > J$,

$$
\begin{aligned}
|a_j - a| &= |a_j - a_{n_{J+1}} + a_{n_{J+1}} - a| \\
&\leq |a_j - a_{n_{J+1}}| + |a_{n_{J+1}} - a| \\
&< \frac{\varepsilon}{2} + \frac{\varepsilon}{2} \\
&= \varepsilon.
\end{aligned}
$$

$\square$

Back to Axioms

In this text we assumed as an axiom that the reals were complete (AoC).[29] Using AoC we proved, in this order, the monotone convergence theorem (MCT) in Theorem 3.27, the Bolzano-Weierstrass theorem (B-W) in Theorem 3.37, and then the Cauchy criterion (CC) in Theorem 3.42. That is,

$$\text{AoC} \Rightarrow \text{MCT} \Rightarrow \text{B-W} \Rightarrow \text{CC}.$$

- In fact, AoC and MCT are equivalent! If we had started by supposing as an axiom that the reals were an ordered field containing $\mathbb{Q}$ which satisfies MCT, then we could have proved AoC and everything else and ended up right where we are now.

- Also, AoC and B-W are equivalent! So by supposing B-W instead of AoC, we again could have developed our entire theory.

- CC is almost enough, but not quite. Surprisingly, with CC you can not prove the Archimedean principle (AP). But if you assume both CC and AP, then you can prove everything.[30]

That is,

$$\text{AoC} \Leftrightarrow \text{MCT} \Leftrightarrow \text{B-W} \Leftrightarrow (\text{CC} + \text{AP}).$$

Neat stuff.

---

[28]For example, one subsequence is $(a_{n_j}) = (a_1, a_3, a_5, a_7, \dots)$, giving $n_1 = 1$, $n_2 = 3$, $n_3 = 5$, $n_4 = 7$, and so on. And so, $n_1 \geq 1$, and $n_2 \geq 2$, and $n_3 \geq 3$, and $n_4 \geq 4$, and so on.

[29]As someone whose mathematical interests are nearly matched by my interest in politics, the post-2018 dueling use of "AOC" has taken a toll on my casual readings.

[30]Suppose you tried to prove that an ordered field with CC (but not AP) has AoC. To prove AoC. you must prove that given a bounded set A, its least upper bound—$\sup(A)$—exists. A reasonable approach that uses CC would be to try to approach it from above. Start with any upper bound b of A and then pick any $a \in A$. Construct a sequence which approaches where we know $\sup(A)$ to be by setting $x_1 = b$, and for any $n > 1$, letting $x_{n+1} = \begin{cases} x_n - \frac{b-a}{2^n} & \text{if this is an upper bound of } A; \\ x_n & \text{otherwise.} \end{cases}$

Try to convince yourself that this sequence should indeed converge to $\sup(A)$. But how would you prove it? Let $\varepsilon > 0$. We must show there exists a point in this sequence that gets within ε of $\sup(A)$. And showing that we can always get closer comes down to showing that there exists an n such that $\frac{1}{2^n} < \frac{\varepsilon}{b-a}$. And there it is—at this step we require the Archimedean principle.

— Notable Exercises —

- The definition of sequence convergence (Definition 3.7) is tough, with lots of qualifiers, there exists/for all statements, and inequalities flying around. And with many more similar definitions on the horizon, it is important to understand it *fully*. Exercise 3.3 probes your understanding of the quantifiers by asking for examples satisfying related definitions, while Exercises 3.5 and 3.7 ask for specific examples about the actual convergence definition, in special circumstances.

- A *subsequential limit* of a sequence (a_n) is any number L for which there is some subsequence (a_{n_k}) converging to L. Some authors work a lot with subsequential limits, whereas I do not. But Exercises 3.6, 3.19 and 3.31 are indirect examples of this concept.

- Exercise 3.14 is a disguised version of a problem whose solutions was famously given over 2200 years ago by Archimedes. It's a challenging problem, but Exercise 3.14 had to somehow be π-related...

- We finish proving the limit laws (Theorem 3.21) in Exercises 3.7 and 3.8. These proofs also contain some important ideas which will be used many more times in this book.

- Exercise 3.28 is Lemma 4.8, and will play a role when we study geometric series in Chapter 4.

- Exercise 3.45 is a (disguised) first example of a *series*, which is the focus of Chapter 4.

— Exercises —

Exercise 3.1. Suppose that a sequence (a_n) converges to 0.001. Prove that finitely many values of a_n are negative.

Exercise 3.2. Give an example satisfying the requested condition or prove that no such example can exist.

(a) A sequence with infinitely many 0s that does not converge to 0.

(b) A sequence with infinitely many 0s that converges to a non-zero number.

(c) A sequence of positive numbers that converges to a negative number.

(d) A sequence of irrational numbers that converges to a rational number.

Exercise 3.3. This problem is to help you get a feel for the quantifiers in the definition of convergence (Definition 3.7). Your task: For each of the following definitions of *Non*verges, give an example a sequence (a_n) and a value a for which

- (a_n) does not converge to a (based on the real definition),

- (a_n) does *Non*verge to a based on the definition given below.

Give a different example for each problem. For each of them, explain why your example works in a few sentences (no need to prove it completely). Your example for *Non*verges-type-4 should <u>not</u> work for *Non*verges-type-3.

(a) Definition 1: The sequence (a_n) *Non*verges-type-1 to a if for all $\varepsilon > 0$ there exists some $n \in \mathbb{N}$ such that $|a_n - a| < \varepsilon$.

(b) Definition 2: The sequence (a_n) *Non*verges-type-2 to a if for all $\varepsilon > 0$ there exists some $N \in \mathbb{N}$ such that, for some $n > N$, we have $|a_n - a| < \varepsilon$.

(c) Definition 3: The sequence (a_n) *Non*verges-type-3 to a if there exists some $\varepsilon > 0$ such that for all $N \in \mathbb{N}$ there exists some $n > N$ such that $|a_n - a| < \varepsilon$.

(d) Definition 4: The sequence (a_n) *Non*verges-type-4 to a if there exists some $\varepsilon > 0$ and there exists some $N \in \mathbb{N}$ such that for some $n > N$ we have $|a_n - a| < \varepsilon$.

Exercise 3.4. Prove the following using the definition of sequence convergence.

(a) Let $a_n = 7 - \dfrac{1}{\sqrt{n}}$. Show that $a_n \to 7$ as $n \to \infty$.

(b) Let $a_n = \dfrac{2n - 2}{5n + 1}$. Show that $\lim\limits_{n \to \infty} a_n = \dfrac{2}{5}$.

(c) Let $a_n = 7 - \dfrac{1}{\sqrt{n + \sqrt{n} + 13}}$. Show that $a_n \to 7$ as $n \to \infty$.

Exercise 3.5. Give an example of a sequence (a_n) where a_n is negative for all n, and yet $a_n \to 0$.

Exercise 3.6.

(a) Consider the sequence

$$\frac{1}{2}, \frac{1}{3}, \frac{2}{3}, \frac{1}{4}, \frac{2}{4}, \frac{3}{4}, \frac{1}{5}, \frac{2}{5}, \frac{3}{5}, \frac{4}{5}, \frac{1}{6}, \frac{2}{6}, \frac{3}{6}, \frac{4}{6}, \frac{5}{6}, \frac{1}{7}, \cdots$$

For which numbers L does the above have a subsequence converging to L?

(b) Does there exist a sequence (a_n) where, for every $L \in \mathbb{R}$, there is exists a subsequence of (a_n) which coverges to L?

Exercise 3.7. Each of the following is an independent question, but for each suppose that the sequence (a_n) has the property that $a_n \in \mathbb{Z}$ for all n.

(a) Is it possible that $a_n \to 3.5$?

(b) If $a_n \neq a_m$ for all $n \neq m$, prove that (a_n) does not converge.

(c) If (a_n) converges, what can be said about this sequence?

Exercise 3.8. Assume that (a_n) converges to some $a \in \mathbb{R}$ and (b_n) converges to some $b \in \mathbb{R}$. Also assume $c \in \mathbb{R}$.

(a) Prove that $(a_n + b_n)$ converges to $a + b$.

(b) Prove that $(c \cdot a_n)$ converges to $c \cdot a$.

Exercise 3.9. Prove that if (a_n) and (b_n) are sequences where $a_n \to \infty$ and $b_n \to \infty$, then $a_n + b_n \to \infty$.

Exercise 3.10. Assume that (a_{2n}) converges to L and (a_{2n-1}) converges to L, then (a_n) also converges to L.

Exercise 3.11. Prove or disprove: If (a_n) converges, then the set $\{a_n : n \in \mathbb{N}\}$ of values the sequence takes has a *maximum*.

Exercise 3.12. Suppose (a_n) is a sequence and $f : \mathbb{N} \to \mathbb{N}$ is a bijection. For each of the following, prove that the statement is true or find a counterexample showing that the statement is false.

(a) If (a_n) diverges to ∞, then $(a_{f(n)})$ diverges to ∞.

(b) If (a_n) converges to L, then $(a_{f(n)})$ converges to L.

(c) If (a_n)'s limit does not exist, then $(a_{f(n)})$'s limit does not exist.

Exercise 3.13. Assume that (a_n) converges to a and (b_n) converges to b. Also assume $c \in \mathbb{R}$.

(a) Prove that $(a_n - b_n)$ converges to $a - b$.

(b) Prove that $\left(\dfrac{a_n}{b_n}\right) \to \dfrac{a}{b}$, provided $b \neq 0$ and each $b_n \neq 0$.

Exercise 3.14. Let $a_0 = 2\sqrt{3}$ and $b_0 = 3$, and define two sequences recursively by

$$a_n = \frac{2a_{n-1} \cdot b_{n-1}}{a_{n-1} + b_{n-1}} \qquad \text{and} \qquad b_n = \sqrt{a_n \cdot b_{n-1}}.$$

Prove that (a_n) is monotonically decreasing and is convergent, and prove that (b_n) is monotonically increasing and convergent. Then prove that they converge to π.

Exercise 3.15. Give an example of two divergent sequences (a_n) and (b_n) for which $(a_n + b_n)$ converges.

Exercise 3.16. Give an example of two divergent sequences (a_n) and (b_n) for which $(a_n \cdot b_n)$ converges.

Exercise 3.17. Is it possible for a sequence (a_n) to converge, a sequence (b_n) to diverge, and for $(a_n + b_n)$ to converge? Prove that it is impossible or give an example showing that it is possible.

Exercise 3.18.

(a) Assume that (a_n) is bounded and (b_n) converges to 0. Prove that $(a_n \cdot b_n)$ also converges to 0.

(b) Give an example of sequences (a_n) and (b_n) such that $b_n \to 0$ but $(a_n \cdot b_n)$ does not converge to 0.

Exercise 3.19. For each item, provide an example (and prove that it works) or prove that no such example exists.

(a) A sequence (a_n) where $6 < a_n < 7$ for all n, and which has a subsequence converging to 6 and also one converging to 7.

(b) A sequence (a_n) such that, for each $k \in \mathbb{N}$, there is a subsequence of (a_n) converging to $\frac{1}{k}$.

(c) A sequence (a_n) such that, for each $k \in \mathbb{N}$, there is a subsequence of (a_n) converging to $\frac{1}{k}$, but there is no subsequence of (a_n) converging to 0.

(d) A sequence (a_n) such that for every real number x, the sequence (a_n) has a subsequence that converges to x.

Exercise 3.20. For each item, provide an example (and prove that it works) or prove that no such example exists.

(a) A bounded sequence that does not converge to $\frac{4}{9}$ but has a subsequence converging to $\frac{4}{9}$.

(b) A monotone sequence that does not converge to $\frac{4}{9}$ but has a subsequence converging to $\frac{4}{9}$.

(c) A sequence with both an increasing subsequence and a decreasing subsequence that does not converge.

(d) A bounded monotone sequence that does not converge.

(e) A sequence that does not converge and has no convergent subsequences.

(f) A bounded sequence with an unbounded subsequence.

Exercise 3.21. Assume a sequence (a_n) has a bounded subsequence. Must (a_n) have a convergent subsequence?

Exercise 3.22. Suppose (a_n) and (b_n) are sequences.

(a) Prove that if $a_n \to L$ and $a_n \leq M$ for all n, then $L \leq M$.

(b) Assume that $a_n \leq b_n$ for all n. Prove that if $a_n \to L$ and $b_n \to M$, then $L \leq M$.

Exercise 3.23.

(a) Prove that if $a_n \to L$, then $|a_n| \to |L|$.

(b) Give an example where $|a_n| \to |L|$ but $a_n \not\to L$.

Exercise 3.24. Suppose a sequence (a_n) has the property that for any $\varepsilon > 0$ there exists some N such that $m, n > N$ implies

$$|a_m - a_n| < \varepsilon.$$

Prove that (a_n) is bounded.

Exercise 3.25. Prove the decreasing case of the monotone convergence theorem (Theorem 3.27).

Exercise 3.26. Prove the infimum case of Proposition 3.29.

Exercise 3.27. Give an example of an unbounded divergent sequence whose terms are all positive and whose limit does not exist. (Recall, there are three types of divergence: (1) diverging to ∞, (2) diverging to $-\infty$, or (3) does not exist. Your sequence should be of this third type.). You do not need to prove your answer.

Exercise 3.28.   Prove that the sequence (r^n) converges to 0 if $r \in (-1, 1)$, converges to 1 if $r = 1$, and diverges otherwise.[31]

Exercise 3.29.   Suppose (a_n) is a sequence for which $a_n \to a$. Define a new sequence by

$$b_n := \frac{a_1 + a_2 + \cdots + a_n}{n}.$$

Prove that $b_n \to a$.

Exercise 3.30.   Let (a_n) be a bounded sequence, and consider a second sequence (b_n) defined by

$$b_n := \sup\{a_n, a_{n+1}, a_{n+2}, \dots\}.$$

Prove that (b_n) converges.

Exercise 3.31.   Give an example of a sequence (a_n) which has:

- A subsequence converging to 1,

- Another subsequence converging to 17,

- And another subsequence converging to $-\pi$.

Give a brief explanation for why your example works.

Exercise 3.32.   Let (a_n) be a sequence of real numbers. Prove that if *every* subsequence of (a_n) converges, then (a_n) converges too.

Exercise 3.33.   Let (a_n) be a sequence of real numbers. Prove that if (a_n) diverges to ∞, then every subsequence of (a_n) diverges to ∞ as well.

Exercise 3.34.   Give an example of two sequences (a_n) and (b_n) which satisfy the following four conditions:

(i) $\lim\limits_{n \to \infty} a_n$ does not exist,   (iii) $(a_n \cdot b_n) \to \infty$, and

(ii) $\lim\limits_{n \to \infty} b_n$ does not exist,   (iv) $(a_n \cdot b_{n+1}) \to -\infty$.

Exercise 3.35.   Give an example of a sequence whose limit does not exist and none of whose subsequences converge.

Exercise 3.36.   Define a sequence (a_n) recursively as follows. Let a_1 and a_2 be a pair of real numbers, and recursively define

$$a_n := \frac{a_{n-1} + a_{n-2}}{2}$$

for all $n \geq 3$. Does (a_n) necessarily converge?

---

[31] This is Lemma 4.8, as it will play a role in Chapter 4.

Exercise 3.37. Define a sequence (a_n) recursively by $a_1 = \sqrt{2}$ and $a_{n+1} = \sqrt{2 + a_n}$ for all $n \geq 1$.

(a) Show that $a_n \leq 2$ for every n.

(b) Show that (a_n) is an monotone increasing sequence, and use this to conclude that (a_n) converges.

(c) Show that $a_n \to 2$.

Exercise 3.38. Give an example of a monotone sequence that is not Cauchy.

Exercise 3.39. Assume (a_n) and (b_n) are Cauchy sequences, and let $c_n = |a_n - b_n|$. Use a triangle inequality argument to prove that (c_n) is Cauchy.

Exercise 3.40. For each of the following, give an example of a sequence with that property or prove that no such sequences exist.

(a) A Cauchy sequences which has an unbounded subsequence.

(b) An unbounded sequence which has a Cauchy subsequence.

Exercise 3.41. Assume A is an uncountable set of real numbers. Prove that there must exist a convergent sequence (a_n) such that $a_n \in A$ for all n, and $a_n \neq a_m$ for all $n \neq m$.

Exercise 3.42. Suppose the sequence (x_n) is the sum of two other sequences, (y_n) and (z_n). That is, $x_n = y_n + z_n$ for all n. If (x_n) is bounded, and (y_n) and (z_n) are both monotone, must (x_n) be convergent? What if (y_n) and (z_n) are also bounded?

Exercise 3.43. Suppose that (a_n) is a sequence of nonnegative real numbers that converges to L. Show that the sequence $(\sqrt{a_n})$ converges to $\sqrt{L}$.

Exercise 3.44. Prove that if (a_n) is a bounded sequence which does not converge, then it must contain two subsequences, both of which converge, but which converge to different values.

Exercise 3.45. Let (a_n) be the sequence where $a_1 = 1$ and, for each $n > 1$, $a_n = a_{n-1} + \frac{1}{n^2}$. In 1734, Leonhard Euler famously proved that (a_n) converges to $\frac{\pi^2}{6}$. Now let (b_n) be the sequence where $b_1 = 1$ and, for each $n > 1$, $b_n = b_{n-1} + \frac{1}{n^3}$. Use Euler's result to prove that (b_n) converges.

— Open Questions —

Question 1. For a natural number input x_0, construct a sequence recursively by

$$x_{n+1} = \begin{cases} \frac{x_n}{2} & \text{if } x_n \text{ is even} \\ 3x_n + 1 & \text{if } x_n \text{ is odd} \end{cases}$$

Is such a sequence always bounded, regardless of what x_0 is? Will this sequence eventually reach 1, regardless of what x_0 is?

Question 2. Consider the sequence where $a_1 = 2$ and, for $n > 1$, a_n is the smallest prime factor of

$$\left(\prod_{i=1}^{n-1} a_i \right) + 1.$$

(This product plays a starring role in Euclid's proof of the infinitude of the primes.) The first 47 elements of this sequence are

2, 3, 7, 43, 13, 53, 5, 6221671, 38709183810571, 139, 2801, 11, 17, 5471, 52662739, 23003, 30693651606209, 37, 1741, 1313797957, 887, 71, 7127, 109, 23, 97, 159227, 643679794963466223081509857, 103, 1079990819, 9539, 3143065813, 29, 3847, 89, 19, 577, 223, 139703, 457, 9649, 61, 4357, 87991098722552272708281251793312351581099392851768893748012603709343, 107, 127, 3313

Does every prime number appear in this sequence? If not, is the problem of testing whether a given prime appears a computable problem?

Chapter 4: Series

4.1 Sequences of Partial Sums

A sum of infinitely many numbers is called a(n infinite) *series*.[1] For example, you have probably seen the series $\displaystyle\sum_{k=1}^{\infty} \frac{1}{2^k}$, which equals 1. Now consider the *sequence*

$$\frac{1}{2}, \frac{3}{4}, \frac{7}{8}, \frac{15}{16}, \ldots$$

This sequence is converging to 1 (by Exercise 1.29 and the monotone convergence theorem). But note that we could rewrite the terms of this sequence like so:

$$\frac{1}{2}, \left(\frac{1}{2}+\frac{1}{4}\right), \left(\frac{1}{2}+\frac{1}{4}+\frac{1}{8}\right), \left(\frac{1}{2}+\frac{1}{4}+\frac{1}{8}+\frac{1}{16}\right), \ldots$$

Of course, even though we wrote each term as a (finite) sum, this is still a sequence of numbers. Now notice that this is the same as

$$\sum_{k=1}^{1} \frac{1}{2^k}, \sum_{k=1}^{2} \frac{1}{2^k}, \sum_{k=1}^{3} \frac{1}{2^k}, \sum_{k=1}^{4} \frac{1}{2^k}, \ldots,$$

so this is still the same sequence. But do you see how this sequence builds up a series by adding one more term each time? Indeed, we will view a *series* as the limit of a *sequence*, in just this way. And when, back in Calc II, you found the value of an infinite series like

$$\sum_{k=1}^{\infty} \frac{1}{2^k},$$

what you were actually doing is finding the limit of a sequence. This is an important point, because we just finished developing a whole chapter's worth of tools and theorems for sequences, and so if you think of a series as a *sequence of partial sums*, then we can use all of our results on sequences to help us understand series.

---

[1] ♫ *Let's get down to business... to defeat... the sums!* ♫

Basic definitions

Definition.

Definition 4.1. Given a(n infinite) series $\displaystyle\sum_{k=1}^{\infty} a_k$,

- The numbers a_k are the *terms* of the series;

- The *sequence of partial sums* is the sequence $\displaystyle\left(\sum_{k=1}^{n} a_k\right)_{n=1}^{\infty}$. That is, it's the sequence (s_n) where[2]

$$s_1 = a_1$$
$$s_2 = a_1 + a_2$$
$$s_3 = a_1 + a_2 + a_3$$
$$s_4 = a_1 + a_2 + a_3 + a_4$$
$$\vdots$$

- The series *converges* to $L \in \mathbb{R}$ (or $\sum_{k=1}^{\infty} a_k = L$) if $s_n \to L$, and *diverges* (to ∞, to $-\infty$, or does not exist) if (s_n) does. Likewise, we say the series is *bounded* or is *monotone* if (s_n) is.

In Chapter 3 we defined what is means for a sequence (s_n) to converge (in terms of ε's and N's) and to diverge (in terms of M's and N's); we can therefore use these and the above to write out an alternative definition for a *series* to converge or diverge. For example, $\sum_{k=1}^{\infty} a_k = \infty$ if, for every $M > 0$ there exists an N such that $\sum_{k=1}^{n} a_k > M$ for all $n > N$.

Example 4.2. Since the sequence

$$\frac{1}{2}, \ \frac{3}{4}, \ \frac{7}{8}, \ \frac{15}{16}, \ \dots, \tag{$\maltese$}$$

converges to 1, we also have that

$$\sum_{k=1}^{\infty} \frac{1}{2^k} = 1,$$

since ($\maltese$) is the sequence of partial sums of this series.[3]   □

---

[2]Remember, to get the first term of a sequence you plug in $n = 1$, to get the second you plug in $n = 2$, and so on. It just so happens that, in this sequence, each term is itself a sum of numbers.

[3]Riddle: If I offered you either $\displaystyle\sum_{k=0}^{\infty} \frac{1}{2^{k+1}}$ dollars or $\displaystyle\sum_{k=0}^{\infty} \frac{k}{2^{k+1}}$ dollars, which would you choose?

> **Corollary.**
>
> **Corollary 4.3** (*Series (limit) laws*). Assume that $\displaystyle\sum_{k=1}^{\infty} a_k = \alpha$ and $\displaystyle\sum_{k=1}^{\infty} b_k = \beta$. Also assume that $c \in \mathbb{R}$. Then
>
> (i) $\displaystyle\sum_{k=1}^{\infty}(a_k + b_k) = \alpha + \beta$,
>
> (ii) $\displaystyle\sum_{k=1}^{\infty}(a_k - b_k) = \alpha - \beta$, and
>
> (iii) $\displaystyle\sum_{k=1}^{\infty} c \cdot a_k = c \cdot \alpha$.

Proof. Let

$$s_n = \sum_{k=1}^{n} a_k \qquad \text{and} \qquad t_n = \sum_{k=1}^{n} b_k.$$

Now, (s_n) is the sequence of partial sums of the series $\sum_{k=1}^{\infty} a_k$ and (t_n) is the sequence of partial sums of the series $\sum_{k=1}^{\infty} b_k$. Therefore, by assumption, we have that

$$s_n \to \alpha \qquad \text{and} \qquad t_n \to \beta.$$

By the sequence limit laws (Theorem 3.21), we know that

- $(s_n + t_n) \to \alpha + \beta$,

- $(s_n - t_n) \to \alpha - \beta$, and

- $(c \cdot s_n) \to c \cdot \alpha$.

Now let $v_n = \sum_{k=1}^{n}(a_k + b_k)$, and since we can rearrange finite sums,

$$\begin{aligned}
s_n + t_n &= (a_1 + a_2 + \cdots + a_n) + (b_1 + b_2 + \cdots + b_n) \\
&= (a_1 + b_1) + (a_2 + b_2) + \cdots + (a_n + b_n) \\
&= v_n,
\end{aligned}$$

Collecting everything,

$$\begin{aligned}
\sum_{k=1}^{\infty}(a_k + b_k) &= \lim_{n \to \infty} \sum_{k=1}^{n}(a_k + b_k) && \text{(Definition of series)}, \\
&= \lim_{n \to \infty}(v_n) && \text{(Definition of } v_n), \\
&= \lim_{n \to \infty}(s_n + t_n) && \text{(Showed above)}, \\
&= \alpha + \beta, && \text{(First bullet point)}
\end{aligned}$$

as desired.

In almost exactly the same way we can prove that

$$\sum_{k=1}^{\infty}(a_k - b_k) = \alpha - \beta \qquad \text{and} \qquad \sum_{k=1}^{\infty} c \cdot a_k = c \cdot \alpha.$$

$\square$

Question 4.4. There were two other limit laws, one for multiplication and one for division. However, we can *not* conclude that $\displaystyle\sum_{k=1}^{\infty} a_k \cdot b_k = \alpha \cdot \beta$. Why not?

Answer. Remember, to determine what any series equals, it all comes back to *sequences of partial sums*. The first sequences involved are

$$(s_n) = \left(\sum_{k=1}^{n} a_k\right) \qquad \text{and} \qquad (t_n) = \left(\sum_{k=1}^{n} b_k\right),$$

and so we have $s_n \to \alpha$ and $t_n \to \beta$. We can then apply limit laws to the limit

$$\lim_{n\to\infty} \left(\sum_{k=1}^{n} a_k\right)\left(\sum_{k=1}^{n} b_k\right),$$

to conclude that

$$\lim_{n\to\infty} \left(\sum_{k=1}^{n} a_k\right)\left(\sum_{k=1}^{n} b_k\right) = \alpha \cdot \beta,$$

but since, in general,

$$\left(\sum_{k=1}^{n} a_k\right)\left(\sum_{k=1}^{n} b_k\right) \neq \sum_{k=1}^{n} a_k \cdot b_k,$$

the limit laws have next-to-nothing to say about

$$\sum_{k=1}^{n} a_k \cdot b_k.$$

$\square$

As a first example of this, note that

$$\left(\sum_{k=1}^{2} k\right)\left(\sum_{k=1}^{2} k^2\right) = (1+2)(1+4) = 15 \neq 9 = (1+8) = \sum_{k=1}^{2} k \cdot k^2.$$

4.2   Series Convergence Tests

This section is a lot of fun because here we begin proving theorems that you learned in calculus. Calculus is truly a paramount accomplishment of the human mind, and yet in just 121 pages we are already set to prove results important enough that just about every STEM major in the world learns them. Let's get after it.

Proposition.

Proposition 4.5 (k^{th}-*term test*).[4]   If $a_k \not\to 0$, then $\displaystyle\sum_{k=1}^{\infty} a_k$ diverges.

By the contrapositive, this is equivalent to saying that if $\sum_{k=1}^{\infty} a_k$ converges, then $a_k \to 0$. And this seems like an easier way to proceed, since the assumption that $a_k \not\to 0$ can mean many things: It can mean that it is converging to something else, diverging to $\pm\infty$, or that its limit simply does not exist; but assuming that $\sum_{k=1}^{\infty} a_k$ converges to 0 has a specific definition, and hence is a more solid place to begin.

Proof. We will prove the contrapositive: If $\displaystyle\sum_{k=1}^{\infty} a_k$ converges, then $a_k \to 0$. To that end, assume that $\displaystyle\sum_{k=1}^{\infty} a_k$ converges and let $s_n = \displaystyle\sum_{k=1}^{n} a_k$. Then we are assuming that the sequence (s_n) converges, and hence (s_n) is Cauchy (by Theorem 3.42). So for any $\varepsilon > 0$ there exists some N such that for all $m, n > N$ we have

$$|s_n - s_m| < \varepsilon.$$

And without loss of generality, let's assume that $n \geq m$. Then, going back to what s_n was defined to be, this means that

$$\left|\sum_{k=1}^{n} a_k - \sum_{k=1}^{m} a_k\right| < \varepsilon, \quad \text{which is equivalent to} \quad \left|\sum_{k=m+1}^{n} a_k\right| < \varepsilon.$$

This holds *for all* values of $m, n > N$. In the special case where $m = n - 1$, this gives

$$\left|\sum_{k=n}^{n} a_k\right| < \varepsilon$$
$$|a_n| < \varepsilon$$
$$|a_n - 0| < \varepsilon.$$

---

[4]Also known as the *divergence test*.

That is, we have shown that for any $\varepsilon > 0$ there exists some N such that for all $n > N$,

$$|a_n - 0| < \varepsilon.$$

That is, by definition, $a_n \to 0$, completing the proof. $\qquad\square$

One class of series has a particularly nice, natural form: the *geometric series*.

> **Definition.**
>
> **Definition 4.6.** A *geometric series* is a series of the form
>
> $$\sum_{k=0}^{\infty} ar^k = a + ar + ar^2 + ar^3 + \ldots,$$
>
> where $a, r \in \mathbb{R}$.

Geometric series are ubiquitous in the real world, because they tend to occur whenever each agent of a system acts independently. Almost all couples decide whether to have another kid based on their own situation, not on the current population. Therefore population growth is geometric. Each atom in a radioactive substance disintegrates independently of the rest, and so radioactive decays are geometric — typically described by their "half life." From bouncing balls in physics to interest rates in economics to Zeno's paradoxes in this text's opening pages, they are perhaps the most natural class of series to study.

Example 4.7. Here are a couple numerical examples of geometric series.

- $\displaystyle\sum_{k=0}^{\infty} 2 \cdot \left(\frac{3}{5}\right)^k$

- $\displaystyle\frac{\pi}{2} + \frac{3\pi^2}{4e} + \frac{9\pi^3}{8e^2} + \frac{27\pi^4}{16e^3} + \ldots$

As is often the case in mathematics, good questions have good answers. Here, we ask when this nice class of series converges — and the answer is likewise nice. The proof will use the following lemma.

> **Lemma.**
>
> **Lemma 4.8** ((r^n) *converges iff* $r \in (-1, 1]$). The sequence (r^n) converges to 0 if $r \in (-1, 1)$, converges to 1 if $r = 1$, and diverges otherwise.

Proof. This is Exercise 3.28. $\qquad\square$

The *geometric series test* now gives the detailed version of the very nice answer.

> ### Proposition.
>
> **Proposition 4.9** (*Geometric series test*). Assume that a and r are non-zero real numbers. Then
>
> $$\sum_{k=0}^{\infty} a \cdot r^k = \begin{cases} \dfrac{a}{1-r} & \text{if} \quad |r| < 1 \\ \text{diverges}^5 & \text{if} \quad |r| \geq 1 \end{cases}$$

Notice that a is always the first term of the series, and r is the *common ratio*.[6]

Proof. The case where $|r| > 1$ follows from Lemma 4.8 and the kth-term test. When $r = 1$, the series is $a + a + a + \ldots$ which clearly diverges, and when $r = -1$ the series is $a - a + a - a + a - \ldots$ whose sequence of partial sums, $(a, 0, a, 0, \ldots)$, is clearly not converging. We therefore turn to the case that $|r| < 1$. Note that

$$(1-r)(1 + r + r^2 + r^3 + \cdots + r^n)$$
$$= 1 + r + r^2 + r^3 + \cdots + r^n$$
$$- r - r^2 - r^3 - \cdots - r^{n-1} - r^{n+1}$$
$$= 1 + 0 + 0 + \cdots + 0 - r^{n+1}$$
$$= 1 - r^{n+1},$$

which, by dividing over the $1 - r$, shows that

$$1 + r + r^2 + r^3 + \cdots + r^n = \frac{1 - r^{n+1}}{1 - r},$$

and hence

$$s_n = a + ar + ar^2 + ar^3 + \cdots + ar^n = \frac{a(1 - r^{n+1})}{1 - r},$$

where s_n is the nth partial sum of $\sum_{k=0}^{\infty} a \cdot r^k$. Thus,

$$\sum_{k=0}^{\infty} a \cdot r^k = \lim_{n \to \infty} s_n$$
$$= \lim_{n \to \infty} \frac{a(1 - r^{n+1})}{1 - r}$$
$$= \frac{a(1 - 0)}{1 - r}. \qquad \text{(Lemma 4.8)}$$
$$= \frac{a}{1 - r}. \qquad \square$$

---

[5] Exercise 4.9 asks you to classify the type of divergence.

[6] **Definition.** The *common ratio* of a geometric series is what you have to multiply any term by to get the next term.

Example 4.10. In Example 4.2, we noted—without proof—that the sequence

$$\frac{1}{2}, \frac{3}{4}, \frac{7}{8}, \frac{15}{16}, \cdots$$

converges to 1. Using Proposition 4.9, we can now prove this.

Note that $\displaystyle\sum_{k=1}^{\infty} \frac{1}{2^k}$ has a first term of $a = 1/2$ and a common ratio of $r = 1/2$. Therefore

$$\sum_{k=1}^{\infty} \frac{1}{2^k} = \frac{1/2}{1 - \frac{1}{2}} = \frac{1/2}{1/2} = 1.$$

$\square$

An alternative approach to finding r and a is this: First write the series in the form given in Proposition 4.9. That would be this:

$$\sum_{k=1}^{\infty} \frac{1}{2^k} = \frac{1}{2} + \frac{1}{4} + \frac{1}{8} + \frac{1}{16} \cdots = \sum_{k=0}^{\infty} \frac{1}{2} \cdot \left(\frac{1}{2}\right)^k.$$

And now it is clear that $a = 1/2$ and $r = 1/2$.

We next prove that if you add up infinitely many non-negative numbers, then two things can not happen. First, the answer can not be negative; this probably seems obvious. Second, the answer can not diverge of the "does not exist" variety; this may seem less obvious. The actual proposition is instead phrased in terms of which two things *can* happen.

Proposition.

Proposition 4.11 ($a_k \geq 0 \Rightarrow \sum a_k$ *converges or* $= \infty$). If $a_k \geq 0$ for all k, then $\displaystyle\sum_{k=1}^{\infty} a_k$ either converges or it diverges to ∞.

The series is either bounded (hence converges) or is not bounded (hence diverges).

Proof. Since each a_k is nonnegative, observe that the sequence of partial sums is monotone increasing. Applying the monotone convergence theorem (Theorem 3.27) then gives the result. $\square$

> **Proposition.**
>
> **Proposition 4.12** (*Comparison test*). Assume $0 \leq a_k \leq b_k$ for all k.
>
> (i) If $\displaystyle\sum_{k=1}^{\infty} b_k$ converges, then $\displaystyle\sum_{k=1}^{\infty} a_k$ converges.
>
> (ii) If $\displaystyle\sum_{k=1}^{\infty} a_k$ diverges, then $\displaystyle\sum_{k=1}^{\infty} b_k$ diverges.

The converse of these statements is not true, and Exercise 4.16 asks for example demonstrating this.

Proof. Let (s_n) be the sequence of partial sums of $\displaystyle\sum_{k=1}^{\infty} a_k$, and let (t_n) be the sequence of partial sums of $\displaystyle\sum_{k=1}^{\infty} b_k$. And here is an observation that will be useful: Note that since $a_k \leq b_k$ for all k,

$$s_n = a_1 + a_2 + \cdots + a_n \leq b_1 + b_2 + \cdots + b_n = t_n$$

for all n. That is,

$$s_n \leq t_n \qquad \text{for all } n. \tag{$\maltese$}$$

To see (i), first note[7] that by Proposition 4.11 that all we need to show is that (s_n) is bounded above. If we can do this we know that $s_n \not\to \infty$, which by that proposition implies that it must converge.

Recall that in Proposition 3.20 we showed that if a sequence converges, then it is bounded. And so, since by assumption $\sum_{k=1}^{\infty} b_k$ converges (meaning that (t_n) converges), we deduce that this series is bounded above by some value M; that is $t_n \leq M$ for all n (this is Definition 3.4). And so by $(\maltese)$, $s_n \leq t_n \leq M$ for all n, showing that (s_n) is also bounded above by M, completing the proof of (i).

The proof of (ii) is similar. Let $M > 0$. Assume that $\sum_{k=1}^{\infty} a_k$ diverges, meaning that (s_n) diverges (and hence by Proposition 4.11 this means it diverges to ∞). And so by the definition of diverging to ∞ (Definition 3.15), there exists some N for which $M < s_n$ for all $n > N$. And so by $(\maltese)$, $M < s_n \leq t_n$ for all $n > N$. Therefore, again by Definition 3.15, (t_n) also diverges to ∞. $\square$

Observe that if, say, $0 \leq a_k \leq b_k$ for all but the first 10 terms, then the conclusion of the theorem would still hold. Indeed, if $\sum b_k$ converges and $0 \leq a_k \leq b_k$ for all k, then the comparison test says that $\sum a_k$ converges. But if you now change the first

---

[7]The rest of the proof is a mile wide and an inch deep. It's almost more memory than math — basically just a piecing together of a bunch of old definitions and propositions.

10 terms of a_k so that they are larger than b_k, there is no way that those ten terms now cause $\sum a_k$ to diverge.

In general, as long as $a_k \leq b_k$ holds for *all but finitely many* terms, the theorem still applies, as the following note says.

> ### Note.
>
> **Note 4.13.** Changing finitely many terms of a sequence or a series does not affect whether or not the sequence or series converges.

Example 4.14. Suppose

$$a_k = \begin{cases} k^2 + 7 & \text{if } k \leq 1,000, \\ \left(\frac{1}{2}\right)^k & \text{if } k > 1,000. \end{cases}$$

Then $\displaystyle\sum_{k=0}^{\infty} a_k$ converges. To see this, simply note that by changing finitely many terms we can reach $\displaystyle\sum_{k=0}^{\infty} \left(\frac{1}{2}\right)^k$, which converges by the geometric series test (where $r = \frac{1}{2}$). Now apply Note 4.13.   □

In the exercises you are asked to prove the ratio test (Exercise 4.17) and the root test (Exercise 4.18), which you might recall from your calculus days. We now move toward proving the series p-test and the alternating series test.

Harmonic Series and the Series p-Test

The *harmonic series* is famous. It is the series

$$1 + \frac{1}{2} + \frac{1}{3} + \frac{1}{4} + \frac{1}{5} + \frac{1}{6} + \frac{1}{7} + \frac{1}{8} + \cdots.$$

The terms are getting smaller and and smaller, and yet:

> ### Proposition.
>
> **Proposition 4.15** ($\sum \frac{1}{k}$ *diverges*). The harmonic series $\displaystyle\sum_{k=1}^{\infty} \frac{1}{k}$ diverges.

Proof. Observe that

$$\sum_{k=1}^{\infty} \frac{1}{k} = 1 + \frac{1}{2} + \frac{1}{3} + \frac{1}{4} + \frac{1}{5} + \frac{1}{6} + \frac{1}{7} + \frac{1}{8} + \dots$$

$$= 1 + \left(\frac{1}{2}\right) + \left(\frac{1}{3} + \frac{1}{4}\right) + \left(\frac{1}{5} + \frac{1}{6} + \frac{1}{7} + \frac{1}{8}\right) + \dots$$

$$\geq 1 + \left(\frac{1}{2}\right) + \left(\frac{1}{4} + \frac{1}{4}\right) + \left(\frac{1}{8} + \frac{1}{8} + \frac{1}{8} + \frac{1}{8}\right) + \dots$$

$$= 1 + \left(\frac{1}{2}\right) + \left(\frac{1}{2}\right) + \left(\frac{1}{2}\right) + \dots.$$

In particular, if s_n is the nth partial sum of the harmonic series, then (s_n) is monotonically increasing and, by the above,

$$s_{2^n} \geq 1 + n \cdot \frac{1}{2}.$$

And since $(1 + n \cdot \frac{1}{2})$ diverges to ∞, by the comparison test (Proposition 4.12) the subsequence (s_{2^n}) diverges to ∞. And for a monotonically increasing sequence, if a subsequence diverges to ∞ — implying that (s_n) is unbounded — the entire sequence, (s_n), is also diverging to ∞ by the monotone convergence theorem (Theorem 3.27). That is, the harmonic series diverges to ∞. $\square$

The above proof is the classic proof that the harmonic series diverges, and it was discovered *in the Middle Ages*, predating almost every other result in this text. It was proved by French philosopher Nicole Oresme around 1350 (who somehow found time to study math while his country fought England in the Hundred Years' War, and the Black Death was becoming perhaps the worst pandemic in human history). Due to the significance of this theorem, though, many later mathematicians worked to find their own proofs of the result.[8]

---

[8]Check out the survey *The Harmonic Series Diverges Again and Again* by Kifowit and Stamps, in which they collect 20 more proofs of the theorem. It includes a 17th century proof by Jacob Bernoulli utilizing partial fractions, an 18th century proof by Leonhard Euler using power series, and many more, from using the integral test to using binomial coefficients. A proof that I particularly like, which is by contradiction but otherwise uses similar ideas to the proof above, is this: Suppose for a contradiction that the harmonic series converges to S. Then

$$S = 1 + \frac{1}{2} + \frac{1}{3} + \frac{1}{4} + \frac{1}{5} + \frac{1}{6} + \frac{1}{7} + \frac{1}{8} + \dots$$

$$= \left(1 + \frac{1}{2}\right) + \left(\frac{1}{3} + \frac{1}{4}\right) + \left(\frac{1}{5} + \frac{1}{6}\right) + \left(\frac{1}{7} + \frac{1}{8}\right) + \dots$$

$$> \left(\frac{1}{2} + \frac{1}{2}\right) + \left(\frac{1}{4} + \frac{1}{4}\right) + \left(\frac{1}{6} + \frac{1}{6}\right) + \left(\frac{1}{8} + \frac{1}{8}\right) + \dots$$

$$= 1 + \frac{1}{2} + \frac{1}{3} + \frac{1}{4} + \frac{1}{5} + \frac{1}{6} + \frac{1}{7} + \frac{1}{8} + \dots$$

$$= S.$$

We've shown $S > S$, a contradiction. $\square$

So we know that $\sum_{k=1}^{\infty} \frac{1}{k} = \infty$. That is, for any $M > 0$, eventually the sum $1 + \frac{1}{2} + \frac{1}{3} + \frac{1}{4} + \ldots$ gets above M. This sum diverges suuuuuuper slowly, though. In fact, in 1968 — way before the age of the supercomputer — J.W. Wrench Jr. showed that $\sum_{k=1}^{n} \frac{1}{k} > 100$ for the first time when

$$n = 15,092,688,622,113,788,323,693,563,264,538,101,449,859,497$$

That is, summing the first 15,092,688,622,113,788,323,693,563,264,538,101,449,859,496 integer reciprocals gives a result that is less than 100. But when you add

$$\frac{1}{15,092,688,622,113,788,323,693,563,264,538,101,449,859,497}$$

to that sum, you get above 100. He showed it in just 6 pages[9] and a few (very weak) computer calculations. Amazing.[10]

So yes, the Harmonic series just *barely* diverges; in a precise sense, it is right on the borderline between convergence and divergence, as the next result shows.

> ### Proposition.
>
> **Proposition 4.16** (*The series p-test*). The series $\sum_{k=1}^{\infty} \frac{1}{k^p}$ converges if and only if $p > 1$.

Assume $k \geq 1$. In the proof we will use the fact that k^p increases if you fix k and increase p, or if you fix p and increase k.

Proof. If $p \leq 1$, then $\frac{1}{k} \leq \frac{1}{k^p}$ for all $k \in \mathbb{N}$. And since $\sum_{k=1}^{\infty} \frac{1}{k}$ diverges, by the comparison test, $\sum_{k=1}^{\infty} \frac{1}{k^p}$ diverges too.

---

[9]He used something called the Euler-Maclaurin formula, which gives an integral expression for how far off the nth partial sum of the harmonic series is from the natural log function.

[10]Nowadays, even WolframAlpha can verify that Wrench was correct. If you have the patience, use WolframAlpha to check his solution on your own. It's pretty amazing to see. Carefully type in "sum of 1/k from k=1 to k=15,092,688,622,113,788,323,693,563,264,538,101,449,859,496", and see what you get. Then change that last '6' to a '7' and see how it changes...

Now assume that $p > 1$. Then

$$\sum_{k=1}^{\infty} \frac{1}{k^p} = 1 + \frac{1}{2^p} + \frac{1}{3^p} + \frac{1}{4^p} + \frac{1}{5^p} + \frac{1}{6^p} + \frac{1}{7^p} + \ldots$$

$$= 1 + \left(\frac{1}{2^p} + \frac{1}{3^p} \right) + \left(\frac{1}{4^p} + \frac{1}{5^p} + \frac{1}{6^p} + \frac{1}{7^p} \right) + \ldots$$

$$< 1 + \left(\frac{1}{2^p} + \frac{1}{2^p} \right) + \left(\frac{1}{4^p} + \frac{1}{4^p} + \frac{1}{4^p} + \frac{1}{4^p} \right) + \ldots$$

$$= 1 + \left(\frac{2}{2^p} \right) + \left(\frac{4}{4^p} \right) + \ldots$$

$$= 1 + \frac{1}{2^{p-1}} + \frac{1}{4^{p-1}} + \ldots$$

$$= 1 + \frac{1}{2^{p-1}} + \frac{1}{2^{2(p-1)}} + \ldots$$

$$= 1 + \frac{1}{2^{p-1}} + \left(\frac{1}{2^{p-1}} \right)^2 + \ldots$$

$$= \sum_{k=0}^{\infty} \left(\frac{1}{2^p} \right)^k.$$

Finally, note that $\sum_{k=0}^{\infty} \left(\frac{1}{2^p} \right)^k$ is a geometric series with $|r| = \frac{1}{2^p} < 1$, so by the geometric series test (Proposition 4.9), this series converges. And the above also shows that, term-by-term, this geometric series is larger than $\sum_{k=1}^{\infty} \frac{1}{k^p}$. So by the comparison test, $\sum_{k=1}^{\infty} \frac{1}{k^p}$ converges too. $\square$

The $p = 2$ case is particularly interesting; that is, the sum $1 + \frac{1}{4} + \frac{1}{9} + \frac{1}{16} + \ldots$. It's called the *Basel Problem* after Basel, Switzerland where both Leonhard Euler and his mentor Johann Bernoulli were born. Bernoulli and many other prominent world-class mathematicians tried and failed to determine the sum. But Euler, at age 24, solved the problem, thrusting himself onto the world stage. It was a significant problem, but the most remarkable part is its incredible answer:

$$\sum_{k=1}^{\infty} \frac{1}{k^2} = \frac{\pi^2}{6}.$$

The number π is defined by circles. What the hell do sums of inverse squares have to do with circles?![11] And why squared? It seems utterly unreasonable that a π

---

[11] As far as I can tell, Euler's original proof provided almost no intuition as to what that π is doing there. For some intuition based on a 2010 paper by Johan Wästlund, watch the excellent 3blue1brown video at `https://youtu.be/d-o3eB9sfls`.

should be involved in the answer, and I hope Euler fell off his seat when he discovered it. I wouldn't say there's too much direct evidence that there's a God. But...I think we need to put this one in God's column.

4.3   Absolute Convergence

Suppose each $a_k \geq 0$. Then $\sum a_k$ may or may not converge, but note that the series $\sum (-1)^k a_k$ is, if anything, *more* likely to converge. The original series, $\sum a_k$, can only converge or diverge to ∞, so changing half of the terms to a negative should only help us cancel more things out and cause the series to converge.

Indeed, we have seen examples that show that the converse to the kth-term test (Proposition 4.5) does not hold. That is, it is possible to have the terms of a series converge to 0 while the series still diverges. If your series is alternating, though, this is not true, as the following theorem states.

Proposition.

Proposition 4.17 (*Alternating series test*).   Assume that (a_k) is a monotonically decreasing sequence and $a_k \to 0$. Then $\displaystyle\sum_{k=1}^{\infty} (-1)^{k+1} a_k$ converges.

Note: Since (a_k) is monotonically decreasing and converging to zero, this implies $a_k \geq 0$ for all k.

Proof. Let (s_n) be the sequence of partial sums of this series. First we will show that the subsequence (s_{2n}) is monotonically increasing. To see this, observe that by grouping the terms of the subsequence (s_{2n}),

$$s_{2n} = (a_1 - a_2) + (a_3 - a_4) + \cdots + (a_{2n-1} - a_{2n}),$$

and simply observe that the fact that (a_n) is monotonically decreasing implies that $(a_i - a_{i+1}) \geq 0$ for all i. But not only is (s_{2n}) monotonically increasing, by grouping the terms differently we can see that it is bounded above by a_1:

$$a_1 \geq a_1 - (a_2 - a_3) - (a_3 - a_4) - \cdots - (a_{2n-2} - a_{2n-1}) - a_{2n}.$$

In summary, (s_{2n}) is a monotonically increasing sequence that is bounded above by a_1, and hence converges by the monotone convergence theorem (Theorem 3.27). Call its limit L.

Through a similar argument to the above, one can show that (s_{2n+1}) is monotonically decreasing and bounded below by $a_1 - a_2$, and hence also converges. And moreover, it must also be converging to L. The reason is, if we call its limit U, then

$$L - U = \lim_{n \to \infty} s_{2n} - \lim_{n \to \infty} s_{2n-1} = \lim_{n \to \infty} (s_{2n} - s_{2n-1}) = \lim_{n \to \infty} a_{2n} = 0.$$

And so $U = L$, as asserted.

What we still need to prove is that if (s_n) is a sequence in which (s_{2n}) and (s_{2n+1}) both converge to L, it must also be the case that (s_n) converges to L. To prove this last claim, let $\varepsilon > 0$ and note that for the first subsequence there is an N_1 for which $|s_n - L| < \varepsilon$ for all even $n > N_1$, and for the second there is an N_2 for which $|s_n - L| < \varepsilon$ for all odd $n > N_2$. Now let $N = \max\{N_1, N_2\}$ and observe that since every $n > N$ was an index in one of the two subsequences, we have that

$$|s_n - L| < \varepsilon.$$

Thus $s_n \to L$. And since s_n is the sequence of partial sums of the series, this means that the series also converges to L, completing the proof. $\qquad\square$

Why is it necessary for the sequence to be monotonically decreasing? If we only assumed that it was an alternating sequence of numbers converging to zero, why is that not enough? The idea is to have a series where the positive terms converge to zero but their sum diverges to ∞; the negative terms converge to zero and their sum also converges, and hence are not big enough to "slow down" the divergence of the positive terms. For example, the series

$$\sum_{k=1}^{\infty} \left(\frac{1}{k} - \frac{1}{k^2} \right)$$

would have this property. In order to get a series $\sum_{k=1}^{\infty} a_k$ which produces this (where the odd-indexed terms are positive and the even-indexed terms are negative), let

$$a_k = \begin{cases} \frac{2}{k+1} & \text{if } k \text{ is odd;} \\ \frac{4}{k^2} & \text{if } k \text{ is even.} \end{cases}$$

Then

$$\sum_{k=1}^{\infty} (-1)^{k-1} a_k = \frac{1}{1} - \frac{1}{1^2} + \frac{1}{2} - \frac{1}{2^2} + \frac{1}{3} - \frac{1}{3^2} + \frac{1}{4} - \frac{1}{4^2} + \dots$$

$$= \left(\frac{1}{1} - \frac{1}{1^2} \right) + \left(\frac{1}{2} - \frac{1}{2^2} \right) + \left(\frac{1}{3} - \frac{1}{3^2} \right) + \left(\frac{1}{4} - \frac{1}{4^2} \right) + \dots$$

$$= \sum_{k=1}^{\infty} \left(\frac{1}{k} - \frac{1}{k^2} \right).$$

Indeed, if your sum alternates between positive and negative, and if either

- The sum of your positive terms diverge to ∞ while the sum of your negative terms converges, *or*

- The sum of your negative terms diverge to $-\infty$ while the sum of your positive terms converges, then your series must diverge. On the other hand, if both the positive terms and the negative terms converge on their own, then your series must converge. One way to think about this is that if you took the absolute value of all the terms, and the resulting sum converges, then the original sum must converge too. We codify this idea in the following proposition.

Proposition.

Proposition 4.18 ($\sum |a_k|$ *converges* $\Rightarrow \sum a_k$ *converges*). If $\displaystyle\sum_{k=1}^{\infty} |a_k|$ converges, then $\displaystyle\sum_{k=1}^{\infty} a_k$ converges too.

Proof. This is Exercise 4.3. $\square$

The converse, though, is not true: By Proposition 4.15, the harmonic series $\displaystyle\sum_{k=1}^{\infty} \frac{1}{k}$ diverges, while the *alternating harmonic series* $\displaystyle\sum_{k=1}^{\infty} (-1)^{k+1} \frac{1}{k}$ converges by Proposition 4.17. Indeed, here are the first few partial sums of this latter series:

$$1 = 1$$
$$1 - \frac{1}{2} = \frac{1}{2} = 0.5$$
$$1 - \frac{1}{2} + \frac{1}{3} = \frac{5}{6} = 0.8333\ldots$$
$$1 - \frac{1}{2} + \frac{1}{3} - \frac{1}{4} = \frac{7}{12} = 0.58333\ldots$$
$$1 - \frac{1}{2} + \frac{1}{3} - \frac{1}{4} + \frac{1}{5} = \frac{23}{60} = 0.78333\ldots$$
$$\vdots$$

It certainly seems to be converging, and to something around 0.7, but just like with the Basel problem, what it is converging to is quite surprising:

Fact.

Fact 4.19.
$$\sum_{k=1}^{\infty} (-1)^{k+1} \frac{1}{k} = \ln(2).$$

Unlike some of the crazy results mentioned throughout this text, this one can be demonstrated just with Calculus II material.[12]

Now, some *alternating series* (that is, series which alternate between positive and negative) are like the alternating harmonic series—they converge, but if you take away the $(-1)^k$ component they won't converge. Other alternating series, though, would converge even without alternating. This distinction is important.

Definition.

Definition 4.20. Consider the series $\displaystyle\sum_{k=1}^{\infty} a_k$.

- If $\displaystyle\sum_{k=1}^{\infty} |a_k|$ converges, then we say $\displaystyle\sum_{k=1}^{\infty} a_k$ *converges absolutely.*

- If $\displaystyle\sum_{k=1}^{\infty} a_k$ converges but $\displaystyle\sum_{k=1}^{\infty} |a_k|$ diverges, then $\displaystyle\sum_{k=1}^{\infty} a_k$ *converges conditionally.*

And with that, we are ready to talk about *rearrangements*, which include what I think is by-far the coolest theorem on this topic.

---

[12]We won't reach this until Chapter 9, but if you recall your Calculus II well enough, you might recall a way to see that this is true. The geometric series test says that (for $x \in (-1, 1)$),

$$1 - x + x^2 - x^3 + x^4 - \cdots = \frac{1}{1+x}.$$

Integrating term-by-term is allowed, and doing so gives the Taylor series for $\ln(x+1)$:

$$x - \frac{x^2}{2} + \frac{x^3}{3} - \frac{x^4}{4} + \frac{x^5}{5} - \cdots = \ln(1+x).$$

Now one can argue that plugging in $x = 1$ is valid. $\square$

But note: The above only argues *that* the fact is true. But *why* is it true?? Why is a logarithm with a base of e so intimately related to inverses of integers? That's a *much* harder question to answer. The following is certainly insufficient, but on the right is a geometric interpretation of how the sum

$$1 - \frac{1}{2} + \frac{1}{3} - \frac{1}{4} + \frac{1}{5} - \frac{1}{6} + \frac{1}{7} - \frac{1}{8} + \dots$$

is equal to the area under the curve of $f(x) = 1/x$ from $x = 1$ to $x = 2$, which by your calculus you recall equals $\int_1^2 \frac{1}{x}\, dx = \ln(2) - \ln(1) = \ln(2)$.

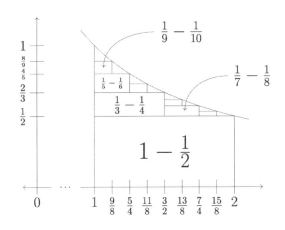

4.4 Rearrangements

> **Definition.**
>
> **Definition 4.21.** A *rearrangement* of a series $\sum_{k=1}^{\infty} a_k$ is a series $\sum_{k=1}^{\infty} b_k$ for which there is a bijection $f : \mathbb{N} \to \mathbb{N}$ such that $b_{f(k)} = a_k$.

It is precise to define a rearrangement in terms of bijections (especially if the series has repeated terms) but, intuitively, a rearrangement of a series is simply the same series where you add up the terms in a different order.

Example 4.22. The alternating harmonic series is

$$1 - \frac{1}{2} + \frac{1}{3} - \frac{1}{4} + \frac{1}{5} - \frac{1}{6} + \frac{1}{7} - \frac{1}{8} + \dots.$$

A rearrangement of this series is

$$\frac{1}{5} - \frac{1}{2} - \frac{1}{152} + \frac{1}{57} - \frac{1}{1847604} - \frac{1}{8} + \dots.$$

A different rearrangement is

$$1 + \frac{1}{3} + \frac{1}{5} - \frac{1}{2} + \frac{1}{7} + \frac{1}{9} + \frac{1}{11} - \frac{1}{4} + \dots.$$

If you have *finitely* many numbers that you are adding/subtracting, then of course the order in which you add/subtract them does not matter — you get the same answer either way. Amazingly, though, if you have *infinitely* many numbers, then it *can* make a difference. It is possible to add up infinitely many numbers and have it equal one thing, but then by simply changing the order that you are adding up the numbers it is possible for the sum to equal something different! The order that you add the numbers can change what it equals! But if that's not freaky enough, it gets worse/better: In some cases, the sum can equal *anything at all*. This is one of my favorite theorems of this entire book, so pay attention.

> **Super Cool Theorem.**
>
> **Theorem 4.23** (*Rearrangement theorem*). If a series $\sum_{k=1}^{\infty} a_k$ converges conditionally, then for any L ($L \in \mathbb{R}$ or $L = \pm\infty$) there exists some rearrangement of $\sum_{k=1}^{\infty} a_k$ which converges to L.

Proof Sketch. Let p_k be the kth positive term of (a_n) and let n_k be the kth negative term of (a_n). Here's the basic idea of the proof:

- Since $\sum_{k=1}^{\infty} a_k$ converges, by the kth-term test we must have $a_k \to 0$.

- Since $\sum_{k=1}^{\infty} a_k$ converges conditionally, we must have:

  - $\displaystyle\sum_{k=1}^{\infty} p_k = \infty$ (while $p_k \to 0$), and

  - $\displaystyle\sum_{k=1}^{\infty} n_k = -\infty$ (while $n_k \to 0$).

  This is the case since, if just one were infinite, then $\sum_{k=1}^{\infty} a_k$ would diverge. And if both were finite, then $\sum_{k=1}^{\infty} a_k$ would converge absolutely.

- Assume you are trying to converge to $L \in \mathbb{R}$ (the case $L = \infty$ is Exercise 4.23), and without loss of generality assume $L > 0$.

- Since $\sum_{k=1}^{\infty} p_k = \infty$, there exists some P_1 such that $\sum_{k=1}^{P_1} p_k > L$. Choose P_1 to be the smallest such P_1 that works.

- Since $\sum_{k=1}^{\infty} n_k = -\infty$, there exists some N_1 such that $\sum_{k=1}^{P_1} p_k + \sum_{k=1}^{N_1} n_k < L$. Choose N_1 to be the smallest such N_1 that works.

- Since $\sum_{k=P_1+1}^{\infty} p_k = \infty$, there exists some $P_2 > P_1$ such that $\sum_{k=1}^{P_1} p_k + \sum_{k=1}^{N_1} n_k + \sum_{P_1+1}^{P_2} p_k > L$. Choose P_2 to be the smallest such P_2 that works.

- Continuing in this way, we construct a sum that gets above L and then below L and then above L and then below L, and so on forever.

- And since we always chose P_i and N_i to be the smallest possible, each time we hopped over L we did so by no more than the size of[13] p_{P_i} or n_{N_i}. And since $p_k \to 0$ and $n_k \to 0$, this "hop over distance" also converges to 0, which forces the sum to converge to L. $\square$

Therefore, there exists a way to rearrange the alternating harmonic series to converge to π, to $-e^{103}$, to LeBron James' phone number, to 8675309, to $-\infty$, and to any other real number you'd like... which I think is just awesome.[14]

---

[13]It looks funny because of the size differences, but what's written next is p with a subscript of P_i, and then n with a subscript of N_i. Meaning the P_i^{th} positive term, or the N_i^{th} negative term.

[14]In fact, even more is true. There's also a rearrangement whose *[footnote continues on next page]*

Back to Zeno

This book began with a pair of paradoxes from Zeno of Elea, which we can at last settle. We begin with Zeno's second paradox—feel free to flip back to the start of the book to recall what it said.

Example 4.24. For simplicity let's assume the arrow is traveling at a constant rate. In particular, let's suppose Zeno's arrow takes t seconds to get halfway to the target, which means it will take $\frac{t}{2}$ seconds to travel the next quarter of the distance, $\frac{t}{4}$ seconds to travel the following eighth of the distance, and so on.

In Zeno's thought experiment, the arrow's first stage was when it traveled from the archer to the point halfway to the target. The arrow's second stage begins at this point, and lasts until the arrow cuts its distance in half again (i.e. when the arrow is three quarters of the way from the archer to the target). And so on.

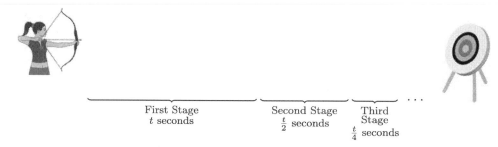

First Stage
t seconds

Second Stage
$\frac{t}{2}$ seconds

Third Stage
$\frac{t}{4}$ seconds

Then by the geometric series test the total amount of time (in seconds) to complete <u>all</u> the stages is

$$\sum_{k=0}^{\infty} \frac{t}{2^k} = \frac{t}{1 - \frac{1}{2}} = 2t.$$

So the the arrow completes all the stages in just 2t seconds. $\quad\square$

What was Zeno's error? He was correct that the arrow must first complete the first stage, and then must complete the second stage, and then must complete the third stage, and so on. And he was correct that each stage takes some finite amount of time. And it seems to make sense that if you have to do one thing at a time, and you have infinitely many things to do, that you will never complete all of those things. This is false though.[15] One *can* do infinitely many things, one after the other, in just a finite amount of time. The reason? Because a positive series can converge. And if each summand represents a unit of time, then the collection of those infinitely many things can indeed converge to a finite time length.

---

limit does not exist. And for those who know what *limit suprema* and *limit infima* are, for any $\alpha, \beta \in \mathbb{R} \cup \{\pm\infty\}$ with $\alpha \leq \beta$, there is a rearrangement whose $\limsup$ is β and $\liminf$ is α.

[15]In math especially, mistakes provide some of the best learning opportunities. And even though my Dean probably won't accept that as sufficient justification to make my real analysis students grade my calculus students' exams, Werner Heisenberg did once say that *"an expert is someone who knows some of the worst mistakes that can be made in his subject, and how to avoid them."* So it is indeed wise to learn from Zeno's error.

Zeno's error in his first paradox was also a faulty assumption that an infinite sum of positive terms must diverge. Just to rub it in, let's work out that one too.

Example 4.25. To make the problem concrete. First, let's be excessively generous[16] and assume the tortoise can move at 1 $^m\!/_s$. Achilles — the greatest warrior of his age — can certainly move quickly. He was easily 10 times faster than even our exceptionally speedy tortoise, so we may assume he can run n $^m\!/_s$, for some $n \geq 10$. (For simplicity, we ignore their acceleration.)[17]

We will look at each stage and determine how long it takes to complete each. We will also keep a *running tally*[18] so that we can find how long it takes to complete all the stages.

- Here, $p_0 = 0$ and $p_1 = 100$.

- So it takes Achilles $\frac{p_1 - p_0}{n} = \frac{100}{n}$ seconds to get from p_0 to p_1.

- Total time this far: $\frac{100}{n}$ seconds.

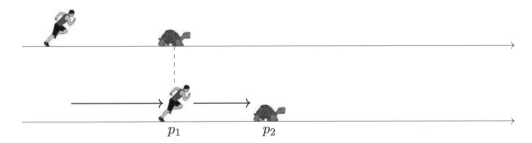

- Here, $p_1 = 100$ and $p_2 = 100 + \frac{100}{n}$.

- So it takes Achilles $\frac{p_2 - p_1}{n} = \frac{100}{n^2}$ seconds to get from p_1 to p_2.

- Total time this far: $\frac{100}{n} + \frac{100}{n^2}$ seconds.

---

[16]The Guinness Book of World Records says the fastest tortoise was Bertie, who clocked in at an impressive 0.28 $^m\!/_s$.

[17]As the famous joke goes: In response to a question about the milk production of cows, the scientist begins, "For simplicity, assume the cow is a perfect sphere and lives in a vacuum."

[18]Ba dum chhh.

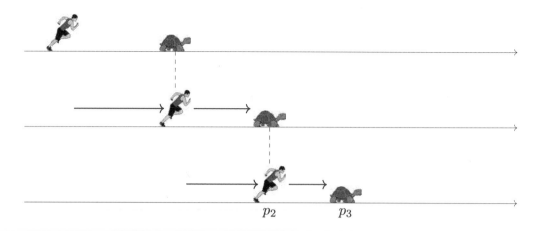

- Here, $p_2 = 100 + \frac{100}{n}$ and $p_3 = 100 + \frac{100}{n} + \frac{100}{n^2}$.

- So it takes Achilles $\frac{p_3 - p_2}{n} = \frac{100}{n^3}$ seconds to get from p_2 to p_3.

- Total time this far: $\frac{100}{n} + \frac{100}{n^2} + \frac{100}{n^3}$.

This pattern will continue. So by the geometric series test (Proposition 4.9), the total amount of time (in seconds) to complete all the stages is

$$\sum_{k=0}^{\infty} \frac{100}{n} \cdot \left(\frac{1}{n}\right)^k = \frac{100/n}{1 - \frac{1}{n}} = \frac{100}{n-1}.$$

And since $n \geq 10$, the two will pass before $\frac{100}{9} < 12$ seconds into the race when the tortoise (and hence also Achilles) is only at the 112 meter mark. For the last 888 meters of the race the tortoise will be eating dust. $\square$

— Notable Exercises —

The exercises contain two more convergence tests that you probably learned in calculus — the ratio and the root tests. These tests are not as important as the others to this text's development of real analysis, but are nevertheless important results with particular utility in determining the convergence of specific series.

- Exercise 4.11 (c) is an important result from number theory. In fact, the converse is essentially true too: If a number is rational, then either it has a terminating decimal expansion or a repeating decimal expansion.

- In Exercise 4.17 we prove the *ratio test*. That is, given a series $\sum_{k=1}^{\infty} a_k$ with $a_k \neq 0$, we will show that if $\lim_{k \to \infty} \left| \dfrac{a_{k+1}}{a_k} \right|$ is less than 1 then the series converges.

- In Exercise 4.18 we prove the *root test*. That is, if $\sum_{k=1}^{\infty} a_k$ is a series where each $a_k \geq 0$ and the limit $\lim_{k \to \infty} (a_k)^{1/k}$ exists, we call this limit ρ. Then this series converges if $\rho < 1$ and diverges if $\rho > 1$. (The test is inconclusive if $\rho = 1$.)

The rearrangement theorem (Theorem 4.23) stated that any conditionally convergent series can be rearranged to converge to anything at all — to any $L \in \mathbb{R}$ or to either ∞ or $-\infty$. After stating the theorem we sketched a proof for the case when $L \in \mathbb{R}$ and $L > 0$.

- In Exercise 4.23 we prove that there exists a rearrangement of this sum which diverges to ∞.

This roughly completes the proof of Theorem 4.23, and in the below we extend this result.

- In Exercise 4.24 we show that there exists a rearrangement of this sum whose limit does not exist.

- In Exercise 4.25 we prove a generalization of Theorem 4.23 called *Riemann's series theorem* in which in a rearrangement can be found whose *limit supremum* and *limit infimum* can be arbitrarily chosen.

- Notice that Theorem 4.23 says nothing about what happens for series which are not conditionally convergent. In Exercise 4.26 we prove that if a series converges absolutely, then every rearrangement converges to the same thing. And in Exercise 4.27 we prove that if each $a_k \geq 0$ and $\sum_{k=1}^{\infty} a_k = \infty$, then any rearrangement of this sum also diverges to ∞.

— Exercises —

Exercise 4.1. Determine whether each of the following converges conditionally, converges absolutely, or diverges. You do not need to prove your answers, but state which of the following tests gives the answers: the kth-term test, the geometric series test, and the alternating series test.

(a) $\displaystyle\sum_{k=1}^{\infty} (-1)^k \frac{1}{\sqrt{k}}$

(b) $\displaystyle\sum_{k=1}^{\infty} (-1)^k \frac{k}{k+7}$

(c) $\displaystyle\sum_{k=1}^{\infty} \frac{1}{(\ln(4))^k}$

(d) $\displaystyle\sum_{k=1}^{\infty} \frac{1}{k!}$

(e) $\displaystyle\sum_{k=1}^{\infty} \left(\sqrt{n+1} - \sqrt{n}\right)$

(f) $\displaystyle\sum_{k=1}^{\infty} \frac{1}{2}$

(g) $\displaystyle\sum_{k=1}^{\infty} (-1)^k \frac{1}{\sqrt[3]{k}}$

(h) $\displaystyle\sum_{k=53}^{\infty} \frac{\ln(k)}{\ln(\ln(k))}$

(i) $\displaystyle\sum_{k=7}^{\infty} \left(\frac{2}{k} + \frac{3}{k^2}\right)$

Exercise 4.2. Does the series $\sum_{k=1}^{\infty} (-1)^k$ converge or diverge? Justify your answer.

Exercise 4.3. Prove Proposition 4.18. That is, prove that if $\displaystyle\sum_{k=1}^{\infty} |a_k|$ converges, then $\displaystyle\sum_{k=1}^{\infty} a_k$ converges too.

Exercise 4.4.

(a) Give an example of a series with nonnegative terms where $\displaystyle\sum_{k=1}^{\infty} a_k$ diverges, but $\displaystyle\sum_{k=1}^{\infty} a_k^2$ converges.

(b) Prove that if $\displaystyle\sum_{k=1}^{\infty} a_k$ converges where each $a_k > 0$, then $\displaystyle\sum_{k=1}^{\infty} a_k^2$ converges.

(c) Show by example that part (b) is not true if we do not insist that each $a_k > 0$.

Exercise 4.5. Give an example of a series where $\displaystyle\sum_{k=1}^{\infty} a_k$ converges, but $\displaystyle\sum_{k=1}^{\infty} a_{2k}$ diverges.

Exercise 4.6. Let (a_k) be a sequence which converges to 0. Prove that the series $\displaystyle\sum_{k=1}^{\infty} (a_k - a_{k+1})$ converges to a_1.

Exercise 4.7. Give an example of a series $\sum\limits_{k=1}^{\infty} a_k$ where

- $\sum\limits_{k=1}^{\infty} a_k$ converges,   • $\sum\limits_{k=1}^{\infty} a_k^2$ diverges,   and   • $\sum\limits_{k=1}^{\infty} a_k^3$ converges.

Exercise 4.8. Find an estimate for $\sum\limits_{k=1}^{\infty} (-1)^k \dfrac{1}{7k}$ that is accurate to 0.01.

Exercise 4.9. The geometric series test (Proposition 4.9) says that a geometric series diverges if $r \geq 1$. Recall that a series can diverge to ∞, to $-\infty$, or can be "does not exist." Which form of divergence is it for a geometric series? Note that your answer may depend on r.

Exercise 4.10. Let $r \in (-1, 1)$. Show directly (without appealing to anything we proved in this chapter) that the sequence of partial sums of the geometric series $\sum\limits_{k=0}^{\infty} r^k$ is Cauchy.

Exercise 4.11.

(a) Find a way to write $77.77777777\ldots$ as a geometric series, and then prove this number is rational by using the geometric series test to write this number as a fraction with integers in the numerator and denominator.

(b) Write $77.77777777\ldots$ as a different geometric series, and use the geometric series test to write this number as a fraction with integers in the numerator and denominator. Are your two fractions the same?

(c) A number q has a *repeating decimal* if the non-integer portion of its decimal expansion is repetitive. For example, $72.578578578578578\ldots$ has a repeating decimal. Prove that if a number q has a repeating decimal, then q is rational.

Exercise 4.12.

(a) Prove that if $\sum\limits_{k=1}^{\infty} a_k$ converges absolutely, and (b_k) is a subsequence of (a_k), then $\sum\limits_{k=1}^{\infty} b_k$ also converges absolutely.

(b) Give an example demonstrating that it is necessary to assume that $\sum\limits_{k=1}^{\infty} a_k$ converges absolutely.

Exercise 4.13.

(a) Prove that if (ka_k) converges to a nonzero real number L, then the series $\sum\limits_{k=1}^{\infty} a_k$ diverges. Give an example to show that the converse is false.

(b) Prove that if $(k^2 a_k)$ converges (to any real number), then the series $\sum\limits_{k=1}^{\infty} a_k$ converges. Give an example to show that the converse is false.

Exercise 4.14.   Prove that if $a_k > 0$ for all k and $\sum\limits_{k=1}^{\infty} a_k^2$ converges, then $\sum\limits_{k=1}^{\infty} \dfrac{a_k}{k}$ converges.

Exercise 4.15.   Prove the *Cauchy condensation test*. That is, suppose that (a_k) is a decreasing sequence for which $a_k \to 0$. Prove that $\sum\limits_{k=1}^{\infty} a_k$ converges if and only if $\sum\limits_{k=1}^{\infty} 2^k a_{2^k}$ converges.

Exercise 4.16.   The converses of (i) and (ii) of the comparison test (Proposition 4.12) are both false. Give a pair of examples demonstrating this.

Exercise 4.17.   Prove the *ratio test* via the following steps. Given a series $\sum\limits_{k=1}^{\infty} a_k$ with $a_k \neq 0$, assume that

$$\lim_{k \to \infty} \left| \frac{a_{k+1}}{a_k} \right| =: r < 1.$$

We will prove that the series converges absolutely.

(a) Let q be such that $r < q < 1$. Explain why there is some N such that $n \geq N$ implies that $|a_{k+1}| \leq |a_k| \cdot q$.

(b) Explain why $\sum\limits_{k=1}^{\infty} |a_N| \cdot q^k$ necessarily converges.

(c) Finally, use part (b) to prove that $\sum\limits_{k=1}^{\infty} |a_k|$ converges.

Exercise 4.18. Prove the *root test* via the following steps. Given a series $\sum_{k=1}^{\infty} a_k$ where each $a_k \geq 0$, assume that the limit $\lim_{k \to \infty} (a_k)^{1/k}$ exists. Call this limit ρ. Then this series converges if $\rho < 1$ and diverges if $\rho > 1$. (The test is inconclusive if $\rho = 1$.)

(a) Suppose $\rho < 1$. Let $\varepsilon = \frac{1-\rho}{2}$ and $\rho_1 = \rho + \varepsilon$. Prove that there is some N for which $(a_n)^{1/n} < \rho_1$ for all $n > N$.

(b) Prove that $\sum_{k=N}^{\infty} a_k$ converges by comparing it to a geometric series. Then conclude that $\sum_{k=1}^{\infty} a_k$ also converges.

(c) Suppose $\rho > 1$. Let $\varepsilon = \frac{\rho-1}{2}$ and $\rho_2 = \rho - \varepsilon$. Prove that there is some N for which $(a_n)^{1/n} > \rho_2$ for all $n > N$.

(d) Use this to argue that $\sum_{k=N}^{\infty} a_k$ diverges by using the kth-term test (Proposition 4.5). Then conclude that $\sum_{k=1}^{\infty} a_k$ also diverges.

Exercise 4.19. Give an example of a divergent series $\sum_{k=1}^{\infty} a_k$ for which $\lim_{k \to \infty} (a_{k+1} - a_k) = 0$.

Exercise 4.20.

(a) Give an example of two divergent series $\sum_{k=1}^{\infty} a_k$ and $\sum_{k=1}^{\infty} b_k$, such that $\sum_{k=1}^{\infty} a_k b_k$ converges.

(b) Give an example of two convergent series $\sum_{k=1}^{\infty} a_k$ and $\sum_{k=1}^{\infty} b_k$, such that $\sum_{k=1}^{\infty} a_k b_k$ diverges.

(c) Prove that if $\sum_{k=1}^{\infty} a_k$ and $\sum_{k=1}^{\infty} b_k$ converge absolutely, then $\sum_{k=1}^{\infty} a_k b_k$ converges absolutely.

Exercise 4.21. Give an example of a divergent series $\sum_{k=1}^{\infty} a_k$ and a convergent series $\sum_{k=1}^{\infty} b_k$ where $a_k \leq b_k$ for all k.

Exercise 4.22. Consider the sum $\sum_{k=1}^{\infty} a_k$, and define s_n to be this series' n^{th} partial sum; that is, $s_n = \sum_{k=1}^{n} a_k$. The series $\sum_{k=1}^{\infty} a_k$ is called *Cesaro summable* if

$$\lim_{n \to \infty} \frac{s_1 + s_2 + \cdots + s_n}{n}$$

converges.

(a) Prove that if $\sum_{k=1}^{\infty} a_k$ converges, then this series is Cesaro summable.

(b) Prove by example that if $\sum_{k=1}^{\infty} a_k$ is Cesaro summable this does *not* imply that $\sum_{k=1}^{\infty} a_k$ converges.

Exercise 4.23. Show that if $\sum_{k=1}^{\infty} a_k$ is conditionally convergent, then there exists a rearrangement of this sum which diverges to ∞.

Exercise 4.24. Show that if $\sum_{k=1}^{\infty} a_k$ is conditionally convergent, then there exists a rearrangement of this sum whose limit does not exist.

Exercise 4.25. Assume that $\sum_{k=1}^{\infty} a_k$ is conditionally convergent. Define the *limit superior* of a sequence (s_n) (Notation: "$\limsup_{n \to \infty} s_n$") to be

$$\limsup_{n \to \infty} s_n = \lim_{n \to \infty} \left(\sup_{m \geq n} s_m \right).$$

And define the *limit inferior* of sequence (s_n) (Notation: "$\liminf_{n \to \infty} s_n$") to be

$$\liminf_{n \to \infty} s_n = \lim_{n \to \infty} \left(\inf_{m \geq n} s_m \right).$$

Prove that for any $\alpha, \beta \in \mathbb{R} \cup \{\pm\infty\}$ with $\alpha \leq \beta$, there is a rearrangement of $\sum_{k=1}^{\infty} a_k$ whose sequence of partial sums, (s_n), have

$$\limsup_{n \to \infty} s_n = \beta \qquad \text{and} \qquad \liminf_{n \to \infty} s_n = \alpha.$$

Exercise 4.26.   Prove that if $\displaystyle\sum_{k=1}^{\infty} a_k$ converges absolutely to L, then any rearrangement of this sum also converges to L.

Exercise 4.27.   Prove that if each $a_k \geq 0$ and $\displaystyle\sum_{k=1}^{\infty} a_k = \infty$, then any rearrangement of this sum also diverges to ∞.

Exercise 4.28.   Prove the *summation by parts* formula. That is, prove that if (a_k) and (b_k) are sequences and $s_n = a_1 + a_2 + \cdots + a_n$, then

$$\sum_{k=j+1}^{n} a_k b_k = s_n b_{n+1} - s_j b_{j+1} + \sum_{k=j+1}^{n} s_k (b_k - b_{k+1}).$$

Exercise 4.29.   There is a neat theorem which, in a small way, relies on the fact that the harmonic series $\displaystyle\sum_{k=1}^{\infty} \frac{1}{k}$ diverges. Here's the theorem: Every positive rational number can be written as a finite sum of distinct numbers of the form $1/n$. For example, the positive rational number $\frac{7243}{4140}$ can be written like this:

$$\frac{7243}{4140} = \frac{1}{1} + \frac{1}{2} + \frac{1}{5} + \frac{1}{21} + \frac{1}{527} + \frac{1}{3,054,492}.$$

Now, take a moment to appreciate how magical this seems, because once you see the algorithm which produces the decomposition, it might lose some of its shine. Here's how I found the above decomposition of $\frac{7243}{4140}$ (which I basically chose at random):

$$\frac{7243}{4140} - 1 = \frac{3103}{4140}$$

$$\frac{3103}{4140} - \frac{1}{2} = \frac{1033}{4140}$$

$$\frac{1033}{4140} < \frac{1}{3} \text{ and } \frac{1}{4}$$

$$\frac{1033}{4140} - \frac{1}{5} = \frac{41}{828}$$

$$\frac{41}{828} < \frac{1}{6}, \frac{1}{7}, \ldots, \frac{1}{20}$$

$$\frac{41}{828} - \frac{1}{21} = \frac{11}{5796}$$

$$\frac{11}{5796} < \frac{1}{22}, \frac{1}{23}, \ldots, \frac{1}{526}$$

$$\frac{11}{5796} - \frac{1}{527} = \frac{1}{3,054,492}.$$

And therefore,
$$\frac{7243}{4140} = \frac{1}{1} + \frac{1}{2} + \frac{1}{5} + \frac{1}{21} + \frac{1}{527} + \frac{1}{3{,}054{,}492}.$$

(a) If the fraction that we start with is much bigger than 1, like $\frac{244{,}406{,}536}{7}$, why is this not an issue? Why will subtracting off 1, and then $1/2$, and then $1/3$, and so on, eventually catch up to it?

(b) Prove that if this operation is performed on a fraction p/q which is less than 1, then the numerator necessarily decreases (in the above example, the numerators decreased at every stage: 7243, 3103, 1033, 41, 11, 1). Note: The algorithm always subtracts off the biggest possible reciprocal. That is, if $\frac{p}{q} < 1$, then we subtract off $\frac{1}{n}$ where $\frac{1}{n+1} < \frac{p}{q} < \frac{1}{n}$.

(c) Explain why this proves the theorem.

— Open Questions —

Question 1. Does $\displaystyle\sum_{k=1}^{\infty} \frac{1}{k^3 \sin^2(k)}$ converge?

Question 2. Does $\displaystyle\sum_{k=1}^{\infty} \frac{(-1)^k k}{p_k}$ converge, where p_k is the kth prime number?

Question 3. Is it true that $\displaystyle\sum_{k=0}^{\infty} \frac{1 + 14k + 76k^2 + 168k^3}{2^{20k}} \binom{2k}{k}^7 = \frac{32}{\pi^3}$?

Question 4. Is $\displaystyle\sum_{k=0}^{\infty} \frac{(-1)^k}{(2k+1)^2}$ irrational?

Question 4. Is $\displaystyle\sum_{k=1}^{\infty} \frac{1}{k^3}$ transcendental?

Question 5. One can show that[19]

$$\sum_{k=1}^{\infty} \left(\frac{1}{k} \times \frac{1}{k+1} \right) = 1.$$

We now reinterpret this algebraic fact geometrically. Notice that the left-hand side is the area of a $\frac{1}{1} \times \frac{1}{2}$ rectangle, plus the area of a $\frac{1}{2} \times \frac{1}{3}$ rectangle, plus the area of a $\frac{1}{3} \times \frac{1}{4}$ rectangle, and so on. The right-hand side is the area of a 1×1 square. This suggests that it is *possible* that you can tile a 1×1 square with this infinite collection of rectangles. Prove or disprove that such a tiling exists.[20]

---

[19] A super neat way to show this is to first use a discrete version of *partial fractions* — something you might remember from calculus as a technique to solve integrals. Indeed,

$$\frac{1}{k(k+1)} = \frac{A}{k} + \frac{B}{k+1} \qquad \Rightarrow \qquad 1 = (A+B)k + A \qquad \Rightarrow \qquad A = 1, \ B = -1.$$

Applying this, the sum then telescopes!

$$\sum_{k=1}^{\infty} \frac{1}{k(k+1)} = \sum_{k=1}^{\infty} \left(\frac{1}{k} + \frac{-1}{k+1} \right) = \lim_{n \to \infty} \left[\left(1 - \frac{1}{2} \right) + \left(\frac{1}{2} - \frac{1}{3} \right) + \left(\frac{1}{3} - \frac{1}{4} \right) + \cdots + \left(\frac{1}{n} - \frac{1}{n+1} \right) \right]$$

$$= \lim_{n \to \infty} \left[1 + \left(-\frac{1}{2} + \frac{1}{2} \right) + \left(-\frac{1}{3} + \frac{1}{3} \right) + \cdots + \left(-\frac{1}{n} + \frac{1}{n} \right) - \frac{1}{n+1} \right]$$

$$= \lim_{n \to \infty} \left[1 - \frac{1}{n+1} \right] = 1 - 0 = 1.$$

[20] It's known this infinite collection of rectangles can be packed in the 1.002×1.002 square. See *Two packing problems* by Vojtech Bálint and *An algorithm for packing squares* by Marc Paulhus.

Chapter 5: The Topology of $\mathbb{R}$

5.1 Open Sets

We are approaching the study of continuous functions.[1] This is a surprisingly subtle task. In fact, there is a whole area of math devoted to the study of sets and continuous functions on those sets; this area is called *topology*. One could easily create a full-year undergraduate math course just on topology, so our treatment, totaling a mere 14 (long-form) pages, will certainly not be comprehensive.

Nevertheless, this brief introduction will aim to do two things. The first is practical: It lays the foundation for our study of continuous functions by introducing special classes of sets which, when we ask for a function's behavior on these sets, will *determine* whether that function is continuous. The second is more philosophical. Although math is traditionally taught by first partitioning it into discrete[2], non-overlapping areas like real analysis, abstract algebra, linear algebra, combinatorics and topology, mathematics is in reality much more interconnected than this, and one's mathematical education should include examples thereof.

This chapter will be focused on three fundamental objects — those of open sets, closed sets and compact sets — which in turn help explain real analysis' place in the sprawling ecosystem of mathematics. We begin with the idea of an *open set*.

Definition.

Definition 5.1. A set $U \subseteq \mathbb{R}$ is *open* if, for every $x \in U$, there is a number $\delta > 0$ such that
$$(x - \delta, x + \delta) \subseteq U.$$

That is, if for all $x \in U$ there exists some δ-*neighborhood of* x, denoted $V_\delta(x)$, such that $V_\delta(x) \subseteq U$.

Note that this is equivalent to saying $U \subseteq \mathbb{R}$ is *open* if, for each $x \in U$, there is a number $\delta > 0$ such that if y satisfies $|x - y| < \delta$, then $y \in U$.

---

[1]Believe it or not, this is a (highly nerdy) joke. After Chapter 6 you may find it mildly amusing.
[2](or continuous)

Example 5.2.

(i) The set $\mathbb{R}$ is open. To see this, note that for any $x \in \mathbb{R}$, if you let $\delta = 17$ then $(x - 17, x + 17) \subseteq \mathbb{R}$.

(ii) The empty set $\emptyset$ is open. To see this, note that, because there are no elements in $\emptyset$, it's true to say "for any $x \in \emptyset$, x is a purple elephant that speaks Spanish." It's vacuously[3] true! You certainly can't disprove it, right? You can't present to me any element in $\emptyset$ that is *not* a purple elephant that speaks Spanish. Likewise, for any $x \in \emptyset$ there exists a δ such that $(x - \delta, x + \delta) \subseteq U$. That is again true, as you certainly cannot find an $x \in \emptyset$ that does not satisfy this.

(iii) The open interval (a, b) is open. To see this, pick any $x \in (a, b)$. Let $\delta = \min\{x - a, b - x\}$. Then $(x - \delta, x + \delta) \subseteq (a, b)$.

Here, $\delta = x - a$

Here, $\delta = b - x$

(iv) The intervals (a, ∞) and $(-\infty, b)$ are open. To see this, pick any $x \in (a, \infty)$ and let $\delta = x - a$; or pick any $x \in (-\infty, b)$ and let $\delta = b - x$.

(v) A non-example: The set $[3, 7]$ is *not* an open set. To see this, note that $3 \in [3, 7]$, but for any $\delta > 0$ we have

$$(3 - \delta, 3 + \delta) \not\subseteq [3, 7].$$

That is, every δ-neighborhood of $x = 3$ is *not* a subset of $[3, 7]$. So we dissatisfied the definition "for all $x \in U$, there is..." by exhibiting a single x for which it fails. Therefore, $[3, 7]$ is *not* open.

$\square$

---

[3]A vacuum in physics is a container in which the air inside has been sucked out, leaving nothing left. Likewise in math, saying something is *vacuously true* means that the set of elements that the statement is referring to is empty; there is therefore nothing to prove, and it's automatically true.

> ## Proposition.
>
> **Proposition 5.3** (*Open sets via arbitrary unions and finite intersections*).
>
> (i) If $\{U_\alpha\}$ is any collection of open sets, then $\bigcup_\alpha U_\alpha$ is also an open set.
>
> (ii) If $\{U_1, U_2, \ldots, U_n\}$ is a finite collection of open sets, then $\bigcap_{k=1}^{n} U_k$ is also an open set.

When you see "$\bigcup_\alpha U_\alpha$" (or "$\bigcup_{\alpha \in S} U_\alpha$" for an *indexing set S*), what this means is that you are taking the union of all the sets in $\{U_\alpha\}$ (or $\{U_\alpha\}_{\alpha \in S}$), whether that is finitely many sets, countably many sets, $|\mathbb{R}|$ many sets, or $|\mathcal{P}(\mathbb{R})|$ many sets. Below is a quick table summarizing what this union would look like in each case.

> ## Note.
>
> | If $\{U_\alpha\}$ contains | Then $\bigcup_\alpha U_\alpha$ is just |
> |---|---|
> | Finitely many sets | $\bigcup_{k=1}^{n} U_k$ |
> | Countably many sets | $\bigcup_{k=1}^{\infty} U_k$ |
> | $|\mathbb{R}|$ many sets | $\bigcup_{x \in \mathbb{R}} U_x$ |
> | $|\mathcal{P}(\mathbb{R})|$ many sets | $\bigcup_{x \in \mathcal{P}(\mathbb{R})} U_x$ |

Proof of Proposition 5.3. For (i), assume that $x \in \bigcup_\alpha U_\alpha$. By the definition of an open set we need to find some δ-neighborhood of x contained inside this union. Since x is in this union, we know $x \in U_{\alpha_0}$ for some α_0. By definition of U_{α_0} being open, this implies that there is some δ-neighborhood $V_\delta(x) \subseteq U_{\alpha_0}$. But since $U_{\alpha_0} \subseteq \bigcup_\alpha U_\alpha$, we have that $V_\delta(x) \subseteq \bigcup_\alpha U_\alpha$, completing part (i).

To prove part (ii), first assume that we have some $x \in \bigcap_{k=1}^{n} U_k$. We need to show that there is a δ-neighborhood[4] of x that is contained inside this intersection. Now, since x is in the intersection, we know $x \in U_k$ for each $k \in \{1, 2, \ldots, n\}$. And since each of these sets is open, that implies that for each k there is a δ_k-neighborhood of x, $V_{\delta_k}(x)$, such that $V_{\delta_k}(x) \subseteq U_k$. Now, these δ_k-neighborhoods may be different sizes, and so if, say, $\delta_1 > \delta_2$, it might not be the case that $V_{\delta_1}(x) \subseteq U_2$. But if we choose $\delta = \min\{\delta_1, \delta_2, \ldots, \delta_n\}$, then $V_\delta(x) \subseteq V_{\delta_k}(x) \subseteq U_k$ for all k, and so $V_\delta(x) \subseteq \bigcap_{k=1}^{n} U_k$, completing the proof. $\qquad\square$

Example 5.4.

(i) We already showed in Example 5.2 that all intervals (a, b) are open, so by Proposition 5.3, also $(a, b) \cup (c, d)$ is open, and $(a, b) \cup (c, d) \cup (e, f)$ is open, and $(a_1, b_1) \cup (a_2, b_2) \cup \ldots$ is open, and any other union of open intervals is open.

(ii) Likewise, $(a, b) \cap (c, d)$ is open. For instance, $(0, 2) \cap (1, 3) = (1, 2)$ is open, as is $(0, 2) \cap (4, 6) = \emptyset$. (But note that such finite intersections are always of the original form: either a single open interval (a, b), or the empty set.)

(iii) We also showed in Example 5.2 that intervals $(-\infty, a)$ and (b, ∞) are open, and hence their union $(-\infty, a) \cup (b, \infty)$ is open, too.

   • Note that this third example actually follows from the first, because $(-\infty, a) \cup (b, \infty)$ is also a union of open intervals:

$$(-\infty, a) \cup (b, \infty) = \bigcup_{n=1}^{\infty} (a - n, a) \cup (b, b + n).$$

(iv) Likewise,

$$\left[(-\infty, a) \cup (b, \infty)\right] \cap \left[(-\infty, c) \cup (d, \infty)\right] = (-\infty, \min\{a, c\}) \cup (\max\{b, d\}, \infty)$$

is open. (But again, this is always of the original form.)

Based on the discussion in the bullet point after (iii), we have not yet seen an example where an open set is anything other than a union of open intervals. This makes even more sense when you combine (1) the fact that every point in an open set is contained inside a δ-neighborhood (which is an open interval) and (2) the fact that if you union together any collection of open sets, then the result is still open. You can therefore imagine that it is true that every open set is a union of open intervals. In fact, even more is true: Every open set is a union of *countably many* disjoint open intervals.

---

[4]Confession: It's difficult for me to do topology without Mister Rogers playing in my head...
♪ *It's a beautiful day in this neighborhood* ♪

> ### Theorem.
>
> **Theorem 5.5** (*Each open set $= \bigcup_{k=1}^{\infty}(a_k, b_k)$*). Every open set is a countable union of disjoint open intervals.

Note that open intervals can be of the form (a, b), $(-\infty, a)$, or (b, ∞).

Proof. Let A be an open set. Then each $x \in A$ is contained inside some open interval $(x - \delta, x + \delta)$. There may be a larger open interval that contains x, though, so let's let I_x be the largest open interval containing x. Formally, $I_x = (\alpha, \beta)$ where

$$\alpha = \inf\{a : (a, x) \subseteq A\} \qquad \text{and} \qquad \beta = \sup\{b : (x, b) \subseteq A\}.$$

Note that, given any pair x and y, either $I_x = I_y$ or $I_x \cap I_y = \emptyset$. The reason for this is that, if instead you had two intervals overlapping but not equal, then you could expand at least one of them, contradicting the fact that we had chosen the *largest* intervals.

We claim that these I_x intervals are the ones we are looking for. To see this, first note that since each $x \in A$ is in $I_x \subseteq A$, the union of all the I_x intervals does indeed equal A. Moreover, we can see that there are only countably many distinct such intervals because by the density of $\mathbb{Q}$ in $\mathbb{R}$, every open interval contains a rational number, so there can not be more intervals than rationals, which is a countable set.[5] $\qquad\square$

5.2 Closed Sets

> ### Definition.
>
> **Definition 5.6.** A set $A \subseteq \mathbb{R}$ is *closed* if A^c is open.

Note that[6] since $(A^c)^c = A$, the above is equivalent to saying that A is closed if there exists some open set B such that $B^c = A$.

Example 5.7.

- The closed interval $[a, b]$ is closed. To see this, note that we have already shown that $(-\infty, a) \cup (b, \infty)$ is open, and its complement is $[a, b]$, and so by definition $[a, b]$ is closed.

- $\mathbb{R}$ is closed. To see this, note that we have already shown that $\emptyset$ is open, and $\emptyset^c = \mathbb{R}$.

---

[5]The details of this final step were asked for in Exercise 2.12.
[6]Recall that if $A \subseteq \mathbb{R}$, then the complement $A^c = \mathbb{R} \setminus A$.

- $\emptyset$ is closed.[7] To see this, note that we have already shown that $\mathbb{R}$ is open, and $\mathbb{R}^c = \emptyset$.

Intuitively, one difference between the open interval $(0, 1)$ and the closed interval $[0, 1]$ is that in the open interval there is a sequence of points from *inside* the set which is converging to something *outside* the set; one example: $\frac{1}{2}, \frac{1}{3}, \frac{1}{4}, \ldots$ is a sequence of numbers from $(0, 1)$ which converges to 0, which is not in $(0, 1)$. The closed interval does not have this property; if a sequence of terms from the set $[0, 1]$ is converging, then we know that it converges to something inside the set. This intuition jives with our definitions of open and closed sets, as the following definition and proposition say.

> ### Definition.
>
> **Definition 5.8.** A point x is a *limit point* of a set A if there is a sequence of points $a_1, a_2, a_3, \ldots$ from $A \setminus \{x\}$ such that $a_n \to x$.

In Exercise 5.7 you're asked to show this: a point x is a limit point of A if and only if every ε-neighborhood of x intersects[8] A at some point other than x.

Example 5.9. The point 0 is a limit point of the interval $(0, 3)$ because, for each $n \in \mathbb{N}$, we see that $a_n = \frac{1}{n+1} \in (0, 1)$, while $a_n \to 0$. The point 0.24 is also a limit point of $[0, 1]$, as it is the limit of the sequence $0.23, 0.239, 0.2399, 0.23999, \ldots$.

Being closed in fact means that you contain your limit points. This is what the next theorem states.

> ### Theorem.
>
> **Theorem 5.10** (*Closed iff contains limit points*). A set A is closed if and only if it contains all of its limit points.

Scratch Work. The "contains all of its limit points" condition can be tough to work with. It's much easier to handle a specific limit point (which, by the definition of limit point, gives a specific sequence converging to that point). Therefore for the forward direction we will proceed by contradiction (giving us a specific sequence from A that converges to an $x \notin A$), and for the reverse direction we will use the contrapositive (meaning that we are in search of just a single sequence from A that converges outside A).

---

[7]Yes, $\mathbb{R}$ and $\emptyset$ are open *and* closed. Yes, this is silly. See the YouTube video "Hitler Learns Topology" (www.youtu.be/SyD4p8_y8Kw) for a dramatic response. These sets are sometimes called *clopen*. In Exercise 5.29 you are asked to prove that these are the only clopen subsets. (For general topologies — beyond the topology of $\mathbb{R}$, which we are studying — there often many more clopen sets.)

[8]Saying a set X *intersects* a set Y means $X \cap Y \neq \emptyset$.

For the forward direction, when we assume for a contradiction that A does not contain all its limit points, this gives us an $x \notin A$ for which there is a sequence (a_n) converging to x, where each $a_n \in A$. Where's the contradiction? Well, converging to x means these points a_n from A are getting closer and closer to x. But since $x \notin A$, that means $x \in A^c$; and since A is closed, A^c is open. And being open means there is a δ-neighborhood of x which is entirely contained inside A^c; i.e., is disjoint from A. That's the contradiction: It's impossible for a sequence (a_n) be getting closer and closer to x, if all these points have to be from A, and hence have to stay outside this δ-neighborhood of x.

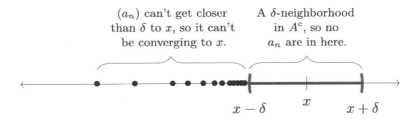

As for the reverse direction, by using the contrapositive we may assume that A is not closed, which is to say that A^c is not open. And being "not open" means that for some $x \in A^c$, every δ-neighborhood of x contains a member of A. So if we let $\delta = 1$, then this 1-neighborhood of x contains a member of A, call it a_1. Likewise, if we let δ be $1/2$, then $1/3$, then $1/4$, and so on, we get an a_2, a_3, a_4, and so on, each from A and contained inside these respective neighborhoods. These elements comprise a sequence from A, which converges to x.

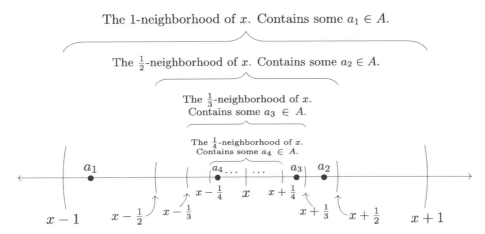

Proof. For the forward direction, suppose that A is closed, and assume for a contradiction that A does not contain all of its limit points. That is, suppose that there is a sequence $a_1, a_2, a_3, \ldots$ from A where $a_n \to x$, but $x \notin A$. Observe that this implies that A^c is open and $x \in A^c$. And so, by openness, there is some δ-neighborhood $(x - \delta, x + \delta) \subseteq A^c$. We now show that this is a contradiction to the claim that $a_n \to x$.

Since $a_n \to x$ where each $a_n \in A$, for all $\varepsilon > 0$ there exists some N such that $|a_n - x| < \varepsilon$ for all $n > N$. In particular, if we let $\varepsilon = \delta$, then such an N exists, which in turn implies that $|a_{N+1} - x| < \delta$. That is, $a_{N+1} \in (x - \delta, x + \delta)$ and $a_{N+1} \in A$. But this is a contradiction, since we showed in the last paragraph that $(x - \delta, x + \delta) \subseteq A^c$, meaning that $a_{N+1} \notin A$.

For the backwards direction, we will prove the contrapositive. Assume that A is not closed; that is, A^c is not open. Being not open means that there is some $x \in A^c$ such that every δ-neighborhood of x contains a point that is not in A^c, meaning that it is in A. In particular, for each $n \in \mathbb{N}$ there is some point $a_n \in A$ such that a_n is in the $\frac{1}{n}$-neighborhood of A. We claim that $a_n \to x$. Note that if we can prove this, then we are done, since we will have constructed a sequence of points from A converging to a point outside of A, proving that A does not contain all of its limit points.

To prove the claim, let $\varepsilon > 0$. By the Archimedean principle there is some N such that $\frac{1}{N} < \varepsilon$. And so for any $n > N$ we have $\frac{1}{n} < \varepsilon$, and hence a_n is in the ε-neighborhood of x. That is,

$$|a_n - x| < \varepsilon$$

for all $n > N$, implying that $a_n \to x$ and completing the proof. $\qquad \square$

The final proposition of this section follows fairly directly from Proposition 5.3 provided you recall the following fact, known as *De Morgan's laws* (whose proof in the finite case was also asked of you in Exercise 1.8).

Fact.

Fact 5.11. De Morgan's laws state that for an arbitrary union or intersection,

$$\left(\bigcup_\alpha U_\alpha \right)^c = \bigcap_\alpha U_\alpha{}^c \qquad \text{and} \qquad \left(\bigcap_\alpha U_\alpha \right)^c = \bigcup_\alpha U_\alpha{}^c.$$

Proposition.

Proposition 5.12 (*Closed sets via finite unions and arbitrary intersections*).

(i) If $\{U_1, U_2, \ldots, U_n\}$ is a collection of closed sets, then $\displaystyle\bigcup_{k=1}^{n} U_k$ is also a closed set.

(ii) If $\{U_\alpha\}$ is a collection of closed sets, then $\displaystyle\bigcap_\alpha U_\alpha$ is also a closed set.

Proof. This is Exercise 5.11. I suggest you make use of De Morgan's laws when you prove it. $\qquad \square$

5.3 Open Covers

> **Definition.**
>
> **Definition 5.13.** Let A be a set.
>
> - The sets $\{U_\alpha\}_{\alpha \in S}$ are a *cover* of A if
>
> $$A \subseteq \bigcup_{\alpha \in S} U_\alpha.$$
>
> - If each U_α is open, then $\{U_\alpha\}_{\alpha \in S}$ is an *open cover* of A.
>
> - If $\{U_\alpha\}_{\alpha \in S}$ has a finite subset $\{U_\alpha\}_{\alpha \in F}$ (meaning $F \subseteq S$) which is still a cover of A, then $\{U_\alpha\}_{\alpha \in F}$ is called a *finite subcover* of A.

This idea may seem odd. It's not at all clear that this should be anything that we should care about. Bear with me, though, because in the grand scheme it is amazing how much this is *exactly* something that we should care about.

Example 5.14.

- An open cover of $(0, 4)$ is

$$\left\{ \left(\frac{1}{k}, 4 - \frac{1}{k} \right) \right\}_{k=1}^{\infty} = \left\{ (1, 3), \left(\frac{1}{2}, 4 - \frac{1}{2} \right), \left(\frac{1}{3}, 4 - \frac{1}{3} \right), \ldots \right\}.$$

Clearly each set in the above is open, as each is an open interval. Moreover, for any $x \in (0, 4)$ note that for large enough k, one of those open intervals will contain x.

- Another open cover of $(0, 4)$ is

$$\{ (-\infty, 2), (1, 3), (1, 2), (0.5, 3.5), (3, 7) \}.$$

Again, all of the above intervals are clearly open sets, as they are open intervals. Note, though, that not all of the above sets are necessary in order to cover $(0, 4)$; we could remove $(1, 2)$, for example. This is fine, though, and in fact it will be a staple of what's to come.

Also note that not all the intervals above are subsets of $(0, 4)$. This is also fine. Oftentimes sets in our covers will go outside the set we are covering.

$\square$

Here is a final example.

Example 5.15. An open cover of $[0, 4]$ is

$$\left\{ \left(\frac{1}{k}, 4 - \frac{1}{k} \right) \right\}_{k=1}^{\infty} \cup \{ (-0.2, 0.2) \} \cup \{ (3.9, 4.1) \}.$$

$\square$

5.4   The Greatest Definition in Mathematics

Without further ado,[9] here it is:

> **Definition.**
>
> **Definition 5.16.** A set A is *compact* if every open cover of A contains a finite subcover of A.

It may seem crazy that this is one of the great definitions in mathematics, but it is.[10] Indeed, one of the reasons it is so significant is that it is not at all obvious. It took great insight to realize that this topological concept is crucial. Indeed, from the topological perspective, compact sets are sets which are *essentially* finite. If your properties only care about what's happening on open sets, and there are essentially only finitely many of them at a time, then things are simple and nice. Real analysis illustrates how messy the infinite is, but compactness illustrates when things are simple(r).

Let's return to our previous two examples.

Example 5.17. The open cover

$$\left\{ \left(\frac{1}{k}, 4 - \frac{1}{k} \right) \right\}_{k=1}^{\infty}$$

of $(0,4)$ has *no* finite subcover (Exercise 5.15). So $(0,4)$ is *not* compact.   □

Example 5.18. The open cover

$$\left\{ \left(\frac{1}{k}, 4 - \frac{1}{k} \right) \right\}_{k=1}^{\infty} \cup \{(-0.2, 0.2)\} \cup \{(3.9, 4.1)\}$$

of $[0,4]$ *does* have a finite subcover. For instance, here's one:

$$\left\{ (-0.2, 0.2), (3.9, 4.1), \left(\frac{1}{11}, 4 - \frac{1}{11} \right) \right\}.$$

So $[0,4]$ *may* be compact (this cover has a finite subcover, but to be compact *every* open cover must have a finite subcover.)   □

In fact, the set $[0,4]$ from Example 5.18 is indeed compact. This is a consequence of a very important theorem called the Heine-Borel theorem.[11]

---

[9]Ok, maybe not the *absolute* greatest, but definitely up there. It's simple to state and it's not obvious at all that it is a property we should care about, but it turns out that it is *exactly* what we should care about. If you have other great definitions in mind, send me an email to let me know.

[10]That said, great definitions do not always come out of nowhere. For an interesting historical look at its development, see *A pedagogical history of compactness* by Manya Raman-Sundström, viewable here: `https://arxiv.org/pdf/1006.4131.pdf`

[11]Fair warning: The proof is quite long and arduous. But *[footnote continues on next page]*

> **Theorem.**
>
> **Theorem 5.19** (*The Heine-Borel theorem*). A set $S \subseteq \mathbb{R}$ is compact if and only if S is closed and bounded.

Proof. This is a long proof, but if you break it down into chunks it is manageable. We will first prove the forward direction.

— Forward direction —

For the forward direction, assume that S is compact. That is, that *every* open cover of S contains a finite subcover of S. To prove that S is closed and bounded, the idea is to pick a specific open cover of S for which the existence of a finite subcover proves the property.

Proof that S is bounded. To prove that S is bounded, we will consider the following open cover. Let $I_n = (-n, n)$; clearly I_n is an open set. Moreover, since

$$\bigcup_{n=1}^{\infty} I_n = \mathbb{R},$$

and $S \subseteq \mathbb{R}$, we see that

$$S \subseteq \bigcup_{n=1}^{\infty} I_n.$$

So we have seen that $\{I_n\}_{n=1}^{\infty}$ is an open cover of S, and since by assumption S is compact this means that there is some finite subcover

$$\{I_{n_1}, I_{n_2}, \ldots, I_{n_k}\}$$

of S. Assume that the sets were written in increasing order so that $n_1 < n_2 < \cdots < n_k$. Notice that this implies that

$$I_{n_1} \subseteq I_{n_2} \subseteq \cdots \subseteq I_{n_k},$$

and hence

$$S \subseteq \bigcup_{\ell=1}^{k} I_{n_\ell} = I_{n_k}.$$

But saying that $S \subseteq I_{n_k}$ precisely means that S is bounded between $-n_k$ and n_k, completing this part of the proof. $\qquad\square_{(S \text{ bounded})}$

---

surprisingly when I surveyed my last real analysis class on what their favorite theorem was throughout the course, quite a few said the Heine-Borel theorem, and they specifically cited the toughness of the proof as a reason! Now, if you don't believe you have that sort of character yet... then this proof is perfect to help you build it up! Also, to read a good discussion of the theorem's history, see *An Analysis of the First Proofs of the Heine-Borel Theorem — Borel's Proof* by Andre, Engdahl and Parker.

<u>Proof that S is closed.</u> Suppose for a contradiction that S is not closed. Then there exists some $x \notin S$ and a sequence $a_1, a_2, a_3, \ldots$ where each $a_k \in S$ and $a_k \to x$.

We again are going to find an open cover of S, and it will turn out that the existence of a finite subcover will give us our contradiction. To this end, for each $n \in \mathbb{N}$ let

$$U_n = \left(-\infty, x - \frac{1}{n}\right) \cup \left(x + \frac{1}{n}, \infty\right).$$

Notice that each U_n is an open set and moreover that

$$\bigcup_{n=1}^{\infty} U_n = \mathbb{R} \setminus \{x\}.$$

And this is great, since all we know about S is that $S \subseteq \mathbb{R}$ and $x \notin S$. So since $\{U_n : n \in \mathbb{N}\}$ covers *everything* except x, certainly it will cover S:

$$S \subseteq \bigcup_{n=1}^{\infty} U_n.$$

Now we use that S is compact. By compactness, this open cover of S must contain a finite subcover of S:

$$\{U_{n_1}, U_{n_2}, \ldots, U_{n_k}\}.$$

And once again if we assume that the subscripts are increasing and hence n_k is the largest, then all of the other U_{n_ℓ} are subsets of U_{n_k}, and hence

$$S \subseteq \bigcup_{\ell=1}^{k} U_{n_\ell} = U_{n_k}.$$

But saying that $S \subseteq U_{n_k}$ means that

$$S \subseteq \left(-\infty, x - \frac{1}{n_k}\right) \cup \left(x + \frac{1}{n_k}, \infty\right).$$

Which in particular implies that

$$S \cap \left(x - \frac{1}{n_k}, x + \frac{1}{n_k}\right) = \emptyset.$$

But this is a contradiction. We had assumed that there was a sequence (a_k) of numbers from S for which $a_k \to x$. But if we let $\varepsilon = \frac{1}{n_k} > 0$, then by the above there are *no* a_k for which

$$|a_k - x| < \varepsilon,$$

clearly contradicting the definition of convergence of a_k to x. $\square_{(S \text{ closed})}$

— Backwards direction —

Now assume that S is closed and bounded. We will show that S is compact. To that end, assume that $\mathcal{F}$ is an open cover of S. We will show that $\mathcal{F}$ contains a finite subcover of S.

For each x, define the set

$$S_x = S \cap (-\infty, x].$$

And let

$$B = \{x : \mathcal{F} \text{ contains a finite subcover of } S_x\}.$$

Intuitively, S_x is the set S up to the point x, and as x increases we are slowly revealing more and more of it. Since S is bounded, both $\inf(S)$ and $\sup(S)$ exist. Moreover, recall that such a set always contains a sequence of points converging to its infimum, and a sequence converging to its supremum (Proposition 3.29). And since S is also closed, this means that those limits — $\inf(S)$ and $\sup(S)$ — must in fact be elements of S.

Let $L = \inf(S)$ and $M = \sup(S)$. In the last paragraph we showed that $L, M \in S$.

Now, returning to the sets S_x and B, note that L being the minimum of S implies that $S_L = \{L\}$. And certainly the cover $\mathcal{F}$ of S contains a finite subcover of $\{L\}$: Since $L \in S$ we know that some open set $U \in \mathcal{F}$ covers $\{L\}$, and then $\{U\}$ is a finite subcover of $\{L\}$. We have therefore shown that $L \in B$. So B is nonempty.

Note that since $S_M = S$, if $M \in B$ then we are done: that would imply that S has a finite subcover and hence is compact. Therefore assume for a contradiction that $M \notin B$. Then $x < M$ for all $x \in B$, implying that B is bounded above. In summary, B is a nonempty subset of $\mathbb{R}$ that is bounded above, and hence by completeness $\sup(B)$ exists. Let $T = \sup(B)$.

We will now show that this leads to a contradiction. In fact, we will consider two cases and show that in either case we can reach a contradiction.

<u>Case 1: $T \in S$.</u> Since $\mathcal{F}$ is a cover of S, we know that the sets in $\mathcal{F}$ collectively cover all of S. In particular, since in this case we are assuming that $T \in S$, there must be some open set $U \in \mathcal{F}$ that covers the number T. That is, $T \in U$ for some open set $U \in \mathcal{F}$. But by the definition of U being an open set, this means that there is some δ-neighborhood around T which is also contained in U:

$$(T - \delta, T + \delta) \subseteq U.$$

In particular, this implies that

$$\left(T - \delta, T + \frac{\delta}{2}\right] \subseteq U.$$

We will now show that T was not in fact $\sup(B)$. In particular, we will show that $(T + \delta/2) \in B$.

Note that since $T - \delta < T = \sup(B)$, this means that $T - \delta \in B$. I.e., there exists a finite subcover $\{F_1, F_2, \ldots, F_k\}$ from $\mathcal{F}$ of the set $S \cap (-\infty, T - \delta]$.

But notice that $\{F_1, F_2, \ldots, F_k, U\}$ is also a finite subset of $\mathcal{F}$ and this set covers $S \cap \left(-\infty, T + \frac{\delta}{2}\right]$. That is, we have found a finite subcover of $S_{T+\delta/2}$, implying that $(T + \delta/2) \in B$, contradicting our choice that $\sup(B) = T$. So in Case 1, we did find a contradiction. $\square_{\text{(Case 1)}}$

Case 2: $T \notin S$. The idea behind this proof is that, since $T \notin S$, we could have increased T by a bit and literally not changed the set. That is, there is some ε for which $S_T = S_{T+\varepsilon}$, and so if S_T can be finitely subcovered, then certainly $S_{T+\varepsilon}$ can be too, since they are literally the same set. This would contradict the claim that $T = \sup(B)$, since $T + \varepsilon$ must then also be in B. The trick is finding a way to increase T by some small amount.

Since by assumption $T \notin S$ and S is closed, this implies that $T \in S^c$ and S^c is open. And to be an element of an open set means that there is some δ-neighborhood

$$(T - \delta, T + \delta) \subseteq S^c.$$

In particular,

$$\left[T - \frac{\delta}{2}, T + \frac{\delta}{2}\right] \cap S = \emptyset.$$

And so

$$S \cap \left(-\infty, T - \frac{\delta}{2}\right] = S \cap \left(-\infty, T + \frac{\delta}{2}\right]. \qquad (\text{☢})$$

But since $T - \delta/2 < T$ and $\sup(B) = T$, we know that there is a finite subcover of $S \cap (-\infty, T - \delta/2]$. But then, by (☢), this is also a finite subcover of

$$S \cap \left(-\infty, T + \frac{\delta}{2}\right].$$

So $(T + \delta/2) \in B$, contradicting our choice that $\sup(B) = T$. $\square_{\text{(Case 2)}}$ $\square$

> **Theorem.**
>
> **Theorem 5.20** (*Heine-Borel, expanded*). Let $A \subseteq \mathbb{R}$. Then the following are equivalent.
>
> (i) A is compact.
>
> (ii) A is closed and bounded.
>
> (iii) If (a_n) is a sequence of numbers in A, then there is a subsequence (a_{n_k}) that converges to a point in A.

Proof. The Heine-Borel theorem (Theorem 5.19) shows that (i) holds if and only if (ii) holds. In Exercise 5.16 you will prove that (ii) holds if and only (iii) holds. $\square$

— Notable Exercises —

- There is a way to introduce topology that is heavily reliant on sequences and limit points. I went a different route in this book, but some authors do it differently. Nevertheless, I do include several exercises dealing with limit points to enforce their importance. These include Exercises 5.7, 5.8, 5.9 and 5.10.

- A snappy solution to Exercise 5.11 utilizes de Morgan's laws (Fact 5.11). In fact, they are so intimately connected that if you didn't know of de Morgan's laws but independently proved this exercise, then this exercise (and the fact that the complement of an open set is a closed set) might lead you to believe that something like de Morgan's law should hold.

  A special case of this exercise was asked of you in Exercise 1.34.

- Exercise 5.22 is similar in flavor to a famous set called the *Cantor set*, which is discussed in Example B.1 of Appendix B.

- Exercises 5.31m and 5.32 introduce an important idea in topology — that of the *interior*, *exterior* and *boundary* of a set.

- Exercise 5.33 introduces another important idea in topology — that of a *connected* set. Parts (a)-(e) get you warmed up with this idea, and part (f) asks you to prove the main theorem about this property.

— Exercises —

Exercise 5.1. For each of the following, determine whether the set is open, whether it is closed, and whether it is compact. (It might be more than one, or none of these.) You do not need to prove your answers.

(a) $\mathbb{Z}$ (c) $\mathbb{R}$ (e) $\mathbb{Q}$

(b) $\{1, \frac{1}{2}, \frac{1}{3}, \frac{1}{4}, \dots\} \cup \{0\}$ (d) $(0, 1) \cup [3, 4]$ (f) $\{17\}$

Exercise 5.2. Determine the set of limits points for each set in Exercise 5.1.

Exercise 5.3. For each of the following, provide a proof or a counterexample.

(a) If A and B are compact, must $A \cup B$ be compact?

(b) If A and B are compact, must $A \cap B$ be compact?

Exercise 5.4.

(a) Prove that if A is closed and B is open, then $A \setminus B$ is closed.

(b) Prove that if A is open and B is closed, then $A \setminus B$ is open.

Exercise 5.5.

(a) Give an example of countably many disjoint open intervals.

(b) Prove that there does not exist a collection of uncountably many disjoint open intervals.

Exercise 5.6. For each of the following, you should state which sets you are choosing and what their intersection/union is, but you do not need to prove your examples work.

(a) Give an example of an infinite collection of open sets whose intersection is *not* open.

(b) Give an example of an infinite collection of closed sets whose union is *not* closed.

(c) Give an example of an infinite collection of compact sets whose union is *not* compact.

Exercise 5.7. Prove that a point x is a limit point of a set A if and only if every ε-neighborhood of x intersects A at some point other than x.

Exercise 5.8. Construct a set A whose set of limit points is $\mathbb{Z}$.

Exercise 5.9. Does there exist a set A whose set of limit points is precisely $\{1, \frac{1}{2}, \frac{1}{3}, \frac{1}{4}, \frac{1}{5}, \ldots\}$?

Exercise 5.10. Show that the set of limit points of a set is closed.

Exercise 5.11. Prove Proposition 5.12. That is, prove the following.

(a) If $\{U_1, U_2, \ldots, U_n\}$ is a collection of closed sets, then $\bigcup\limits_{k=1}^{n} U_k$ is also a closed set.

(b) If $\{U_\alpha\}$ is a collection of closed sets, then $\bigcap\limits_{\alpha} U_\alpha$ is also a closed set.

Exercise 5.12. For each of the following tasks, give an example as requested or prove that one does not exist.

(a) A nonempty open set that is a subset of $\mathbb{Q}$.

(b) A nonempty closed set that is a subset of $\mathbb{Q}$.

(c) Two nonempty disjoint open sets whose union is $\mathbb{R}$.

(d) An infinite set with no limit points.

(e) A bounded infinite set with no limit points

(f) An infinite union of compact sets that is not compact.

(g) An infinite intersection of compact sets that is not compact.

Exercise 5.13.

(a) If finitely many points are removed from an open set, must the set still be open?

(b) If countably many points are removed from an open set, must the set still be open?

(c) If uncountably many points are removed from an open set, must the set still be open?

Exercise 5.14. Let A be a closed set, let x be a point from A, and let $B = A \backslash \{x\}$. Give necessary and sufficient conditions on A and x for B to be a closed set. Prove that your conditions works.

Exercise 5.15. Prove that

$$\left\{ \left(\frac{1}{k}, 4 - \frac{1}{k} \right) \right\}_{k=1}^{\infty}$$

is an open cover of of $(0, 4)$, but that this cover has *no* finite subcover. This implies that $(0, 4)$ is *not* compact.

Exercise 5.16. Let $A \subseteq \mathbb{R}$. Prove that A is closed and bounded (i.e. compact) if and only if every sequence of numbers from A has a subsequence that converges to a point in A.

Exercise 5.17. Let A be compact and $U \subseteq A$ be closed. Prove that U is compact.

Exercise 5.18. For each of the following, prove it to be true or provide a counterexample

(a) If A is compact and B is bounded, must $A \cap B$ be compact?

(b) If A is compact and B is closed, must $A \cap B$ be compact?

(c) If A is compact and B is bounded, must $A \cup B$ be compact?

(d) If A is compact and B is closed, must $A \cup B$ be compact?

Exercise 5.19. Prove that if $\{U_\alpha\}$ is a collection of compact sets, then $\bigcap_\alpha U_\alpha$ is also a compact set.

Exercise 5.20. For sets A and B, define $A + B = \{a + b : a \in A, b \in B\}$.

(a) Prove that if A and B are open, then $A + B$ is open.

(b) Prove that if A and B are compact, then $A + B$ is compact.

(c) It is *not* true that the sum of closed sets must be closed. Provide an example to demonstrate this.

Exercise 5.21. For sets A and B, define $A \cdot B = \{a \cdot b : a \in A, b \in B\}$.

(a) If A and B are open, must $A \cdot B$ be open?

(b) If A and B are compact, must $A \cdot B$ be compact?

(c) If A and B are closed, must $A \cdot B$ be closed?

Exercise 5.22.   Let A be the set of numbers in $[0,1]$ whose decimal expansions use only the numbers 2, 5 and 8. For example, $0.5822585582 \in A$, and also $0.2525252525\ldots \in A$. Prove that A is a closed set.

Exercise 5.23.   One open cover of the set $[2,10]$ is the collection

$$\left\{ \bigcup_{n=1}^{\infty} \left(2 + \frac{1}{n^2}, 10 - \frac{1}{n^2} \right) \right\} \cup \{(0,3), (9.82, 10.1)\}.$$

Since $[2,10]$ is compact, we know that this open cover must have a finite subcover. Give an example of such a subcover of this cover. You do *not* need to prove your answer.

Exercise 5.24.   A set A is said to have the *intersecting closedness property* if it satisfies this condition: If $\mathcal{S}$ is any collection of closed sets for which the intersection of any finite number of sets from $\mathcal{S}$ contains an element of A, then the intersection of every set in $\mathcal{S}$ also contains an element of A.

Prove that A has the intersecting closedness property if and only if A is compact.

Exercise 5.25.   Call a subset A of real numbers *closed-cover-compact* if every closed cover of A (that is, a cover consisting of closed sets) has a finite subcover. Which sets A are closed-cover-compact?

Exercise 5.26.

(a) Prove that if A is compact, then $\sup(A)$ exists and $\sup(A) \in A$. Does the same hold for the infimum?

(b) Give an example of a set which contains its supremum and its infimum, but is not compact.

(c) If a set contains its supremum, its infimum, and all of its limit points, must it be compact?

Exercise 5.27.   Suppose $U_1, \ldots, U_n$ is a finite open cover of a compact set A. Note that this implies that $B \subseteq U_1 \cup U_2 \cup \cdots \cup U_n$. Is it possible that the union $B = U_1 \cup U_2 \cup \cdots \cup U_n$?

Exercise 5.28.   For which compact sets A does there exist an $m \in \mathbb{N}$ such that every open cover of A contains a subcover containing at most m open sets?

Exercise 5.29.   Call a subset A of real numbers *clopen* if it is both open and closed. Prove that the only clopen sets are $\emptyset$ and $\mathbb{R}$.

Exercise 5.30.   Give two examples of sets which are neither open nor closed. Have one example be a bounded set and the other be an unbounded set.

Definitions.

- The *interior* of a set A, denoted $\text{Int}(A)$, is the set of points x such that there is a neighborhood of x that is a subset of A.

- The *exterior* of a set A, denoted $\text{Ext}(A)$, is the set of points x such that there is a neighborhood of x that is a subset of A^c.

- The *boundary* of a set A, denoted ∂A, is the set of points X such that every neighborhood of x contains points from both A and A^c.

Exercise 5.31.

(a) Find the interior, exterior and boundary for each of the following sets: $(0, 1)$, $[0, 1]$, $\mathbb{R}$ and $\mathbb{Q}$.

(b) Prove that for any set A, we have $\mathbb{R} = \text{Int}(A) \cup \partial A \cup \text{Ext}(A)$, and this is a disjoint union.

Exercise 5.32.   Prove that the only sets with empty boundary are $\mathbb{R}$ and $\emptyset$.

Definition.   A set A is said to be *connected* if there do <u>not</u> exist two open sets U and V such that

(i) $U \cap V = \emptyset$,

(ii) $U \cap A \neq \emptyset$ and $V \cap A \neq \emptyset$, and

(iii) $(U \cap A) \cup (V \cap A) = A$.

Exercise 5.33.

(a) Explain intuitively what it means for a set A to be connected.

(b) Give an example of a connected set and of a set that is not connected. You do not need to prove your answers.

(c) Give an example of a set A which is not connected, but $A \cup \{4\}$ is connected.

(d) Prove that $\{1, 2, 3, 4, 5\}$ is not connected.

(e) Prove that $\mathbb{Q}$ is not connected.

(f) Give an example of a set A which is not connected, but there exists some $x_0 \in \mathbb{R}$ such that $A \cup \{x_0\}$ is connected.

(g) Prove that a set of real numbers with more than one element is connected if and only if it is an interval.

— Open Question —

In the area of topology called *knot theory*,[12] a piece of rope that is tied at the ends is called a *knot*. If you took an untangled piece of rope and tied the ends, what you get is called the *unknot*.

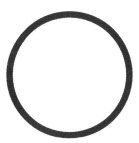

If the rope is tangled up before the ends are tied, though, it may or may not be possible to untangle it to resemble the unknot. The below, for instance, is the *trefoil knot*, and as you can see it is impossible to untangle it to make it look like the unknot.

Given a knot, if you can untangle it to resemble the unknot, then you essentially had (a messy version of) the unknot the whole time. Therefore if your knot can be completely untangled, then we say that your knot *is* the unknot. Now here's the open question: If knots are all drawn on the paper like the one above, does there exist a polynomial-time algorithm to determine whether a knot is the unknot?

---

[12] A mathematician walks into a bar with a dog and a cow. The bartender says, "Hey, you can't bring wild animals in here!"

The mathematician replies, "You don't understand, these are very special animals."

"How so?"

"They're knot theorists," he says. And, seeing the bartender's look of disbelief, he adds, "here, I'll prove it to you — ask them a question."

The bartender turns the dog, "Name a knot invariant."

"Arf! Arf!" barks the dog.

Scowling, the bartender faces the cow. "Name a topological invariant."

"Mu! Mu!" responds the cow.

Irate, the bartender throws all three out. The dog, bothered at being yelled at, turns to the mathematician and asks, "Should have said the Jones polynomial?"

Chapter 6: Continuity

6.1 Approaching Continuity

I made a meme describing your discrete approach towards understanding continuity.[1]

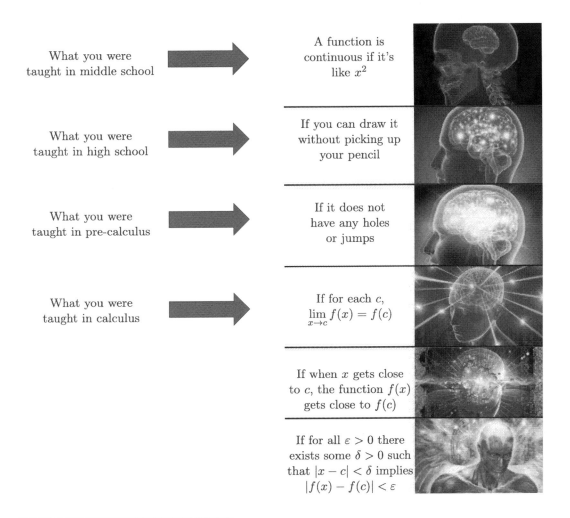

---

[1]Here's to hoping that memes have a longer half-life than I fear!

6.2   Weird Examples

What should it mean to say a function is continuous? Or to say that the function is continuous at the point $x = c$?

When you were young you might have just imagined some basic examples, which were probably all polynomials. At some point you may have been told that it means that you can draw it without picking up your pencil. Then that there weren't any holes or gaps in the graph — that might have been where you were when you got your high school diploma. (And at this point, you must admit that all of these definitions are imprecise and informal; they do not provide the rigor that mathematicians demand.)

When taking calculus you might have learned that it's in fact a localized property, and f being continuous at some point c means that as x gets close to c, you need to have $f(x)$ get close to $f(c)$. This was then formalized by saying that f is continuous at c if[2]

$$\lim_{x \to c} f(x) = f(c).$$

This is now getting pretty good, although you were likely given little to no explanation for why this is how it is. In fact, at this point we have not even defined what it means to take a limit of a function. But intuitively it should mean something like this: As x gets *near* c, $f(x)$ will be getting *near* $f(c)$.

We have two goals for the next few sections:

- Understand what is meant by "$\lim_{x \to c} f(x)$."

- Understand what is meant by "$f(x)$ is continuous at c."

In 15 pages time we will have ironed these out. There is still some preliminary work to go, but for now let's just look at some interesting examples that highlight the complexities and subtleties of these goals.

Important Examples

Question 6.1. Does there exist a function which is continuous at one point but not continuous at another?

Of course! For example, here is the graph of such an example, where we have a jump discontinuity at the point c_1, and so it is discontinuous there, while at every other point (such as at c_2) the function is continuous.

---

[2]The notation "$\lim_{x \to c} f(x)$" will be formally defined soon.

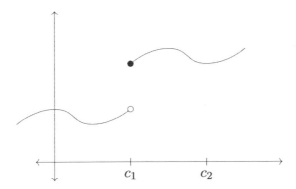

As the above example illustrates, since a function can be continuous at one point while being discontinuous at another, the property of being continuous must be point-centric.

Question 6.2. Do there exist functions which are continuous nowhere? That is, can we find a function f which is not continuous at any point c?

The first example is known as the *Dirichlet's function*. Dirichlet's idea was to consider a piecewise function f in which $f(x)$ was equal to 1 whenever x was a rational number, and $f(x) = 0$ whenever x was irrational. Here's that:

Example 6.3. Define the *Dirichlet function* to be

$$f(x) = \begin{cases} 1 & \text{if } x \in \mathbb{Q} \\ 0 & \text{if } x \notin \mathbb{Q}. \end{cases}$$

Recall that the irrationals are dense in the rationals, and the rationals are dense in the irrationals. Therefore the graph looks pretty wacky:

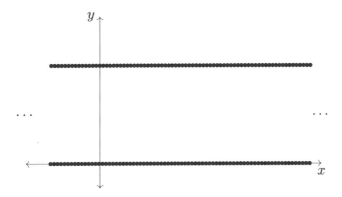

Pick any point, like $c = 3$. Is f continuous at c? Based on our intuition of "f is continuous at c if x being close to c implies that $f(x)$ is close to $f(c)$," it is not, right? Since $c = 3$ is rational, $f(c) = 1$. But you can find irrational numbers x as close to c as you want, and for those numbers $f(x) = 0$, which is not near $f(c) = 1$.

In calculus, we talked about continuity in terms of limits. So what in this situation should $\lim_{x \to 3} f(x)$ be equal to? At this point in this book, the only limits we know how to take are limits of sequences. So, for example, $(a_n) = \left(3 - \frac{1}{n}\right)$ is a sequence converging to 3. And for each $n \in \mathbb{N}$, note that $\left(3 - \frac{1}{n}\right) \in \mathbb{Q}$. So each term of this sequence has $f(a_n) = 1$, and so for this sequence we do have $f(a_n) \to f(c)$, since all this is saying is that a sequence of 1s is converging to 1. So this suggests that maybe $\lim_{x \to 3} f(x) = 1$.

However, not all sequences behave like this. For instance, note that, for each $n \in \mathbb{N}$, $b_n := 3 + \frac{\sqrt{2}}{n}$ is irrational, and also $b_n \to 3$. So we have a sequence (b_n) which is converging to 3, but $(f(b_n))$ is just a sequence of 0s, and hence converges to 0, which is not $f(3)$.

(Worse yet, you could find a third sequence (c_n) for which $\lim_{n \to \infty} f(c_n)$ does not even exist, by having the terms alternate between being rational and irrational.)

So we can not in general find $\lim_{x \to 3} f(x)$ by simply finding a sequence (a_n) converging to 3 and applying f to it and checking what $f(a_n)$ is converging to, because it might depend on which sequence we pick! Indeed, our eventual definition of these functional limits will label the above limit as "does not exist."

And of course, there was nothing special about $c = 3$. By the same reasoning, any $c \in \mathbb{R}$ will have the property that $\lim_{x \to c} f(x)$ does not exist. We are therefore going to say that this function f is *continuous nowhere*. $\qquad \square$

Question 6.4. Can a function be continuous at only one point? Or does being continuous somewhere mean that there is a region around that point on which the function is also continuous? (Think about these before jumping to the answer.)

The next example is similar, but highlights an important point: just because a function is not "smooth"-looking, does not mean that it is discontinuous everywhere. This destroys the "continuous means you can draw it without picking up your pencil" claim.

Example 6.5. Define the *modified Dirichlet function* to be

$$g(x) = \begin{cases} x & \text{if } x \in \mathbb{Q} \\ 0 & \text{if } x \notin \mathbb{Q}. \end{cases}$$

Again, recall that the irrationals are dense in the rationals, and the rationals are dense in the irrationals. Therefore the graph again looks pretty wacky:

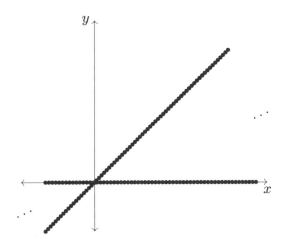

Pick any $c \neq 0$. Like with the Dirichlet function, there is a gap between the dots on the x-axis near c, and those on the line $y = x$ near c. And if (a_n) is a sequence of rational numbers converging to c and b_n is a sequence of irrational numbers converging to c, then

$$\lim_{n \to \infty} g(a_n) = c \neq 0 = \lim_{n \to \infty} g(b_n).$$

Therefore, just like with the Dirichlet function, g is not going to be continuous at any such c.

What about at $c = 0$? Well, here things seem to be different. We can no longer construct two sequences converging to 0 which have different limits when we apply g to them; all such sequences also converge to 0. Sure, the function is still jumping all over the place, so it doesn't look anything like what we imagine it means to be continuous at a point, but it does seem to satisfy our guiding principle of "$f(x)$ is continuous at c if x being close to c implies that $f(x)$ is close to $f(c)$." Indeed, as x gets close to 0, all the $g(x)$-values are close to (or equal to) $g(c) = 0$. When we eventually write down a precise definition, it will agree that g is continuous at 0 (and discontinuous at every other point).

So yes, this will be an example of a function which is continuous at only one point![3]   □

Question 6.6. Can a function be continuous at every irrational number and discontinuous at every rational number?

Now we are asking for all sorts of craziness. If this is possible, then you should stop trusting all of your prior intuition about continuity. Well, get ready, because the next example does just this, and is called *Thomae's function*.

---

[3]Make sure you take a moment to appreciate how remarkably, wonderfully weird this is.

Example 6.7. Define *Thomae's function*[4] to be

$$h(x) = \begin{cases} 1 & \text{if } x = 0 \\ 1/n & \text{if } x \in \mathbb{Q} \text{ and } x = m/n \text{ in lowest terms with } n > 0 \\ 0 & \text{if } x \notin \mathbb{Q}. \end{cases}$$

Notice that at any rational point c, we have $h(c) > 0$; meanwhile, there are irrational numbers x arbitrarily close to c which have $h(x) = 0$. So h will indeed be discontinuous at every rational number.

At any irrational number c, though, we can show that h is actually continuous. We will prove this in detail later, but the basic idea is this: Let $\varepsilon > 0$, and find an N for which $\frac{1}{N} < \varepsilon$. We can show that, in some region around c, all x-values have $h(x) < \frac{1}{N} < \varepsilon$. To see this, note that there are only finitely many rational numbers which have a denominator of N or smaller and are also within, say, distance 1 of c. Find the one that is closest to c, and suppose it has distance $\delta > 0$ from c. Then, all rational numbers in $(c - \delta, c + \delta)$ have denominators that are larger than N. So within this range, all function values $h(x)$ are smaller than $1/N$. Awesome!

And, as you can imagine, as N gets large this forces all the function values to be converging to 0 in these shrinking regions around c, kind of like what happened at $c = 0$ in the modified dirichlet function. So yes, this will be an example of a function which is discontinuous on $\mathbb{Q}$, and continuous on $\mathbb{R} \setminus \mathbb{Q}$![5]   □

This function's graph looks like this:[6]

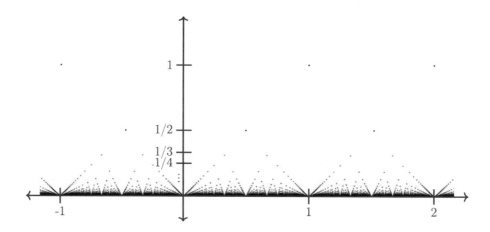

So yes, with these functions in mind, we certainly are in need of a careful definition of continuity. But first, as we have discussed, we need a careful definition of a functional limit.

---

[4]AKA the *popcorn function*, AKA the *raindrop function*, AKA the *countable cloud function*, AKA the *ruler function*, AKA the *Riemann function*, AKA the *Stars over Babylon*.

[5]Make sure you take a moment to appreciate how remarkably, wonderfully weird this is.

[6]So cool that it made the cover.

6.3   Functional Limits

Continuing with our discussion, we think about "$\lim\limits_{x \to c} f(x) = L$" as meaning "as x gets arbitrarily close to c, $f(x)$ is getting arbitrarily close to L." We will state this in a fairly general way. Rather than requiring that f be a function from $\mathbb{R}$ to $\mathbb{R}$, we will allow the domain to be some subset $A \subseteq \mathbb{R}$. Moreover, c does not necessarily have to be a member of A; if c is a limit point of A, then we can still pick x-values from A that approach it, and everything goes through just fine.

Definition.

Definition 6.8. Let $f : A \to \mathbb{R}$ and let c be a limit point of A. Then we say that

$$\lim_{x \to c} f(x) = L$$

if for all $\varepsilon > 0$ there exists some $\delta > 0$ such that for every $x \in A$ for which $0 < |x - c| < \delta$, we have

$$|f(x) - L| < \varepsilon.$$

We also say $\lim\limits_{x \to c} f(x)$ *converges* to L in such a situation.

This definition is both important and confusing, so some comments are in order.

<u>Comment 1.</u> The reason we require that $x \in A$ is just so that $f(x)$ makes sense. Remember, $f : A \to \mathbb{R}$, so if $x \notin A$ then $f(x)$ is not even defined. (Note that we never take $f(c)$, so that's why it's ok if $c \notin A$.)

<u>Comment 2.</u> The reason why we require $0 < |x - c|$ (i.e. $x \neq c$) is because limits only care about what happens as you *approach* the point c, not what actually happens *at* the point c. Indeed, $f(c)$ could be changed to anything at all — *or could not even exist* — and the limit would be unaffected.

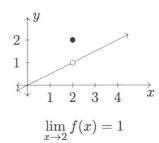

$$\lim_{x \to 2} f(x) = 1$$

<u>Comment 3.</u> When the context is clear, for convenience we do not always explicitly state the conditions $x \in A$ and $0 < |x - c|$.

Comment 4. Below is the important picture to keep in mind, even though as we will and have seen, it is overly simplistic.[7]

Suppose f, c, and L are as pictured below, and we are interested in showing that $\lim_{x \to c} f(x) = L$.

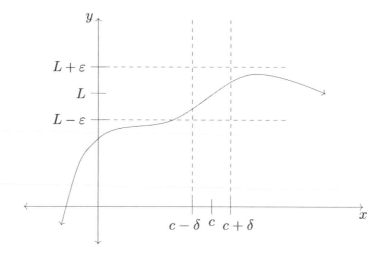

The picture shows that if x has the property that $0 < |x - c| < \delta$ (meaning that the x-values are between the vertical dashed lines), then $f(x)$ has the property that $|f(x) - L| < \varepsilon$ (meaning that the graph of $f(x)$ has y-values inside of the horizontal dashed lines). Indeed, there is another way to think about this using the neighborhood terminology from topology.

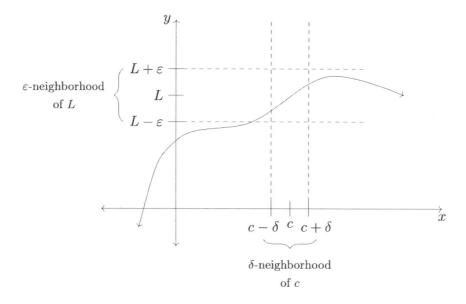

So we could restate the definition: $\lim_{x \to c} f(x) = L$ if for all $\varepsilon > 0$ there exists a $\delta > 0$ such that if x is in the δ-neighborhood of c, then $f(x)$ is in the ε-neighborhood of L.

---

[7] *"Art is a lie that enables us to realize the truth."* –Pablo Picasso

<u>Comment 5.</u> To show something is *not* continuous, you must prove the negation of the definition. The definition of f being continuous at c is this:

> For all $\varepsilon > 0$ there exists some $\delta > 0$ such that for all $x \in A$ for which $0 < |x - c| < \delta$, we have $|f(x) - L| < \varepsilon$.

By negating the above, here is what it means for f to *not* be continuous at c:

> There exists $\varepsilon > 0$ such that for all $\delta > 0$ there exists some $x \in A$ satisfying $0 < |x - c| < \delta$ for which $|f(x) - L| \geq \varepsilon$.

That is, we must find just a single ε which fails. The (still simplistic) picture for $\lim_{x \to c} f(x) \neq L$ is this:

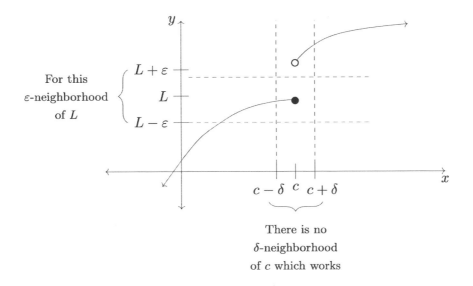

For this ε-neighborhood of L

There is no δ-neighborhood of c which works

For that particular ε, there is no possible $\delta > 0$ that we could choose to have the limit criterion hold. In particular, for any $\delta > 0$ that we choose, the x-values between c and $c + \delta$ are guaranteed to have $|f(x) - L| \geq \varepsilon$. And since we found a single ε for which the criterion fails, f is *not* continuous at c.

And with that, we conclude our comments on Definition 6.8, the definition of a limit.[8]

---

[8] Just like with the definition of sequence convergence, this is a definition that you should aim to *fully* understand. And so I will echo my earlier advice: If you don't fully understand Definition 6.8, then go brew a pot of tea, find a comfortable Thinking Chair, and ponder it carefully. Don't move on until that pot is emptied.

Here's an update on your brain expansion status:[9]

A function is continuous if it's like x^2	
If you can draw it without picking up your pencil	
If it does not have any holes or jump	
If for each c, $\lim_{x \to c} f(x) = f(c)$	
If when x gets close to a, the function $f(x)$ gets close to $f(a)$	
If for all $\varepsilon > 0$ there exists some $\delta > 0$ such that $|x - c| < \delta$ implies $|f(x) - f(c)| < \varepsilon$	

Where you were at the start of the chapter ⟹ (If when x gets close to a, the function $f(x)$ gets close to $f(a)$)

Where you are now ⟹ (If for all $\varepsilon > 0$ there exists some $\delta > 0$ such that $|x - c| < \delta$ implies $|f(x) - f(c)| < \varepsilon$)

---

[9]I'm really doubling down on this meme still being a thing...

Examples

Let's now do some examples of this definition.

Example 6.9. Let $f(x) = 5x + 2$. Prove $\lim\limits_{x \to 3} f(x) = 17$.

Scratch Work. Let $\varepsilon > 0$. Recall, $|x - 3| < \delta$ is what we will assume, for whichever $\delta > 0$ that we choose. So in this scratch work we will assume that our conclusion holds,

$$|f(x) - L| < \varepsilon,$$

and then we unwind this until we get just an $|x - 3|$ on one side, which then tells us which δ will make it work.

$$
\begin{aligned}
\text{WTS:} \quad & |f(x) - L| < \varepsilon \\
& |5x + 2 - 17| < \varepsilon \\
& |5x - 15| < \varepsilon \\
& 5 \cdot |x - 3| < \varepsilon \\
& |x - 3| < \frac{\varepsilon}{5} \quad \checkmark \\
& \text{So let } \delta = \frac{\varepsilon}{5}
\end{aligned}
$$

We were able to get down to just $|x - 3|$ on the left, which is the thing that we can control. So if we choose δ to be $\varepsilon/5$, the above can be redone to provide a proof that $\lim\limits_{x \to 3} f(x) = 17$.

Solution. Let $\varepsilon > 0$. And let $\delta = \frac{\varepsilon}{5} > 0$. Then for any x where $0 < |x - 3| < \delta$,

$$
\begin{aligned}
|f(x) - L| &= |5x + 2 - 17| \\
&= |5x - 15| \\
&= 5 \cdot |x - 3| \\
&< 5 \cdot \delta \\
&= 5 \cdot \frac{\varepsilon}{5} \\
&= \varepsilon.
\end{aligned}
$$

$\square$

The next example is that of a rational function—a polynomial divided by another polynomial.

Example 6.10. Let $f : \mathbb{R} \setminus \{2\} \to \mathbb{R}$ where $f(x) = \dfrac{3x^2 - 12}{x - 2}$. Prove $\lim\limits_{x \to 2} f(x) = 12$.

Scratch Work. Let $\varepsilon > 0$. Recall, $|x - 2| < \delta$ is what we will assume, for whichever $\delta > 0$ that we choose. So in our scratch work we will assume that our conclusion holds,

$$|f(x) - L| < \varepsilon,$$

and then we unwind this until we get just an $|x - 2|$ on one side, which then tells us which δ will make it work. Here's that:

$$
\begin{aligned}
\text{WTS:} \quad & |f(x) - L| < \varepsilon \\
& \left| \frac{3x^2 - 12}{x - 2} - 12 \right| < \varepsilon \\
& \left| \frac{3(x - 2)(x + 2)}{x - 2} - 12 \right| < \varepsilon \\
& |3(x + 2) - 12| < \varepsilon \\
& 3 \cdot |(x + 2) - 4| < \varepsilon \\
& 3 \cdot |x - 2| < \varepsilon \\
& |x - 2| < \frac{\varepsilon}{3} \quad \checkmark \\
& \text{So let } \delta = \frac{\varepsilon}{3}.
\end{aligned}
$$

We were able to get down to just $|x - 2|$ on the left, which is the thing that we can control. So if we choose δ to be $\varepsilon/3$, the above can be redone to provide a proof that $\lim\limits_{x \to 3} f(x) = 12$.

Solution. Let $\varepsilon > 0$. And let $\delta = \frac{\varepsilon}{3} > 0$. Then for any x where $0 < |x - 2| < \delta$,

$$
\begin{aligned}
|f(x) - L| &= \left| \frac{3x^2 - 12}{x - 2} - 12 \right| \\
&= \left| \frac{3(x - 2)(x + 2)}{x - 2} - 12 \right| \\
&= |3(x + 2) - 12| \\
&= 3 \cdot |(x + 2) - 4| \\
&= 3 \cdot |x - 2| \\
&< 3 \cdot \delta \\
&= 3 \cdot \frac{\varepsilon}{3} \\
&= \varepsilon. \qquad \qquad \square
\end{aligned}
$$

6.4 Properties of Functional Limits

> **Proposition.**
>
> **Proposition 6.11** (*Functional limits are unique*). A limit $\lim_{x \to c} f(x)$ can converge to at most one value.

Proof. Suppose $A \subseteq \mathbb{R}$, c is a limit point of A, and $f : A \to \mathbb{R}$. Suppose

$$\lim_{x \to c} f(x) = L_1 \qquad \text{and} \qquad \lim_{x \to c} f(x) = L_2.$$

We will show that $|L_1 - L_2| < \varepsilon$ for all $\varepsilon > 0$, which would imply that $L_1 = L_2$, as desired.

To this end, let $\varepsilon > 0$. Since $\lim_{x \to c} f(x) = L_1$, there exists some $\delta_1 > 0$ such that $0 < |x - c| < \delta_1$ implies

$$|f(x) - L_1| < \frac{\varepsilon}{2}.$$

Likewise, since $\lim_{x \to c} f(x) = L_2$, there exists some $\delta_2 > 0$ such that $0 < |x - c| < \delta_2$ implies

$$|f(x) - L_2| < \frac{\varepsilon}{2}.$$

Let $\delta = \min\{\delta_1, \delta_2\}$, and notice that now $0 < |x - c| < \delta$ implies both $0 < |x - c| < \delta_1$ and $0 < |x - c| < \delta_2$. And so, for all x where $0 < |x - c| < \delta$, we have

$$
\begin{aligned}
|L_1 - L_2| &= |L_1 - f(x) + f(x) - L_2| \\
&\leq |L_1 - f(x)| + |f(x) - L_2| \qquad \text{(Triangle inequality)} \\
&= |f(x) - L_1| + |f(x) - L_2| \\
&< \frac{\varepsilon}{2} + \frac{\varepsilon}{2} \\
&= \varepsilon.
\end{aligned}
$$

$\square$

In our discussion in Section 6.2 on the three interesting examples, we mentioned that one reason a functional limit does not exist is because there are two sequences a_n and b_n, both of which converge to c, for which $\lim_{n \to \infty} f(a_n) \neq \lim_{n \to \infty} f(b_n)$. We will show now, though, that if all such sequences *do* converge to the same thing, then the functional limit *does* exist, and equals that common value.

> **Theorem.**
>
> **Theorem 6.12** ($\lim\limits_{x \to c} f(x) = L$ *iff every such* $f(a_n) \to L$). Assume that $A \subseteq \mathbb{R}$, $f : A \to \mathbb{R}$, and c is a limit point of A. Then $\lim\limits_{x \to c} f(x) = L$ if and only if, for every sequence a_n from A for which each $a_n \neq c$ and $a_n \to c$, we have $f(a_n) \to L$.

Proof Idea. For the forward direction, we are assuming that if x is close to a (say, $|x - a| < \delta$), then $f(x)$ is close to $f(a)$ (say, $|f(x) - f(a)| < \varepsilon$). What we aim to show is that $a_n \to c$ implies that $f(a_n) \to L$. Broadly, this is done like this:

$$
\begin{aligned}
a_n \to c \quad &\Longrightarrow \quad a_n \text{ is close to } c \text{ for } n > N \qquad (\text{Say, } |a_n - c| < \delta, \text{ for } n > N) \\
&\Longrightarrow \quad f(a_n) \text{ is close to } L \text{ for } n > N \qquad (\text{By continuity of } f) \\
&\Longrightarrow \quad f(a_n) \to L.
\end{aligned}
$$

For the reverse direction, we are assuming that every sequence a_n for which $a_n \to c$ has the property that $f(a_n) \to L$, and we are trying to show that $\lim\limits_{x \to c} f(x) = L$. The assumption about "every sequence..." is a little awkward to apply. However, if we use a proof by contradiction or by contraposition, then we will be assuming that $\lim\limits_{x \to c} f(x) \neq L$, and we will be aiming to find a *single* sequence (a_n) which does *not* have the asserted property. This is a clearer path.

Proof. Throughout the proof, assume that $x \neq c$ and $a_n \neq c$ for all n.

First we will prove the forward direction. Recall that the definition of "$\lim\limits_{x \to c} f(x) = L$" is that for any $\varepsilon > 0$ there is some $\delta > 0$ such that if $0 < |x - c| < \delta$, then $|f(x) - L| < \varepsilon$. Let a_n be an arbitrary sequence from A converging to c. We will now show that $f(a_n)$ is converging to L.

Let $\varepsilon > 0$. We wish to show that there exists some N for for which $|f(a_n) - L| < \varepsilon$ for all $n > N$. To see this, simply note that since $a_n \to c$ and $\delta > 0$, by definition[10] there exists some N for which $|a_n - c| < \delta$ for all $n > N$. But by the last paragraph, this then implies that $|f(a_n) - L| < \varepsilon$, as desired.

We will now prove the reverse direction. Assume that for every sequence a_n from A for which each $a_n \neq c$ and $a_n \to c$, we have $f(a_n) \to L$. And assume for a contradiction that $\lim\limits_{x \to c} f(x) \neq L$. That is, we are assuming this:

> This is *not* true: For all $\varepsilon > 0$ there exists some $\delta > 0$ such that for all $x \in A$ for which $0 < |x - c| < \delta$, we have $|f(x) - L| < \varepsilon$.

Therefore, by negating the above,

> This *is* true: There exists $\varepsilon > 0$ such that for all $\delta > 0$ there exists some $x \in A$ satisfying $0 < |x - c| < \delta$ for which $|f(x) - L| \geq \varepsilon$.

---

[10] Here, δ is taking the role of ε in the definition of a convergent sequence (Definition 3.7).

In particular, for this $\varepsilon > 0$, by letting $\delta_n = \frac{1}{n}$ we see that there exists some $x_n \in A$ satisfying $0 < |x_n - c| < \frac{1}{n}$ for which $|f(x_n) - L| \geq \varepsilon$. Note that $|x_n - c| < \frac{1}{n}$ implies that $x_n \to c$. So we have found a sequence (x_n) converging to c for which $|f(x_n) - L| \geq \varepsilon$ for all n, and hence $f(x_n) \nrightarrow L$. This contradicts our original assumption, completing the proof. $\qquad\square$

Note.

Note 6.13. Theorem 6.12 gives a criterion to determine that a functional limit does not exist: If you can find two such sequences a_n and b_n which both converge to c but for which

$$\lim_{n \to \infty} f(a_n) \neq \lim_{n \to \infty} f(b_n),$$

then we can conclude that $\lim_{x \to c} f(x)$ does not exist.

With Theorem 6.12, any result that holds for all sequences converging to a point c will also apply to functional limits! Indeed, this theorem quickly gives us functional analogues to the sequence limit laws (Theorem 3.21) and the sequence squeeze theorem (Theorem 3.23).

Corollary.

Corollary 6.14 (*Func-y limit laws*). Let f and g be functions from $A \subseteq \mathbb{R}$ to $\mathbb{R}$ and let c be a limit point of A. Lastly, assume that

$$\lim_{x \to c} f(x) = L \qquad \text{and} \qquad \lim_{x \to c} g(x) = M.$$

Then,

(i) $\lim_{x \to c} [k \cdot f(x)] = k \cdot L$ for any $k \in \mathbb{R}$,

(ii) $\lim_{x \to c} [f(x) + g(x)] = L + M$,

(iii) $\lim_{x \to c} [f(x) \cdot g(x)] = L \cdot M$, and

(iv) $\lim_{x \to c} \dfrac{f(x)}{g(x)} = \dfrac{L}{M}$, provided $M \neq 0$ and $g(x) \neq 0$ for any $x \in A$.

Proof. By the sequence limit laws (Theorem 3.21), each of the asserted properties holds for all sequences $a_n \to c$. Therefore by Theorem 6.12 they also hold for the functional limit. $\qquad\square$

> **Corollary.**
>
> **Corollary 6.15** (*Func-y squeeze theorem*). Let f, g and h be functions from $A \subseteq \mathbb{R}$ to $\mathbb{R}$ and let c be a limit point of A, and suppose that
>
> $$f(x) \leq g(x) \leq h(x)$$
>
> for all $x \in A$. Lastly, assume that
>
> $$\lim_{x \to c} f(x) = L \qquad \text{and} \qquad \lim_{x \to c} h(x) = L.$$
>
> Then
>
> $$\lim_{x \to c} g(x) = L.$$

Proof. By the sequence squeeze theorem (Theorem 3.23) the result holds for all sequences $a_n \to c$. So by Theorem 6.12, it also holds for the functional limit. $\square$

6.5 Continuity

We have already mentioned the definition of continuity, but let's state it here formally. One way to state it is this: A function is continuous at a point c (where c is a limit point of $A \setminus \{c\}$) if $\lim_{x \to c} f(x) = f(c)$. This definition is sometimes hard to apply; furthermore it doesn't really explain, at a nuts-and-bolts level, the reason why a function is continuous at a point. So we are going to unwind this functional-limit definition, and write it in terms of epsilons and deltas. This will additionally clarify the case where c is an isolated point.

> **Definition.**
>
> **Definition 6.16.** A function $f : A \to \mathbb{R}$ is *continuous at a point* $c \in A$ if for all $\varepsilon > 0$ there exists some $\delta > 0$ such that for all $x \in A$ where $|x - c| < \delta$ we have
>
> $$|f(x) - f(c)| < \varepsilon.$$
>
> If f is continuous at every point in its domain, then f is called *continuous*.

There are only small differences between the above and Definition 6.8 where $L = f(c)$. First, Definition 6.8 assumes $0 < |x - c|$, but in the case that $c \in A$ (which Definition 6.16 assumes), no issues arise. Definition 6.8 also assumes that c is a limit point of A, but if it's not, then observe that f is guaranteed to be continuous at c.

With functional limits, we proved that its ε-δ definition was equivalent to a topological and sequential definition. The same holds with continuity, as the next theorem shows.

> ### Theorem.
>
> **Theorem 6.17** (*Continuity topologically and sequentially*). Let $f : A \to \mathbb{R}$ and $c \in A$. Then the following are equivalent.
>
> (i) f is continuous at c,
>
> (ii) For all $\varepsilon > 0$ there exists some $\delta > 0$ such that for all $x \in A$ where $|x - c| < \delta$ we have $|f(x) - f(c)| < \varepsilon$.
>
> (iii) For any ε-neighborhood of $f(c)$, denoted $V_\varepsilon(f(c))$, there exists some δ-neighborhood of c, denoted $V_\delta(c)$, with the property that for any $x \in A$ for which $x \in V_\delta(c)$, we have $f(x) \in V_\varepsilon(f(c))$.
>
> (iv) For all sequences (a_n) from A which converge to c, we have $f(a_n) \to f(c)$.
>
> (v) $\lim\limits_{x \to c} f(x) = f(c)$ if c is a limit point of A.

Proof Sketch. The fact that (i) is true if and only if (ii) is simply Definition 6.16. And (ii) being equivalent to (iii) is again just due to a definition: Applying Definition 3.14 we can simply replace the absolute values with ε- or δ-neighborhoods.

The proof that (iv) is equivalent to (ii) is essentially identical to that of Theorem 6.12. The proof that (v) is equivalent to (iv) is due to Theorem 6.12. $\qquad\square$

Poincaré once said that "mathematics is the art of giving the same name to different things." This is certainly true for most of Theorem 6.17. Most lines in this theorem are simply rephrasing the original definition using other terms that we have learned. The exception is (iv) — this is substantially different, and it will be useful to keep it in mind as an equivalent condition. Indeed, this sequential formulation is often easiest to work with, especially to show that a function is *not* continuous. In fact, we will now formally note this.

> ### Note.
>
> **Note 6.18.** By Theorem 6.17 part (iv), if you can find a single sequence (a_n) from the domain for which $a_n \to c$ while $f(a_n) \nrightarrow f(c)$, then you may conclude that f is not continuous at c.
>
> In particular, if you can find two sequences (a_n) and (b_n) from the domain, both of which converge to c, but $\lim\limits_{n \to \infty} f(a_n) \neq \lim\limits_{n \to \infty} f(b_n)$, then at least one of these is not equal to $f(c)$ and so f is not continuous at c.

Indeed, using this note we now officially see why our argument worked for the Dirichlet function.

Example 6.19. Our argument in Example 6.3 that the Dirichlet function is continuous nowhere now works. We showed that every c has a pair of sequences converging to it with different limits when you apply f. By Note 6.18, that's enough. □

Next are more results which are easily obtained from previous results.

> ### Proposition.
>
> **Proposition 6.20** (*Continuity limit laws*). Let $f : A \to \mathbb{R}$ and $g : A \to \mathbb{R}$ be continuous at c, and let $c \in A$. Then,
>
> (i) $k \cdot f(x)$ is continuous at c, for any $k \in \mathbb{R}$;
>
> (ii) $f(x) + g(x)$ is continuous at c;
>
> (iii) $f(x) \cdot g(x)$ is continuous at c; and
>
> (iv) $\dfrac{f(x)}{g(x)}$ is continuous at c, provided $g(x) \neq 0$ for all $x \in A$.

Proof. By Theorem 6.17 part (v), we can rephrase each of these in terms of a functional limit, and after doing so we exactly have the statement of Corollary 6.14, and hence this completes the proof. □

Compounding Examples

Limit laws are great because they allow you to use simple examples you know to be continuous to build more complicated examples with that property. So let's take a page to see how far these limit laws can take us. We begin with one of the simplest continuous functions, $f(x) = x$, and then we repeatedly apply Proposition 6.20.

Example 6.21. The function $f : \mathbb{R} \to \mathbb{R}$ given by $f(x) = x$ is continuous. Here's the proof of that: Let c be any real number. Let $\varepsilon > 0$, and then let $\delta = \varepsilon$. Then $|x - c| < \delta$ implies $|x - c| < \varepsilon$. And so, by the definition of f, this is equivalent to $|f(x) - f(c)| < \varepsilon$. And hence by the definition of a continuous function, f is continuous at c. And since c was arbitrary, f is continuous. □

Example 6.22. By the previous example and Proposition 6.20, this then implies that every polynomial is continuous. Indeed, since by n applications of Proposition 6.20 part (iii) where each function is $f(x) = x$, this implies that $f(x) = x^n$ is also continuous. Then by (i) we see that for any $a_n \in \mathbb{R}$ we have the monomial $f(x) = a_n x^n$ being continuous. Now apply (ii) for different functions of this form, we get a polynomial

$$f(x) = a_n x^n + a_{n-1} x^{n-1} + \cdots + a_1 x + a_0,$$

which is then continuous.   □

If I gave you the limit of a polynomial (from $\mathbb{R}$ to $\mathbb{R}$), like $\lim\limits_{x \to 2}(x^3 - 4x^2 + 1)$, how would you find the limit? From calculus you probably are dying to just plug the limit in. Well by the above, now you can.

Example 6.23. Define a *rational function* to be a polynomial divided by a polynomial, with some domain for which the denominator is never zero. We see that rational functions are always continuous, since by the previous example we know that any two polynomials are continuous, and by Proposition 6.20 part (iv) we then know that their quotient is as well, as we have already noted that the domain is properly restricted so that the denominator is never zero.   □

This approach has now ran dry, though. Any additional applications of Proposition 6.20 will continue to yield rational functions. We next investigate another way to obtain new continuous functions from known continuous functions: compositions.

Recall that given two functions f and g, that $f \circ g$ is notation for the composition function. That is,

$$(f \circ g)(x) = f(g(x)).$$

It might look funny that we have parentheses around the "$f \circ g$" and around the "x", but the ones around the "$f \circ g$" are there to be clear that the "$f \circ g$" is its own thing, so those parentheses are just grouping them together — it is a function and we should be thinking of it as a unit. The parentheses around the "x" are the usual ones around the argument of a function, like $f(x)$ or $h(x)$.

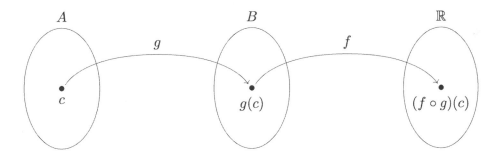

Proposition.

Proposition 6.24 (*f, g continuous* $\Rightarrow$ *f $\circ$ g is too.*). Suppose that $A, B \subseteq \mathbb{R}$, $g : A \to B$ and $f : B \to \mathbb{R}$. If g is continuous at $c \in A$, and f is continuous at $g(c) \in B$, then $f \circ g : A \to \mathbb{R}$ is continuous at c.

Proof. We will use the sequential formulation of continuity, as per Theorem 6.17 part (iv). Consider an arbitrary sequence (a_n) from A converging to c. We aim to show that $(f \circ g)(a_n)$ converges to $(f \circ g)(c)$.

Since each a_n is in g's domain, and $a_n \to c$, and g is continuous at c, by Theorem 6.17 part (iv) we know that $g(a_n) \to g(c)$.

Now, note that since g is a function from A to B, each $g(a_n)$ is an element of B. So $g(a_n)$ is a sequence in B! Likewise, since $c \in A$, $g(c) \in B$. Now, the last paragraph concluded that $g(a_n) \to g(c)$, so we can again apply Theorem 6.17 part (iv). We have that each $g(a_n)$ is in f's domain, and $g(a_n) \to g(c)$, and f is continuous. So by Theorem 6.17 part (iv), we know that $f(g(a_n)) \to f(g(c))$.

We have shown that for an arbitrary sequence a_n in $(f \circ g)$'s domain, we have that $(f \circ g)(a_n)$ converges to $(f \circ g)(c)$, as desired. $\square$

Now here are a few more standard functions. In Exercise 6.6 you are asked to check that they are continuous.

Fact.

Fact 6.25.

- $f : \mathbb{R} \to \mathbb{R}$, $f(x) = e^x$ is continuous.

- $f : (0, \infty) \to \mathbb{R}$, $f(x) = \log(x)$ is continuous

- $f : \mathbb{R} \to \mathbb{R}$, $f(x) = \sin(x)$ is continuous.

- $f : \mathbb{R} \to \mathbb{R}$, $f(x) = \cos(x)$ is continuous.

6.6   Topological Continuity

Recall that not every functions has an inverse. For example, $f : \mathbb{R} \to \mathbb{R}$ given by $f(x) = x^2$ does not have an inverse—each input of a function needs to give exactly one output, but if $f^{-1}(x)$ were a function, then is $f^{-1}(4)$ equal to 2 or -2? So we can not simply define the general inverse for an arbitrary function. We can, however, always define the inverse of a *set*.

Definition.

Definition 6.26. Let $X, Y \subseteq \mathbb{R}$ and $f : X \to Y$. For $B \subseteq Y$, define the *inverse image* (or *pre-image*)

$$f^{-1}(B) = \{x \in X : f(x) \in B\}.$$

So the inverse of a set is itself a set.

Example 6.27.

- If $f : \mathbb{R} \to \mathbb{R}$ is given by $f(x) = 5x$ and B is the open interval $(1, 20)$, then

$$f^{-1}(B) = (1/5, 4).$$

- If $f : \mathbb{R} \to \mathbb{R}$ is again given by $f(x) = 5x$, then

$$f^{-1}(\mathbb{R}) = \mathbb{R}.$$

- If $f : \mathbb{R} \to \mathbb{R}$ is given by $f(x) = x^2$ and B is the open interval $(1, 49)$, then

$$f^{-1}(B) = (-7, -1) \cup (1, 7).$$

- If $f : \mathbb{R} \to \mathbb{R}$ is again given by $f(x) = x^2$, then

$$f^{-1}((1, 4) \cup (25, \infty)) = (-\infty, -5) \cup (-2, -1) \cup (1, 2) \cup (5, \infty).$$

$\square$

Note that in all of these examples, we were taking the inverse image of an open set, and what we got out was itself an open set. This is not an accident, and it is *almost* a general property; there is just one way to prevent this from holding, which the next example illustrates.

Example 6.28.

- If $f : [2, 10] \to \mathbb{R}$ is given by $f(x) = 5x$ and B is the open interval $(1, 20)$, then

$$f^{-1}(B) = [2, 4).$$

You see? It would have been an open set, but was just forced to be restricted because of the domain. Another way we could have written the above is that

$$f^{-1}(B) = (1/5, 4) \cap [2, 10].$$

- Likewise, with the same f as above,

$$f^{-1}(\mathbb{R}) = \mathbb{R} \cap [2, 10].$$

$\square$

What these examples suggest is that if $f : X \to \mathbb{R}$ is continuous and B is open, then $f^{-1}(B)$ is either itself an open set, or is equal to $A \cap X$, where A is some open set. This is further supported if you recall the fact from your intro-to-proofs class that $f^{-1}(\bigcup_\alpha B_\alpha) = \bigcup_\alpha f^{-1}(B_\alpha)$ (also asked of you in Exercise 6.11).

Indeed, as the next theorem states, this is an if-and-only-if condition.

> **Theorem.**
>
> **Theorem 6.29** (*f continuous* $\Leftrightarrow$ *B open implies* $f^{-1}(B) = open \cap X$). Let $f : X \to \mathbb{R}$. Then f is continuous if and only if for every open set B, we have $f^{-1}(B) = A \cap X$ for some open set A.

Proof Sketch. For the forward direction, we begin with the following picture, where we have a domain set X, a codomain set $\mathbb{R}$ (each drawn as a blob), a function f going between them, a subset $B \subseteq \mathbb{R}$, and its inverse image under f:

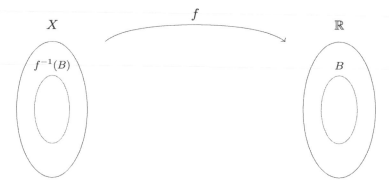

We want to show that $f^{-1}(B) = A \cap X$ where A is open. As you'll see, getting $f^{-1}(B)$ to be in X will be easy. The thing we need to show is that it behaves like an open set. That is, we want to show that if $x_0 \in f^{-1}(B)$, then there is some open neighborhood around x_0 that is also contained inside $f^{-1}(B)$. That is, we want to be able to draw a circle like this:

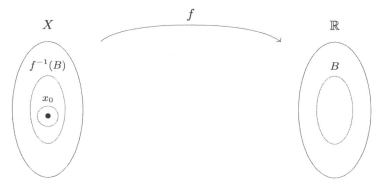

How do we do it? Well, we don't know anything about $f^{-1}(B)$, but we do know something about B — it's open. So any point over there we *can* find a region around (which is contained inside B). Seems like a start, so let's use that. To do it, let's take x_0 and apply f to it. What we get, $f(x_0)$, will be a point in B, and hence subject to B's openness property.

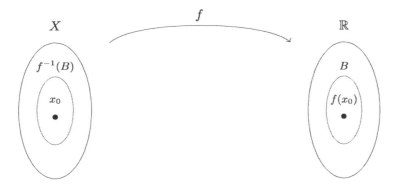

Since B is open we can find some open neighborhood around $f(x_0)$ which is contained inside B:

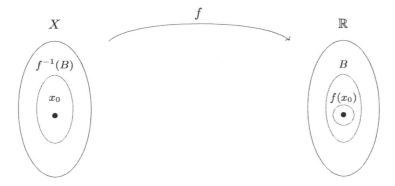

How do we use this to get an open set around x_0? The idea is to use the fact that f is continuous, which we haven't used up to this point. Recall the topological definition of continuity (but calling the domain X) from Theorem 6.17 part (iii):

> For $f : X \to \mathbb{R}$ and $c \in X$, f is continuous at c if and only if, for any ε-neighborhood $V_\varepsilon(f(c))$ of $f(c)$, there exists a δ-neighborhood $V_\delta(c)$ of c with the property that for any $x \in X$ for which $x \in V_\delta(c)$, we have $f(x) \in V_\varepsilon(f(c))$.

This is almost precisely what we need.

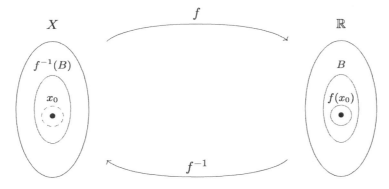

With a few details omitted, this is the big idea. Now let's prove it formally.

Proof **of Theorem 6.29.**   First assume that f is continuous and let B be an open set in $\mathbb{R}$. We aim to show that $f^{-1}(B) = A \cap X$ for some open set A. If we were only trying to show that $f^{-1}(B)$ equaled some open set A, then (by the definition of an open set) what we would want to show is that for any $x_0 \in f^{-1}(B)$, there is some open interval $(x_0 - \delta, x_0 + \delta)$ which is contained entirely inside of A. But since we are trying to show that $f^{-1}(B) = A \cap X$ for some open set A, we only care about this property holding for $x \in X$. So we just have to amend the openness property like so:

> Want to show: For any $x_0 \in f^{-1}(B)$, there is some δ-neighborhood $V_\delta(x_0)$ of x_0 with the property that for any $x \in X$ where $x \in V_\delta(x_0)$, we have $x \in f^{-1}(B)$.

Now, to prove this, we will use the fact that f is continuous and B is open. Since $x_0 \in f^{-1}(B)$, we know that $f(x_0) \in B$. So we have a point inside of an open set. By the definition of openness, there must exist some $\varepsilon > 0$ such that $(f(x_0) - \varepsilon, f(x_0) + \varepsilon) \subseteq B$. That is, we have an ε-neighborhood $V_\varepsilon(f(x_0))$ of $f(x_0)$ inside of B.

Now we use that f is continuous. By the topological definition of continuity (Theorem 6.17 part (iii)), there does indeed exist some δ-neighborhood $V_\delta(x_0)$ of x_0 with the property that for any $x \in X$ where $x \in V_\delta(x_0)$, we have $f(x) \in V_\varepsilon(f(x_0))$; i.e., for all these x, we have $x \in f^{-1}(V_\varepsilon(f(x_0))) \subseteq f^{-1}(B)$, as desired.

To prove the reverse direction, assume that for every open set B, $f^{-1}(B) = A \cap X$ for some open set A. We aim to show that f is continuous, and we will do so by showing that it satisfies the topological definition that we have been using. Choose any $x_0 \in X$ and $\varepsilon > 0$. Since $V_\varepsilon(f(x_0))$ is an open set, the hypothesis gives that

$$f^{-1}\left(V_\varepsilon(f(x_0))\right) = A \cap X$$

for some open set A. And of course, $x_0 \in A \cap X$. Since A is open and $x_0 \in A$, there must exist some δ-neighborhood $V_\delta(x_0)$ of x_0 contained inside A. Therefore, if $x \in X$ and $x \in V_\delta(x_0)$, then $x \in A \cap X$, and hence $f(x) \in V_\varepsilon(x_0)$, as desired. $\square$

Note that we now have another characterization of continuity. The conditions in Theorem 6.17 were all local conditions that demonstrated continuity at a point, while this is a broader condition guaranteeing continuity at *every* point — meaning the entire function is continuous.

6.7   The Extreme Value Theorem

This next section begins with one more connection between continuity and one of the fundamental objects of topology: compact sets.

> **Theorem.**
>
> **Theorem 6.30** (*The continuous image of a compact set is compact*). Suppose $f : X \to \mathbb{R}$ is continuous. If $A \subseteq X$ is compact, then $f(A)$ is compact.

Proof Sketch. The idea behind the proof is contained in the following picture.

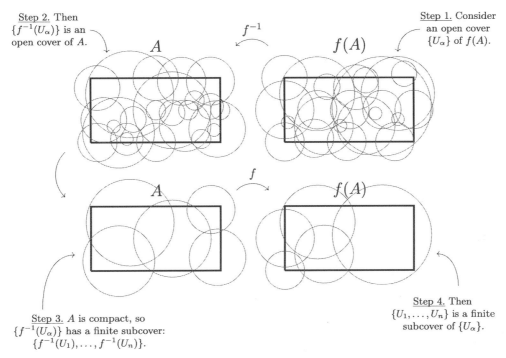

Step 2. Then $\{f^{-1}(U_\alpha)\}$ is an open cover of A.

f^{-1}

A

$f(A)$

Step 1. Consider an open cover $\{U_\alpha\}$ of $f(A)$.

f

A

$f(A)$

Step 3. A is compact, so $\{f^{-1}(U_\alpha)\}$ has a finite subcover: $\{f^{-1}(U_1), \ldots, f^{-1}(U_n)\}$.

Step 4. Then $\{U_1, \ldots, U_n\}$ is a finite subcover of $\{U_\alpha\}$.

The only additional justification that is needed is in Step 2. Theorem 6.29 does not technically guarantee that $f^{-1}(U_\alpha)$ is open, but that $f^{-1}(U_\alpha) = V_\alpha \cap X$ for some open set V_α. So technically our open cover of A will be $\{V_\alpha\}$ — but the above sketch will still essentially work.

Proof. Suppose A is compact. We aim to show that $f(A)$ is also compact, which we will show via the definition. To that end, let $\{U_\alpha\}$ be an open cover of $f(A)$. We will find a finite subcover of this cover, which will complete the proof.

By Theorem 6.29, each $f^{-1}(U_\alpha) = V_\alpha \cap X$ for some open set V_α. In particular, note that $f^{-1}(U_\alpha) \subseteq V_\alpha$.

We will show that $\{V_\alpha\}$ is an open cover of A. By the above, each V_α is open. To see that $\{V_\alpha\}$ is a cover of A, we will instead show that $\{f^{-1}(U_\alpha)\}$ is a cover. This is sufficient since we already noted that each $f^{-1}(U_\alpha) \subseteq V_\alpha$. Consider any $x_0 \in A$;

we will show that some $f^{-1}(U_{\alpha_0})$ covers x_0. Consider $f(x_0)$. Since $f(x_0) \in f(A)$ and $\{U_\alpha\}$ is a cover of $f(A)$, there exists some U_{α_0} which covers $f(x_0)$; that is, $f(x_0) \in U_{\alpha_0}$. But then $x_0 \in f^{-1}(U_{\alpha_0})$, as desired.

So indeed, $\{f^{-1}(U_\alpha)\}$ covers A, and hence $\{V_\alpha\}$ is an open cover of A. And hence, by the compactness of A, there must exist some finite subcover $\{V_1, V_2, \ldots, V_n\}$ of A. Recall that each V_i has a corresponding set $f^{-1}(U_i)$ for which $f^{-1}(U_i) = V_i \cap X$. We claim $\{U_1, U_2, \ldots, U_n\}$ is a finite subcover of $f(A)$. Doing so would complete the proof.

The sets $U_1, U_2, \ldots, U_n$ are from the open cover $\{U_\alpha\}$, so clearly these are open. So all we need to show is that they do indeed cover $f(A)$. To this end, choose any $y \in f(A)$. To be in $f(A)$ means that there is some $x \in A$ such that $f(x) = y$. We showed above that $f^{-1}(U_1), f^{-1}(U_2), \ldots, f^{-1}(U_n)$ covers A, and so since $x \in A$ this in particular means that there is some $f^{-1}(U_k)$ that covers x. But $x \in f^{-1}(U_k)$ implies that $y = f(x) \in U_k$. So indeed, given an arbitrary $y \in f(A)$, one of the open sets $U_1, U_2, \ldots, U_k$ did cover it. So our open cover does have a finite subcover, completing the proof. $\qquad\square$

This has a nice corollary. Notice that $f : \mathbb{R} \to \mathbb{R}$ given by $f(x) = x$ is a continuous function that is also *unbounded*; meaning, the range of f, which is $\mathbb{R}$, is an unbounded set. You see, if the domain is an unbounded set (like $\mathbb{R}$), then it is easy to have a range that is also unbounded (like $\mathbb{R}$). That said, you don't *need* the domain to be unbounded in order to get an unbounded range. For instance, $f : (0,1) \to \mathbb{R}$ given by $f(x) = \frac{1}{x}$ is a continuous function on the bounded set $(0,1)$, but yet f is unbounded.

However, if $f(x)$ is continuous on $[0,1]$, then we *can* guarantee that f is *bounded*; meaning, the range of f is a bounded set. And on any closed interval, this same property will hold. But not just any closed set — for example, $\mathbb{R}$ is a closed set, so the first example in the last paragraph shows that guaranteeing a closed domain is not sufficient. You also need the domain to be bounded. But that is enough — if the domain is closed and bounded, then any continuous f will be bounded. Said differently, if f is a continuous function on a *compact* set, then f is bounded.

> ### Corollary.
>
> **Corollary 6.31** (*Continuous on compact set $\Rightarrow$ bounded*). A continuous function on a compact set is bounded.

Proof. If A is compact and $f : A \to \mathbb{R}$ is continuous, then by Theorem 6.29, $f(A)$ is also compact. And hence by the Heine-Borel theorem (Theorem 5.19), $f(A)$ is closed and bounded. So f is bounded. $\qquad\square$

Here's another corollary, and you know it's important because it has a name. It's called the extreme value theorem.[11]

---

[11] It's a *corollary* because it follows from a previous result. But due to its importance it is known as the extreme value *theorem*.

> ### Theorem.
>
> **Theorem 6.32** (*The extreme value theorem*). A continuous function on a compact set attains its supremum and infimum.

By "attains" its supremum we mean that not only does the supremum exist (call this supremum M), but there is in fact some point a_0 for which $f(a_0) = M$.

Proof. Let A be a compact set; so A is closed and bounded. By Theorem 6.30, $f(A)$ is also closed and bounded. Since it is bounded above and below, the supremum and infimum of $f(A)$ exist. Let $M = \sup\{f(a) : a \in A\}$ and $L = \inf\{f(a) : a \in A\}$ (that is, $M = \sup(f(A))$ and $L = \inf(f(A))$).

There then exists a sequence $f(a_n)$ from $f(A)$ converging to M, and a sequence $f(b_n)$ from $f(A)$ converging to L (Proposition 3.29). Since $f(A)$ is closed, they both converge to some $f(a_0)$ and $f(b_0)$ inside the set $f(A)$. And so f achieves its supremum and infimum. $\square$

Because this theorem is so important, let's do a few examples to make sure we understand it.

Example 6.33. If $f : [-2, 1] \to \mathbb{R}$ is given by $f(x) = x^2$, then $f([-2, 1])$'s supremum is 4 and its infimum is 0. And moreover, these are achieved. That is, we can find points in the domain whose images equal these values. In particular, $-2 \in [-2, 1]$ and $f(-2) = 4$. And $0 \in [-2, 1]$ and $f(0) = 0$. $\square$

The two assumptions in the extreme value theorem were that f is continuous and f's domain is compact. Now we will give two examples to show that both assumptions are necessary.

Example 6.34. If $f : (-2, 1] \to \mathbb{R}$ is given by $f(x) = x^2$, then its supremum is 4 and its infimum is 0. The supremum is *not* achieved, though. That is, there is no $c \in (-2, 1]$ where $f(c) = 4$. This shows that the assumption that the domain be compact is indeed necessary. $\square$

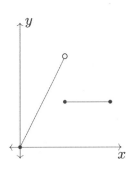

Example 6.35. Suppose $f : [0, 4] \to \mathbb{R}$ is given by

$$f(x) = \begin{cases} 2x & \text{if } x \in [0, 2), \\ 2 & \text{if } x \in [2, 4]. \end{cases}$$

Note that here the supremum is 4 but the supremum is *not* achieved. That is, there is no $c \in [2, 4]$ where $f(c) = 4$. This shows that the assumption that the function be continuous is indeed necessary. $\square$

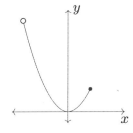

6.8   The Intermediate Value Theorem

Suppose you hike up a mountain. You leave base camp at 7am and you reach the summit at 2pm. You camp there for the night and the next morning you leave at 7am and hike back down the mountain, taking the same trail down that you took up.

In this scenario, the magic of real analysis says that there must have been a moment when you were at the *exact* same spot on the trail at the *exact* same time on the two days. And I do mean *exactly*: perhaps both times you were 57.254% of the way up the mountain when the time was exactly 11:32 and 42.4 seconds.

How could we prove such a thing? I'm glad you asked. We begin with a lemma.

Lemma.

Lemma 6.36 (*f continuous, $f(c) > 0 \Rightarrow f(x) > 0 \ \forall \ |x - c| < \delta$*). If f is continuous and $f(c) > 0$, then there exists some $\delta > 0$ such that $f(x) > 0$ for all x in $(c - \delta, c + \delta)$.

Likewise, if f is continuous and $f(c) < 0$, then there exists some $\delta > 0$ such that $f(x) < 0$ for all x in $(c - \delta, c + \delta)$.

Proof Sketch. Since f is continuous at c, every $\varepsilon > 0$ has a $\delta > 0$ for which

$$|x - c| < \delta \quad \Longrightarrow \quad |f(x) - f(c)| < \varepsilon.$$

That is,

$$x \in (c - \delta, c + \delta) \quad \Longrightarrow \quad f(c) - \varepsilon < f(x) < f(c) + \varepsilon,$$

which is already looking like what we want to show. And remember, we have complete control over ε! So, for example, if $f(c) > 0$, then we could choose ε to be anything less than $f(c)$, such as $f(c)/2$. Then by the left inequality above,

$$f(c) - \varepsilon < f(x),$$

will give us what we want. Below is the familiar picture of a continuous function, where we set $\varepsilon = f(c)/2$. Notice how the portion of the graph between $c - \delta$ and $c + \delta$ is entirely above $f(c) - \varepsilon$, which in turn is larger than 0, as desired.

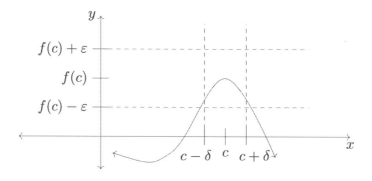

Proof. Assume we are in the first situation and let $\varepsilon = \dfrac{f(c)}{2} > 0$. Then since f is continuous there exists some $\delta > 0$ such that $|x - c| < \delta$ implies $|f(x) - f(c)| < \varepsilon$. With a little algebra, we can show that this implies that $f(x) > 0$. To that end, first we unpack the the inequality $|f(x) - f(c)| < \varepsilon$:

$$|f(x) - f(c)| < \varepsilon$$
$$-\varepsilon < f(x) - f(c) < \varepsilon$$
$$f(c) - \varepsilon < f(x) < f(c) + \varepsilon.$$

And now the left inequality, combined with $\varepsilon = \frac{f(c)}{2}$, implies that

$$f(c) - \frac{f(c)}{2} < f(x)$$
$$0 < \frac{f(c)}{2} < f(x),$$

as desired.

In the second situation let $\varepsilon = \dfrac{-f(c)}{2} > 0$, and apply the same reasoning. □

Proposition.

Proposition 6.37 (*f continuous*, $f(a) > 0 > f(b) \Rightarrow \exists\, c \in (a, b)$, $f(c) = 0$). If f is continuous on $[a, b]$ and $f(a)$ and $f(b)$ have different signs, then there is some $c \in (a, b)$ for which $f(c) = 0$.

Proof. Without loss of generality, assume that $f(a)$ is positive and $f(b)$ is negative. Let

$$A = \{t : f(x) > 0 \text{ for all } x \in [a, t]\}.$$

Since $f(a) > 0$, we know $a \in A$ and hence $A \neq \emptyset$. Moreover, since $f(b) < 0$, we know $b \notin A$ and hence A is bounded above by b. By completeness, these imply that $\sup(A)$ exists. Let $c = \sup(A)$, and note that $c \in (a, b)$. If $f(c) = 0$ then we are done, so assume for a contradiction that $f(c) \neq 0$.

If $f(c) > 0$, then by Lemma 6.36 there exists some $\delta > 0$ for which $f(x) > 0$ for all $x \in (c - \delta, c + \delta)$; in particular, $f(x) > 0$ for all $x \in (c - \delta, c + \delta/2]$. And certainly $(c - \delta) \in A$ since otherwise $c - \delta$ would have been an upper bound of A, contradicting our claim that c was the *least* upper bound of A. And so by definition, $(c - \delta) \in A$ means that $f(x) > 0$ for all $x \in [a, c - \delta]$. But at the start of this paragraph we also showed that $f(x) > 0$ for all $x \in (c - \delta, c + \delta/2]$. Putting these together, we see that $f(x) > 0$ for all $x \in [a, c + \delta/2]$, which means that $(c + \delta/2) \in A$, contradicting our claim that c is an upper bound on A.

By very similar reasoning, if $f(c) < 0$ then there is some $\delta > 0$ for which

$f(c - \delta/2) < 0$, contradicting c being the *least* upper bound of A. This completes the proof. $\qquad\square$

With that, we are now ready for this section's titular theorem.

Theorem.

Theorem 6.38 (*The intermediate value theorem*). If f is continuous on $[a, b]$ and α is any number between $f(a)$ and $f(b)$, then there is some $c \in (a, b)$ for which $f(c) = \alpha$.

Note that "α is any number between $f(a)$ and $f(b)$" means either "$f(a) < \alpha < f(b)$" or "$f(b) < \alpha < f(a)$." But we don't know whether $f(a)$ or $f(b)$ is larger, so it's easier to say it the way we did.

Proof Sketch. By assumption, either $f(a) < \alpha < f(b)$ or $f(b) < \alpha < f(a)$; let's assume it's the former.

Now, by Proposition 6.37, which just proved, if $\alpha = 0$, then we are done. So what do we do if $\alpha \neq 0$? Well, that proposition is still a powerful tool, so we still want to make use of it. How do we do so? We create a new function! If you let $g(x) = f(x) - \alpha$ (intuitively, g is the same as f, just scooted down by α), then "$f(a) < \alpha < f(b)$" turns into "$g(a) < 0 < g(b)$," and we are able to apply Proposition 6.37. This gives us a c such that $g(c) = 0$, which corresponds to $f(c) = \alpha$, as desired.

This idea of proving the "zero case" before proving the general case will be used several more times in this book.

Proof. If $f(a) = f(b)$, then there are no numbers between $f(a)$ and $f(b)$, so there is nothing to show. So assume that $f(a) \neq f(b)$. Without loss of generality, we may assume $f(a) > f(b)$. So $f(a) > \alpha > f(b)$.

Define $g(x) = f(x) - \alpha$ and note that g is continuous on $[a, b]$. Then note that since $f(a) > \alpha$, we have $g(a) > 0$. And since $f(b) < \alpha$, we have $g(b) < 0$. So by Proposition 6.37, there is some $c \in (a, b)$ such that

$$g(c) = 0$$

That is,

$$f(c) - \alpha = 0$$
$$f(c) = \alpha.$$

So we have found a $c \in (a, b)$ such that $f(c) = \alpha$, as desired. $\qquad\square$

Interestingly, in some sense the converse to Theorem 6.38 is not true. That is, suppose you know that a function f has this property: Given any pair of points

a and b, and any value α between $f(a)$ and $f(b)$, there exists some $c \in (a, b)$ for which $f(c) = \alpha$; this is called the *intermediate value property*. Must such an f be continuous? Surprisingly, the answer is 'no.' The below, called the *topologist's sine curve*, is such an example.

$$f(x) = \begin{cases} \sin(1/x) & \text{for } x \neq 0, \\ 0 & \text{for } x = 0. \end{cases}$$

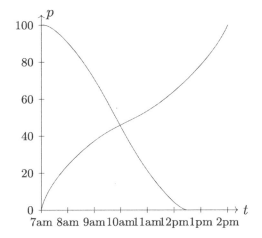

The Hiker Problem

At the start of this section we mentioned the problem of the hiker. We wish to show that if a hiker leaves base camp at 7am and travels up a mountain trail, camps at the summit that night, and then the next day departs the summit at 7am and travels back down that same trail, then at some point the hiker was at the *exact* same point on the trail at the *exact* same time both days. We can graph their two journeys on the same graph (with a 'time' axis vs an 'progress' axis where 100 means you are at the summit and 0 means you are at base camp).

As you can imagine, these continuous functions will have to intersect somewhere, and when they do, both their times and altitudes will match.[12]   That said, the intermediate value theorem can not directly be applied to prove this. Rather, it follows from a more general result that you are asked to prove in Exercise 6.46.

---

[12]There's an intuitive way to think about this. Instead of thinking about the same person hiking on different days, imagine that a hiker and his twin were hiking the same mountain on the same day. One starts at base camp and hikes up the mountain, while the other starts at the summit and hikes down the mountain. It's intuitively clear that at some point these two will pass each other on the trail, and that point is the point we are looking for.

6.9   Uniform Continuity

Consider $f : \mathbb{R} \to \mathbb{R}$ where $f(x) = 5x$. To prove f is continuous at each $c \in \mathbb{R}$, you would let $\varepsilon > 0$ and (from your scratch work) you would let $\delta = \varepsilon/5 > 0$. Then, for all $x \in \mathbb{R}$ such that $|x - c| < \delta$,

$$
\begin{aligned}
|f(x) - f(c)| &= |5x - 5c| \\
&= 5 \cdot |x - c| \\
&< 5 \cdot \delta \\
&= 5 \cdot \frac{\varepsilon}{5} \\
&= \varepsilon.
\end{aligned}
$$

Likewise, to prove that $f(x) = 10x$ is continuous at each $x \in \mathbb{R}$, you would let $\delta = \frac{\varepsilon}{10}$. Note that steeper slopes are requiring smaller δ's.

Now consider the function $f(x) = x^2$ on the domain $(0, \infty)$. We know this is a continuous function: Let $\varepsilon > 0$. For any $c \in (0, \infty)$ we can find some $\delta > 0$ such that, if $|x - c| < \delta$, we have $|f(x) - f(c)| < \varepsilon$. In particular, if you wanted to write a proof that f is continuous, you might begin with the following scratch work:

$$
\begin{aligned}
|f(x) - f(c)| &< \varepsilon \\
\left|x^2 - c^2\right| &< \varepsilon \\
|x - c| \cdot |x + c| &< \varepsilon \\
|x - c| &< \frac{\varepsilon}{|x + c|} \\
\delta &< \frac{\varepsilon}{|x + c|}
\end{aligned}
$$

If we insist $\delta < 1$, then $|x - c| < 1$ implies $|x + c| < 2c + 1$. So by choosing

$$
\delta < \frac{\varepsilon}{2c + 1}
$$

we can guarantee the previous line holds.

Then, to write a formal proof that $f(x) = x^2$ is continuous on $(0, \infty)$, you would let $\varepsilon > 0$, $c > 0$ and $\delta = \min\{1, \frac{\varepsilon}{2c+1}\}$. Then for any $x \in (0, \infty)$ for which $|x - c| < \delta$, we have

$$
|f(x) - f(c)| = |x^2 - c^2| = |x - c| \cdot |x + c| < \delta \cdot |x + c| \le \frac{\varepsilon}{2c + 1} \cdot |x + c|.
$$

Note that $|x - c| < 1$ implies $|x + c| \le (c + 1) + c = 2c + 1$. Combining this with the above we have

$$
|f(x) - f(c)| < \frac{\varepsilon}{2c + 1} \cdot |x + c| \le \frac{\varepsilon}{2c + 1} \cdot (2c + 1) = \varepsilon.
$$

Notice how δ depends on c: We had set $\delta = \min\{1, \frac{\varepsilon}{2c+1}\}$, and so as c gets larger and larger, δ in turn gets smaller and smaller. Remember that for a fixed $\varepsilon > 0$, the steeper the function is, the smaller the δ you need. So the fact that δ is doing this makes sense, since larger points of $f(x) = x^2$ give steeper slopes.

The important point from this is that for $f(x) = x^2$ with domain $(0, \infty)$, we certainly can not hope, for a fixed ε, to choose just a single δ and have it work for all of the c-values.

For some curves, though, we can pick just a single δ, like the $f(x) = 5x$ and $10x$ functions from earlier. Consider now $f(x) = x^2$ on $[1, 4]$. The steepest this curve gets is at the very end, where its slope is about 8 (we haven't covered derivatives yet, but I'm sure you remember that the slope at the end is $f'(4) = 2(4) = 8$).

So if the slope is never more than 8, then for any $\varepsilon > 0$, setting $\delta = \varepsilon/8$ will work. Indeed, pick any $c \in [1, 4]$, let $\varepsilon > 0$, and let $\delta = \varepsilon/8 > 0$. Then for any $x \in [1, 4]$ for which $|x - c| < \delta$, we have

$$\begin{aligned}
|f(x) - f(c)| &= |x^2 - c^2| \\
&= |x - c| \cdot |x + c| \\
&< \delta \cdot 8 \qquad\qquad (x, c \in [1, 4] \Rightarrow |x + c| \leq 8) \\
&= \frac{\varepsilon}{8} \cdot 8 \\
&= \varepsilon.
\end{aligned}$$

So given any ε, a single δ worked no matter which c we were at. (Indeed, when x and c are close, the "$|x + c|$" being about $2x$ gives reason to see why the derivative of $f(x) = x^2$ is going to be $f'(x) = 2x$.) Let's summarize these observations.

- Let $\varepsilon > 0$. Let $f(x) = x^2$ on $(0, \infty)$ and let $c \in (0, \infty)$. It is true that f is continuous at c, but in the proof of this it is impossible to find a single δ which works no matter which c you choose. The reason is that for c-values getting larger and larger, it requires an increasingly smaller δ. In fact, the δ would be approaching 0. (Since δ depends on both on ε and c, this is sometimes written "$\delta = \delta(\varepsilon, c)$," as δ is in fact a function of ε and c.)

- Let $\varepsilon > 0$. Let $f(x) = x^2$ on $[1, 4]$ and let $c \in [1, 4]$. It is true that f is continuous at c, and moreover, this time, in the proof of this it *is* possible to find a single δ which works no matter which c you choose. Indeed, $\delta = \varepsilon/8$ worked. (Since δ depends just on ε, this is sometimes written "$\delta = \delta(\varepsilon)$," as δ is only a function of ε.)

We are going to distinguish between these cases. We will call functions in the second case *uniformly continuous*. Note also from this that the domain is super important. Functions always have domains, and saying a function is uniformly continuous means that it only has to satisfy the condition for the chosen domain.

The first way to state the definition of uniform continuity is this:

> Let $f : A \to \mathbb{R}$. We say f is *uniformly continuous* if for all $\varepsilon > 0$ there exists some $\delta > 0$ such that for all $c \in A$ and for every $x \in A$ for which $|x - c| < \delta$, we have $|f(x) - f(c)| < \varepsilon$.

But note that since c is completely unanchored, there is no difference between that x and that c; they are just any two points from A. Therefore a second way to state it is this:

Definition.

Definition 6.39. Let $f : A \to \mathbb{R}$. We say f is *uniformly continuous* if for all $\varepsilon > 0$ there exists some $\delta > 0$ such that, for all $x, y \in A$,

$$|x - y| < \delta \qquad \text{implies} \qquad |f(x) - f(y)| < \varepsilon.$$

Indeed,[13] below is a table of continuous functions (on a domain), and the δ that you would get if you tried to prove it was continuous at c. For each, we noted whether it was uniformly continuous or not.

$f(x)$ on a domain	δ	Uniformly continuous?
x^2 on $(0, \infty)$	$\min\left\{1, \frac{\varepsilon}{2c+1}\right\}$	No
x^2 on $[0, 4]$	$\min\left\{1, \frac{\varepsilon}{2c+1}\right\}$	Yes[14]
$\sqrt{x}$ on $(0, \infty)$	ε^2	Yes
e^x on $(0, \infty)$	$\min\left\{1, \frac{\varepsilon}{e^c(e-1)}\right\}$	No
$5x + 2$ on $\mathbb{R}$	$\varepsilon/5$	Yes
$1/x$ on $(0, \infty)$	$\min\left\{\frac{c^2\varepsilon}{2}, \frac{\varepsilon}{2}\right\}$	No
$1/x$ on $(3, \infty)$	$\min\left\{\frac{c^2\varepsilon}{2}, \frac{\varepsilon}{2}\right\}$	Yes
$\sin(x)$ on $\mathbb{R}$	ε	Yes
$\cos(x)$ on $\mathbb{R}$	ε	Yes

---

[13]Unrelated side note: if you look at the definition, it almost looks like a continuous, point-centric version of being Cauchy.

[14]The first two rows have the same δ but one is uniformly continuous while the other is not. What's the difference? It is in the domain. For the first, c can be arbitrarily big, and so $\frac{\varepsilon}{2c+1}$ will get arbitrarily small. For the second, $c \in [0, 4]$. If the biggest c is 4, then the smallest δ is $\frac{\varepsilon}{2(4)+1} = \frac{\varepsilon}{9}$. So for any $\varepsilon > 0$, letting $\delta = \frac{\varepsilon}{9}$ will work, regardless of which c you chose.

So being uniformly continuous is a strictly stronger condition than being continuous. How do we identify when a continuous function is in fact uniformly continuous? Well, here's one nice result: If your continuous function is on a *compact* set, then you are guaranteed that it is uniformly continuous.

> **Proposition.**
>
> **Proposition 6.40** (*Continuous f on compact $A \Rightarrow f$ uniformly continuous*). If $f : A \to \mathbb{R}$ is continuous and A is compact, then f is uniformly continuous on A.

Proof Idea. How do we combine compactness and continuity? First, we will let $\varepsilon > 0$. Since f is continuous—meaning that at every point c, f is continuous at c—for every c there exists a $\delta(c)$-neighborhood[15] around c for which x being in this $\delta(c)$-neighborhood implies $f(x)$ is in the ε-neighborhood of $f(c)$.

Now, every point c has a $\delta(c)$ corresponding to it. To show that f is uniformly continuous, we must prove that there is a single δ that works for *every* possible c. To cut down the infinitely many δ's (one for each $c \in A$) to a finite number, we use compactness! Each δ-neighborhood is an open interval, and these open intervals collectively cover all of the compact set A, and so by the definition of compactness (Definition 5.16), this open cover contains a *finite* subcover. And this is important, because once we have a finite collection of δ's,[16] we are then guaranteed that one of these is the smallest δ.

Pre-proof trick. There is a trick that we have used a number of times, but because the proof below uses it twice, I thought I'd just highlight it now. Recall that if $|a - b| < \beta/2$ and $|b - c| < \beta/2$, then $|a - c| < \beta$. This follows quickly from applying the triangle inequality: $|a - c| = |a - b + b - c| \leq |a - b| + |b - c| = \frac{\beta}{2} + \frac{\beta}{2} = \beta$.

Proof. Suppose that f is continuous and A is compact. Let $\varepsilon > 0$. For each $c \in A$, since f is continuous at c, there exists some $\delta(c) > 0$ for which $x \in A$ and $|x - c| < \delta(c)$ imply that $|f(x) - f(c)| < \varepsilon/2$. Note then that the collection

$$\left\{ \left(c - \frac{\delta(c)}{2}, c + \frac{\delta(c)}{2} \right) \ : \ c \in A \right\}$$

of open intervals is an open cover of A, and hence by the compactness of A there

---

[15] I am writing "$\delta(c)$" like one writes "$f(x)$." Each point c has a corresponding δ; since δ changes depending on c, δ is indeed a function of c.

[16] To grammar nazis: I am aware that the contraction does not pluralize. However, in math where we use letters as variables, "δs" can be misread as a product between 'δ' and 's'. Occasionally I add an apostrophe before the 's' when pluralizing a variable in order to avoid mathematical confusion, even if it is grammatically problematic.

Also to grammar nazis: Apologies for all the mistakes that I have no idea I'm making. Please email me all my typos and errors.

Also to grammar nazis: Good job learning real analysis!

exists some finite subcover,

$$\left\{ \left(c_1 - \frac{\delta(c_1)}{2}, c_1 + \frac{\delta(c_1)}{2} \right), \left(c_2 - \frac{\delta(c_2)}{2}, c_2 + \frac{\delta(c_2)}{2} \right), \ldots, \left(c_n - \frac{\delta(c_n)}{2}, c_n + \frac{\delta(c_n)}{2} \right) \right\},$$

of A.

Pick any $x, y \in A$. By the definition of uniform continuity we aim to find a $\delta > 0$ such that if $|x - y| < \delta$ we have $|f(x) - f(y)| < \varepsilon$. Set

$$\delta = \min \left\{ \frac{\delta(c_1)}{2}, \frac{\delta(c_2)}{2}, \ldots, \frac{\delta(c_n)}{2} \right\}.$$

Since $x \in A$ and we found a finite subcover of A, we know that

$$x \in \left(c_k - \frac{\delta(c_k)}{2}, c_k + \frac{\delta(c_k)}{2} \right),$$

for some $k \in \{1, 2, \ldots, n\}$. Therefore $|c_k - x| < \frac{\delta(c_k)}{2}$. And also, since $|x - y| < \delta \leq \frac{\delta(c_k)}{2}$, by the pre-proof trick we see that

$$|c_k - y| < \delta(c_k).$$

That is, we have seen that both

$$|x - c_k| < \delta(c_k) \qquad \text{and} \qquad |y - c_k| < \delta(c_k).$$

And therefore by the original definition of $\delta(c_k)$, we see that

$$|f(x) - f(c_k)| < \varepsilon/2 \qquad \text{and} \qquad |f(y) - f(c_k)| < \varepsilon/2.$$

Applying the pre-proof trick one more time, this implies that

$$|f(x) - f(y)| < \varepsilon,$$

completing the proof.   □

Note.

Note 6.41. Proposition 6.40 in particular implies that any function that is continuous on a closed interval $[a, b]$ is uniformly continuous on that interval. This is the form of the proposition that is most commonly used.

— Notable Exercises —

- In Exercise 6.6 you will prove the continuity of some fundamental functions. The sine and cosine functions, in conjunction with the material of Chapter 9, are particularly important in the branch of analysis called *Fourier analysis*. A teaser straight from the 1822 mouth of Joseph Fourier: "There is no function $f(x)$, or part of a function, which cannot be expressed by a trigonometric series." That certainly caught the math world's attention!

- Exercise 6.24 asks you to prove that a peculiar function satisfies the intermediate value property, and Exercise 6.24 asks you to prove that this function is continuous at 0. This function, and others, are called *topologist's sine curves*, and will make a splash in Chapter 7.

- It is often useful in mathematics to decompose something down into simpler parts. In number theory we talk about prime factorizations, and in abstract algebra we talk about the extent to which general ℕ-like algebraic structures can be uniquely factored. In group theory in particular, an important question is which groups can be expressed has a direct product of cyclic groups. As we just mentioned in the above bullet point, Fourier claimed that all functions can be decomposed in some way into sums of trig functions. In Chapter 9 we will see similar decompositions into sums of polynomials. Exercises 6.30 and 6.31 — although a fairly tame form of this — are yet more useful decomposition (of a function, in this case) into a form which is sometimes easier to handle.

- Theorem 6.30 says that if f is continuous and A is compact, then $f(A)$ is compact. Exercise 6.33 shows that even more is true in the case that A is the closed interval $[a, b]$.

- Make sure to appreciate what Exercise 6.35 is having you prove. It's a neat result.

- Exercise 6.46 can be used to prove the hiker problem that we discussed at the beginning and end of Section 6.8. Here's a hint to solve that exercise: That exercise asks you to prove the existence of a point with a special property. We have two theorems in this chapter whose conclusions are about the existence of a point with a special property (the intermediate value theorem and the extreme value theorem). And as my own undergraduate real analysis professor Jamie Radcliffe used to tell us, "*use your theorems!*"

- There are a number of important *fixed point* results in mathematics. A first version of this type of result is given in Exercise 6.48, and a famous version is given in Exercise 6.56.

— Exercises —

Exercise 6.1. Use the definition of the functional limit to prove the following:

(a) $\lim_{x \to -2} (4x + 3) = -5$.

(c) $\lim_{x \to 1} \left(\dfrac{x^3}{x - 1} - \dfrac{1}{x - 1} \right) = 3$.

(b) $\lim_{x \to 0} x^2 = 0$.

(d) $\lim_{x \to 2} x^3 = 8$.

Exercise 6.2. Let $f(x) = \dfrac{|x - 3|}{x - 3} + 4$. Prove that $\lim_{x \to 3} f(x)$ does not exist.

Exercise 6.3. This is a *false* statement:

> If $|f(x) - L| < \varepsilon$ for all x for which $0 < |x - a| < \delta$, then $|f(x) - L| < \varepsilon/2$ for all x for which $0 < |x - a| < \delta/2$.

Prove that this is false by finding a function f and numbers a and L which are a counterexample. (Make sure you also state f's domain and codomain.)

Exercise 6.4. For each of the following, give an example or prove that no such example exists.

(a) If $\lim_{x \to a} f(x)$ and $\lim_{x \to a} g(x)$ do not exist, can $\lim_{x \to a} [f(x) + g(x)]$ exist?

(b) If $\lim_{x \to a} f(x)$ and $\lim_{x \to a} g(x)$ do not exist, can $\lim_{x \to a} f(x)g(x)$ exist?

(c) If $\lim_{x \to a} f(x)$ and $\lim_{x \to a} [f(x) + g(x)]$ exist, must $\lim_{x \to a} g(x)$ exist?

(d) If $\lim_{x \to a} f(x)$ exists and $\lim_{x \to a} [f(x) + g(x)]$ does not exist, can $\lim_{x \to a} g(x)$ exist?

(e) If $\lim_{x \to a} f(x)$ and $\lim_{x \to a} f(x)g(x)$ exist, must $\lim_{x \to a} g(x)$ exist?

Exercise 6.5. Let $f : \mathbb{R} \to \mathbb{R}$ be given by $f(x) = x^2 + 2x + 1$. Use the ε-δ definition of continuity to prove that $f(x)$ is continuous at $x = 1$.

Exercise 6.6. Use the ε-δ definition of continuity to prove that each of the following functions is continuous on its domain.

(a) $f : \mathbb{R} \to \mathbb{R}$, $f(x) = |x|$.

(d) $s : (0, \infty) \to \mathbb{R}$, $h(x) = \log(x)$.

(b) $g : (0, \infty) \to \mathbb{R}$, $f(x) = \sqrt{x}$.

(e) $t : \mathbb{R} \to \mathbb{R}$, $s(x) = \sin(x)$.

(c) $h : \mathbb{R} \to \mathbb{R}$, $g(x) = e^x$.

(f) $v : \mathbb{R} \to \mathbb{R}$, $t(x) = \cos(x)$.

Exercise 6.7. Let $a, b \in \mathbb{R}$, and let $f : \mathbb{R} \to \mathbb{R}$ be given by $f(x) = ax + b$. Use the ε-δ definition of continuity to prove that f is continuous.

Exercise 6.8. Sketch the graph of a function $f : \mathbb{R} \to \mathbb{R}$ which has the property that $\lim_{x \to 3} f(x) = 5$, but f is not continuous at 3. Include in your picture the ε showing clearly why f does not satisfy the definition of continuity.

Exercise 6.9. Assume that $f : \mathbb{R} \to \mathbb{R}$ continuous at c, and $f(c) > 0$. Prove that there exists some $\delta > 0$ such that $f(x) > 0$ for all x in the δ-neighborhood of c.

Exercise 6.10. Let $g_1(x) : \mathbb{R} \to \mathbb{R}$ be defined as

$$g_1(x) = \begin{cases} x & \text{if } x \in \mathbb{Q} \\ 0 & \text{if } x \notin \mathbb{Q}. \end{cases}$$

Prove that g_1 is continuous at c if and only if $c = 0$.

Exercise 6.11. Suppose that $f : X \to Y$ and $\{B_\alpha\}$ is a collection of subsets of Y. Prove that for $f^{-1}(\bigcup_\alpha B_\alpha) = \bigcup_\alpha f^{-1}(B_\alpha)$.

Exercise 6.12. Let $f : \mathbb{R} \to \mathbb{R}$. Prove that $\lim_{x \to 0^+} f(1/x) = \lim_{x \to \infty} f(x)$. That is, prove that if one converges to some L, then the other also converges to L; if one diverges to ∞ or to $-\infty$ or whose limit does not exist, then the other does the same.

Exercise 6.13. Define

$$f(x) = \begin{cases} 1 & \text{if } x \geq 0 \\ -1 & \text{if } x < 0. \end{cases}$$

Prove that f is discontinuous at $c = 0$.

Exercise 6.14. Let $g(x) = x^2$ and let f be *Dirichlet function*:

$$f(x) = \begin{cases} 1 & \text{if } x \in \mathbb{Q} \\ 0 & \text{if } x \notin \mathbb{Q}. \end{cases}$$

Determine the following.

(a) For which x is $f(x) \leq x$?

(b) For which x is $f(x) \leq g(x)$?

(c) What is $g(f(x)) - f(x)$?

Exercise 6.15. Suppose $f : \mathbb{R} \to \mathbb{R}$ is a continuous function with the property that $f(x) = 0$ for every $x \in \mathbb{Q}$. Prove that $f(x) = 0$ for every $x \in \mathbb{R}$.

Exercise 6.16. For each of the following sets A, construct a continuous function $f : A \to \mathbb{R}$ that is unbounded on A. Why these do not contradict Corollary 6.31?

(a) $A = \mathbb{Z}$,   (b) $A = (0, 1)$,   (c) $A = [1, 2] \cap \mathbb{Q}$.

Exercise 6.17. Let

$$k(x) = \begin{cases} x & \text{if } x \in \mathbb{Q} \\ -x & \text{if } x \notin \mathbb{Q}. \end{cases}$$

Prove that $\lim_{x \to 0} k(x)$ exists, but for any $a \neq 0$, $\lim_{x \to a} k(x)$ does not exist.

Definition 6.42. Let I be an interval, $f : I \to \mathbb{R}$ and $c \in I$. Define the restriction $f_L : (-\infty, c] \cap I \to \mathbb{R}$ by $f_L(x) = f(x)$. If $\lim_{x \to c} f_L(x)$ exists, then we say that the *left-hand limit* of f at c exists, and

$$\lim_{x \to c^-} f(x) := \lim_{x \to c} f_L(x).$$

Likewise, define the restriction $f_R : [c, \infty) \cap I \to \mathbb{R}$ by $f_R(x) = f(x)$. If $\lim_{x \to c} f_R(x)$ exists, then we say that the *right-hand limit* of f at c exists, and

$$\lim_{x \to c^+} f(x) := \lim_{x \to c} f_R(x).$$

Exercise 6.18. Prove that $\lim_{x \to c} f(x)$ exists if and only if $\lim_{x \to c^-} f(x) = \lim_{x \to c^+} f(x)$.

Exercise 6.19. Let h be *Thomae's function*:

$$h(x) = \begin{cases} 1 & \text{if } x = 0 \\ 1/n & \text{if } x \in \mathbb{Q} \text{ and } x = m/n \text{ in lowest terms with } n > 0 \\ 0 & \text{if } x \notin \mathbb{Q}. \end{cases}$$

Prove that h is continuous at every irrational number and discontinuous at every rational number.

Exercise 6.20. Suppose that, for each $n \in \mathbb{N}$, A_n is a finite set of numbers from $[0, 1]$. Further, assume that $A_n \cap A_m = \emptyset$ whenever $n \neq m$. Define $f : [0, 1] \to \mathbb{R}$ to be

$$f(x) = \begin{cases} 1/n & \text{if } x \in A_n \\ 0 & \text{if } x \notin A_n \text{ for any } n. \end{cases}$$

Prove that $\lim_{x \to a} f(x) = 0$ for all $a \in [0, 1]$.

Exercise 6.21. Suppose that a function f satisfies $|f(x)| \leq |x|$ for all x. Show that f is continuous at 0.

Exercise 6.22. Suppose that a function g is continuous at 0 and $g(0) = 0$, and f is some function which satisfies $|f(x)| \leq |g(x)|$ for all x. Show that f is also continuous at 0.

Exercise 6.23. Suppose f is continuous on $[a, b]$ and $f(a) = f(b)$. Let

$$m = \inf\{f([a, b])\} \qquad \text{and} \qquad M = \sup\{f([a, b])\}.$$

Prove that if $m < L < M$, then there exist *two* values $x_1, x_2 \in [a, b]$ such that $f(x_1) = f(x_2) = L$.

Definition. A function f has the *intermediate value property* if, for any L between $f(a)$ and $f(b)$ (say, $a < b$), there exists some $c \in [a, b]$ such that $f(c) = L$.

Note: By the intermediate value theorem, if f is continuous, then f has the intermediate value property.

Exercise 6.24. Prove that $g : [0, 1] \to \mathbb{R}$ has the intermediate value property, where

$$g(x) = \begin{cases} x \sin(1/x) & \text{if } x \neq 0, \\ 0 & \text{if } x = 0. \end{cases}$$

Exercise 6.25. Prove that $g : \mathbb{R} \to \mathbb{R}$ is continuous at 0, where

$$g(x) = \begin{cases} x \sin(1/x) & \text{if } x \neq 0, \\ 0 & \text{if } x = 0. \end{cases}$$

Exercise 6.26.

(a) Give an example of a function $f : \mathbb{R} \to \mathbb{R}$ which is discontinuous at $1, \frac{1}{2}, \frac{1}{3}, \frac{1}{4}, \ldots$, but is continuous at every other point.

(b) Give an example of a function $f : \mathbb{R} \to \mathbb{R}$ which is discontinuous at $1, \frac{1}{2}, \frac{1}{3}, \frac{1}{4}, \ldots$, and at 0, but is continuous at every other point.

Exercise 6.27. Give an example of a function on the interval $[0, 1]$ that does not attain its supremum.

Exercise 6.28.

(a) Prove that if $f : A \to \mathbb{R}$ is continuous, then $|f|$ is continuous.

(b) Give an example where $|f|$ is continuous, but f is not continuous.

Exercise 6.29. Suppose $f : \mathbb{R} \to \mathbb{R}$ is continuous, and $a, b \in \mathbb{R}$.

(a) Prove that $g(x) := f(x + b)$ is continuous.

(b) Prove that $h(x) := f(ax + b)$ is continuous.

Exercise 6.30. A function E is said to be *even* if $E(-x) = E(x)$ for all x. A function O is said to be *odd* if $O(-x) = -O(x)$ for all x. Prove that if $f : \mathbb{R} \to \mathbb{R}$ is continuous, then $f(x) = E(x) + O(x)$ for some even function $E(x)$ and odd function $O(x)$.

Exercise 6.31. Let f be continuous. Prove that f may be written as $f(x) = g(x) - h(x)$, for some continuous functions g are h which are also nonnegative (meaning $g(x) \geq 0$ and $h(x) \geq 0$ for all x).

Exercise 6.32. How many continuous functions f are there which satisfy $(f(x))^2 = x^2$ for all x?

Exercise 6.33. Prove that if $f : X \to \mathbb{R}$ is continuous and $[a, b] \subseteq X$, then $f([a, b]) = [c, d]$ for some $c, d \in \mathbb{R}$.

Exercise 6.34. In the below, by "finite open interval" we mean an interval (a, b) where $a, b \in \mathbb{R}$ and $a < b$. And by "finite closed interval" we mean an interval $[a, b]$ where $a, b \in \mathbb{R}$ and $a < b$.

(a) Let $f : A \to B$ be a continuous function where $f(A) = B$. Is it possible for A to be a finite open interval while B is a finite closed interval? Either provide an example showing it is possible, or prove that it is not possible.

(b) Let $f : A \to B$ be a continuous function where $f(A) = B$. Is it possible for A to be a finite closed interval while B is a finite open interval? Either provide an example showing it is possible, or prove that it is not possible.

Exercise 6.35. Let $f : [0, 1] \to \mathbb{R}$ be continuous with $f(0) = f(1)$. Show that there must exist $x, y \in [0, 1]$ which are of distance $1/2$ apart (i.e. $|x - y| = 1/2$) for which $f(x) = f(y)$.

Exercise 6.36. Let S be a dense subset of $\mathbb{R}$, and assume that f and g are continuous functions on $\mathbb{R}$. Prove that if $f(x) = g(x)$ for every $x \in S$, then $f(x) = g(x)$ for all $x \in \mathbb{R}$.

Exercise 6.37. For each of the following, prove that there is a number x satisfying the expression.

(a) $x^{83} + x^{17} + \dfrac{242}{1 + x^4 + \sin^2(x)} = 201$.

(b) $\cos(x) = x$

Exercise 6.38.   Give an alternative proof of the intermediate value theorem (Theorem 6.38) using the result of Exercise 1.34.

> **Definition.**   A function $f : \mathbb{R} \to \mathbb{R}$ has a *removable discontinuity at c* if $\lim_{x \to c} f(x)$ exists, but $\lim_{x \to c} f(x) \neq f(c)$.

Exercise 6.39.

(a) Draw a picture of a function which has has a removable discontinuity at 2.

(b) Does Thomae's function have any removable discontinuities? If so, at which points?

(c) Suppose $f : \mathbb{R} \to \mathbb{R}$ has a removable discontinuity at c, and $g : \mathbb{R} \to \mathbb{R}$ is a function for which

$$g(x) = \begin{cases} f(x) & \text{if } x \neq c \\ \lim_{x \to c} f(x) & \text{if } x = c. \end{cases}$$

Prove that g is continuous at c.

(d) Does there exist a function $f : \mathbb{R} \to \mathbb{R}$ which has a removable discontinuity at every point? You do not need to prove your answer, but if you think the answer is 'yes', then give an example demonstrating this, and if you think the answer is 'no', then give a brief sketch of why you think so.

Exercise 6.40.

(a) Prove that if $f : A \to \mathbb{R}$ is uniformly continuous and A is bounded, then $f(A)$ is bounded.

(b) Give an example of an unbounded continuous function on $(0, 1)$.

(c) Give an example of a bounded continuous function on $(0, 1)$ that is not uniformly continuous. You do not need to prove your answer.

Exercise 6.41.   Prove that if $f : \mathbb{Z} \to \mathbb{R}$, then f is uniformly continuous.

Exercise 6.42.   Prove that the function $f(x) = \sqrt{x}$ is uniformly continuous on $[0, \infty)$.

Exercise 6.43.   Let $f : [a, b] \to \mathbb{R}$ be continuous and suppose that $f(x) > 0$ for all x. Show that there is some $L > 0$ such that $f(x) \geq L$ for all $x \in [a, b]$.

Exercise 6.44.   Prove that there does not exist a continuous function which maps the closed interval $[0, 1]$ onto the open interval $(0, 1)$.

Exercise 6.45.

(a) Give an example of a continuous function $f : (0, 1) \to \mathbb{R}$ and a Cauchy sequence (a_n) where $(f(a_n))$ is *not* a Cauchy sequence.

(b) Prove that if f is uniformly continuous and (a_n) is a Cauchy sequence, then $(f(a_n))$ is a Cauchy sequence.

Exercise 6.46. Let f and g be continuous function on $[a, b]$, and suppose that $f(a) < g(a)$ while $f(b) > g(b)$. Prove that $f(c) = g(c)$ for some $c \in [a, b]$.

Exercise 6.47. At the end of Section 6.8 we discussed the hiker problem, and in this exercise you prove a generalization of it. Show that if the hiker hikes up the mountain from time s_1 to s_2 on one day, and hikes down the mountain from time t_1 to t_2 the next day, and $[s_1, s_2] \cap [t_1, t_2] \neq \emptyset$, then there exists some time at which the hunter was at the exact same spot on the mountain on both days.

Exercise 6.48. Let $f : [0, 1] \to [0, 1]$ be a continuous function. Prove that $f(x) = x$ for some x.

Exercise 6.49. Suppose that $f : \mathbb{R} \to \mathbb{R}$ is a function that is continuous at every rational number. Prove that there is an irrational number at which f is continuous.

Exercise 6.50. How many continuous functions f are there where $(f(x))^2 = x^2$ for all x?

Exercise 6.51.

(a) Give an example of a continuous function defined on $\mathbb{R}$ which obtains every value exactly once. (That is, for each $b \in \mathbb{R}$ there is exactly one $a \in \mathbb{R}$ for which $f(a) = b$.)

(b) Prove that there does not exist a continuous function defined on $\mathbb{R}$ which obtains every value exactly *twice*.

(c) Give an example of a continuous function which obtains every value exactly three times. (If your function's graph is simple and clear, just sketching its graph is sufficient.)

Exercise 6.52. Is it possible for a continuous function $f : \mathbb{R} \to \mathbb{R}$ to have the property that $f(\mathbb{R}) = \mathbb{Q}$?

Exercise 6.53. Prove that if $f : \mathbb{R} \to \mathbb{R}$ is continuous and A is connected (See definition preceding Exercise 5.33 in Chapter 5), then $f(A)$ is connected.

Exercise 6.54. Give an example of a function $f : \mathbb{R} \to \mathbb{R}$ which is

- Monotonically increasing (meaning $f(x) \leq f(y)$ for all $x \leq y$),

- Discontinuous at every $c \in \mathbb{Z}$,

- Continuous at every $c \in \mathbb{R} \setminus \mathbb{Z}$, and

- Bounded.

You do *not* need to prove that your example works. Note: If you can not find an example with all 4 conditions holding, for partial credit try to find one with the first three holding.

Exercise 6.55. Assume that $f : \mathbb{R} \to \mathbb{R}$ and $g : \mathbb{R} \to \mathbb{R}$ are both uniformly continuous functions.

(a) Prove that $f + g$ is uniformly continuous.

(b) If f and g are both bounded, prove that fg is uniformly continuous.

(c) Give an example showing that if g is not bounded, then it is possible for fg to not be uniformly continuous.

(d) Prove that $g \circ f$ is uniformly continuous.

Exercise 6.56. In this exercise you will prove the *Banach fixed-point theorem*. Let $f : \mathbb{R} \to \mathbb{R}$, and assume there exists some $C < 1$ such that

$$|f(x) - f(y)| \leq C|x - y|$$

for all $x, y \in \mathbb{R}$ (such functions are called *Lipschitz*).

(a) Prove that f is continuous on $\mathbb{R}$.

(b) Pick any $x_1 \in \mathbb{R}$ and consider the sequence where, for $n > 1$, we have $x_n = f(x_{n-1})$. That is, this is the sequence

$$(x_1, f(x_1), f(f(x_1)), f(f(f(x_1))), \dots).$$

Prove that this sequence converges by showing it is Cauchy.

(c) Let $x = \lim_{n \to \infty} x_n$. Prove that x is a fixed point of f. That is, prove that $f(x) = x$.

(d) Prove that this x is the only fixed point of f.

Exercise 6.57. Prove that if $f : [a, b] \to \mathbb{R}$ is a one-to-one and continuous function, then its inverse f^{-1} is a continuous function.

Exercise 6.58. Fix a real number $b > 1$.

(a) What do you think 2^π should mean?

(b) Prove that if r is a rational number, then

$$b^r = \sup\{b^t : t \in \mathbb{Q}, \, t < r\}.$$

It is therefore consistent with our previous definition to define, for any $x \in \mathbb{R}$,

$$b^x = \sup\{b^t : t \in \mathbb{Q}, \, t < x\}.$$

(c) Prove that for all $x, y \in \mathbb{R}$ we have $b^{x+y} = b^x \cdot b^y$.

(d) Prove that the function $\exp_b : \mathbb{R} \to \mathbb{R}$ defined by $\exp_b(x) = b^x$ is continuous and strictly increasing.

(e) Prove that the range of $\exp_b$ is the set $\{y \in \mathbb{R}\} \, y > 0$.

Exercise 6.59.

(a) Define a function $f : A \to B$ to be *monotone increasing* if for all $x, y \in A$ for which $x < y$, we have $f(x) \leq f(y)$. Is it possible for a monotone increasing function to have uncountably many discontinuities?

(b) Define a function $f : A \to B$ to be *strictly monotone increasing* if for all $x, y \in A$ for which $x < y$, we have $f(x) < f(y)$. Prove that if $f : [a, b] \to \mathbb{R}$ is continuous, injective (i.e., 1-to-1) and $f(a) < f(b)$, then f is strictly monotone increasing.

Exercise 6.60. A clock has been made with perfect precision: the hour and minute hands move continuously around the clock, and you are able to measure their positions with perfect accuracy. There's only one problem: the clocksmith accidentally put on identically-looking hands for the minute hand and the hour hand. Is it always possible to determine the time? If not, how many times a day can you not determine the time?

Chapter 7: Differentiation

Speedometers

Two millennia after Zeno's perplexing paradoxes, Galileo Galilei—perhaps the first true practitioner of the scientific method—noticed something interesting. He let a ball roll down a long incline and recorded how far it traveled in each 1 second interval. He noticed that if it traveled, say, 1 foot in the first second, then it would travel 3 feet the next second, then 5 feet in the third second, then 7, then 9, and so on. The odd numbers were somehow fundamental to rolling and falling objects.

After Galileo's[1] death, Isaac Newton and others were trying to solve some fundamental physics problems (such as Galileo's) which, like Zeno's paradoxes, required a deeper understanding of the infinite. Here's a modern version of such a problem, that also contains Galileo's mystery: If a car is driving in a straight line and you know the position of the car at every moment, can you determine the velocity of the car at every moment? Let's consider a specific example.

Example 7.1. Suppose you are sitting in your car at a red light. The light turns green and you accelerate (which of course changes your position/location). The position graph is below. The t-axis represents the amount of time (in seconds) since the light turned green, and the d-axis represents the distance (in feet) that the car has traveled since that moment.

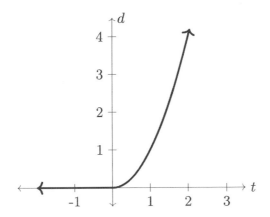

The right-hand side of the graph is the function $d(t) = t^2$. In other words, after 1 second the car has traveled $d(1) = 1$ foot. After 2 seconds it has traveled $d(2) = 4$

---

[1] Move over Cher and Elvis, Galilieo's the real OG at being known by just his first name.

feet. After 3 seconds it has gone 9 feet, and so on. As you can see, the car is speeding up: in the first second it traveled 1 foot, in the second second it traveled $4 - 1 = 3$ feet, and in the third second it traveled $9 - 4 = 5$ feet, and so on.

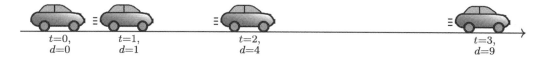

Here's the question we want to answer: At a fixed time t_0, what is the speed of the car at that moment? This is called the car's *instantaneous velocity*. If you're the driver you can tell your speed by looking at the speedometer. Thus we are asking whether we can determine a function which reports exactly what the car's speedometer does.

This is not an easy question, and a full course of high school physics is likely insufficient to find an answer for even some simple distance functions. Indeed, it took perhaps the greatest scientist in history — Isaac Newton — to find the answer; and in doing so, differential calculus was born.[2]

7.1   Graphical Interpretations of Velocity

One of the principal tenets of calculus is that geometry can be used to solve algebra problems. This is good news — the geometry tends to be pretty basic and visual, so not only is it helpful to solve the problems but also it helps us all see (and thus understand) what's going on. To solve the problem in Example 7.1 we will do this. We begin by reminding you of the geometry that is needed.

Secant Lines

A *secant line* is a line segment from one point on a curve to another. Like this:

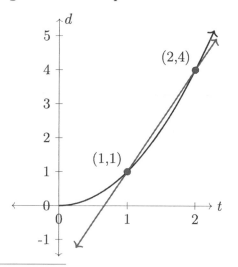

---

[2]Although these car examples are a fun and carefree introduction to derivatives, with prudence we must now turn to a serious Public Service Announcement: *Never drink and derive.*

Every line has a slope, and, just like all lines, the slope of the secant line from a point (x_1, y_1) to another point (x_2, y_2) is "change in y over change in x":

$$\text{slope of secant} = \frac{y_2 - y_1}{x_2 - x_1}.$$

If these points are on the curve of the function d, then $y_1 = d(x_1)$ and $y_2 = d(x_2)$, so the same formula can be written as

$$\text{slope of secant} = \frac{d(x_2) - d(x_1)}{x_2 - x_1}.$$

Average Velocity

Returning to Example 7.1, how would we, for instance, determine the car's velocity after one second? Remember that this is something that you do in your car all the time, as your speedometer is constantly telling you your velocity. We are simply asking what we would see if we looked at the speedometer after 1 second of driving. Well, the one equation you might remember from high school — and a formula Newton knew — was that of *average velocity*:

$$\text{average velocity} = \frac{\text{change in position}}{\text{change in time}}.$$

So, for example, if a car travels 120 miles in 2 hours, then its average velocity is $120m/2h = 60m/h$. If Usain Bolt jogs[3] 100 meters in 10 seconds, then his average velocity is 10m/s.

Now, you might be tempted to just say "the instantaneous velocity at $t = 1$ seconds must just be the average velocity from $t = 1$ to $t = 1$." Sadly this doesn't work out. From $t = 1$ to $t = 1$ no time has passed and no change in position has occurred, which would give

$$\text{average velocity} = \frac{\text{change in position}}{\text{change in time}} = \frac{0}{0} = \text{undefined}.$$

And in this case we would agree with Zeno that something is amiss, and we can not simply apply this formula.

Nevertheless, we will be able to use this formula in a different way. Notice that the formula looks a lot like the formula for the slope of a secant line. . . [4]

---

[3] Yes, *jogs*.

[4] This observation is also more straightforward today than it was back then. Today, every high school student has had years and years of working with graphs of functions. Today when someone mentions the function "$y = x^2$" its graph immediately enters your mind, but until the 17th century this was not the case. Indeed, the pervading Greek perspective considered it blasphemous to even think about numbers on a continuum, or to compare quantities with non-matching units, which happens when you put time on one axis and distance traveled on the other. Graphing functions was a true intellectual leap.

Average Velocity = Slope of Secant Line

If we want to estimate a car's velocity at time $t = 1$ second, one thing we could do is find the average velocity of the car from $t = 1$ second to $t = 2$ seconds — the car's velocity won't change tooooo much over just one second, so the average velocity over this time period should be a decent guess for its speed at $t = 1$.

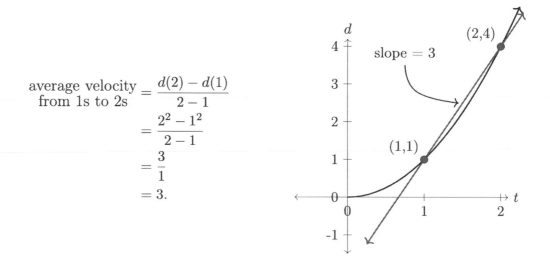

$$\begin{aligned}\text{average velocity} \atop \text{from 1s to 2s} &= \frac{d(2) - d(1)}{2 - 1} \\ &= \frac{2^2 - 1^2}{2 - 1} \\ &= \frac{3}{1} \\ &= 3.\end{aligned}$$

The average velocity from $t = 1$ to $t = 2$ is 3 feet per second; note that the average velocity calculation matches the slope of the corresponding secant line. Thus, 3 feet per second is a decent guess for the velocity at $t = 1$, but let's try to improve it. If we cut down the time range to just $t = 1$ to $t = 1.5$, then the car would have less time to change its velocity, and thus we would have a better estimate. Doing this calculation gives

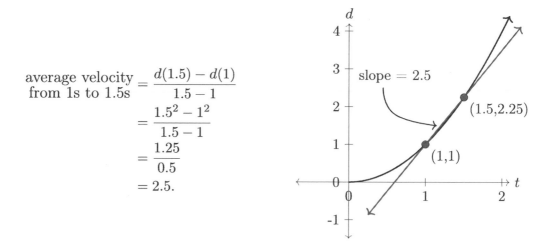

$$\begin{aligned}\text{average velocity} \atop \text{from 1s to 1.5s} &= \frac{d(1.5) - d(1)}{1.5 - 1} \\ &= \frac{1.5^2 - 1^2}{1.5 - 1} \\ &= \frac{1.25}{0.5} \\ &= 2.5.\end{aligned}$$

The average velocity from $t = 1$ to $t = 1.5$ is 2.5 feet per second; our last guess of 3 ft/s was clearly not perfect. We must be getting closer to the true velocity, but why stop here? Let's keep going. An even better estimate for the velocity would be the average velocity from $t = 1$ to $t = 1.25$.

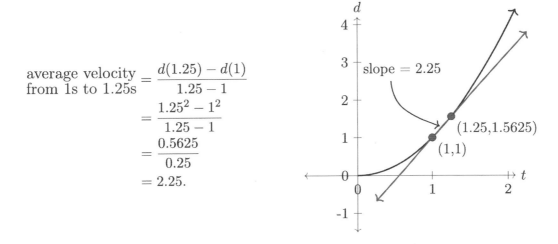

$$\text{average velocity from 1s to 1.25s} = \frac{d(1.25) - d(1)}{1.25 - 1}$$
$$= \frac{1.25^2 - 1^2}{1.25 - 1}$$
$$= \frac{0.5625}{0.25}$$
$$= 2.25.$$

The average velocity from $t = 1$ to $t = 1.25$ is 2.25 ft/s, again with the formula matching the secant's slope. So 2.25 ft/s is an even better estimate, but let's not stop now.

Instantaneous Velocity = Slope of Tangent Line

If we continue to do this for smaller and smaller gaps, the average velocities look like the left-hand side below.

From	To	Average velocity
t=1	t=2	3 ft/s
t=1	t=1.5	2.5 ft/s
t=1	t=1.25	2.25 ft/s
t=1	t=1.1	2.1 ft/s
t=1	t=1.01	2.01 ft/s
t=1	t=1.001	2.001 ft/s
t=1	t=1.0001	2.0001 ft/s
$\vdots$	$\vdots$	$\vdots$
t=1	t=1	Inst. Vel. = 2 ft/s?

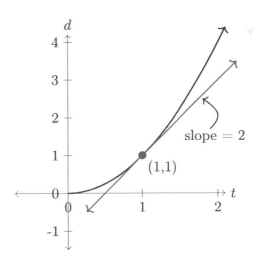

As the time interval gets smaller and smaller, the average velocity approaches 2 feet per second. And, indeed, as you would probably guess, the car's instantaneous velocity at $t = 1$ second turns out to be exactly 2 feet per second.

And what about the right-hand side of the above? The secant lines approach *the tangent line*, and the slopes of the secant lines approach the slope of the tangent line.

If $f(x)$ is a differentiable function at x_0, then the rate of change of f at a point (x_0, y_0) is going to equal to the slope of the tangent line to f at (x_0, y_0). In our particular case, this means that the velocity at a time t_0 is equal to the slope of the tangent line to $d(t)$ at the point $(t_0, d(t_0))$.

Moreover — and we will talk later about how important this is — we would get the same conclusion if instead of beginning at $t = 1$ and ending a little past it, we instead started a little before $t = 1$ and ended at it, as you can see below.

From	To	Average Velocity
t=0	t=1	1 ft/s
t=0.5	t=1	1.5 ft/s
t=0.75	t=1	1.75 ft/s
t=0.9	t=1	1.9 ft/s
t=0.99	t=1	1.99 ft/s
t=0.999	t=1	1.999 ft/s
t=0.9999	t=1	1.9999 ft/s
$\vdots$	$\vdots$	$\vdots$
t=1	t=1	Inst. Vel. = 2 ft/s?

7.2   The Derivative

The limits of these average-velocity and slope-of-secant formulas are exactly how the *derivative* is defined.[5]

Definition.

Definition 7.2. Let I be an interval, $f : I \to \mathbb{R}$, and $c \in I$. We say f is *differentiable* at c if
$$\lim_{x \to c} \frac{f(x) - f(c)}{x - c}$$
exists. That is, this limit equals some real number.

If C is the collection of points at which f is differentiable, then the *derivative* of f is the function $f' : C \to \mathbb{R}$ where

$$f'(c) = \lim_{x \to c} \frac{f(x) - f(c)}{x - c}.$$

---

[5]In the below, an *interval* is any set of the form $(a, b), (a, b], [a, b), [a, b], (a, \infty), [a, \infty), (-\infty, b)$ or $(-\infty, b]$.

Notation 7.3. Given a function f, we write $\frac{d}{dx} f(x)$ for its derivative. That is,

$$\frac{d}{dx} f(x) = f'(x).$$

Examples

Example 7.4. Let I be an interval and $f : I \to \mathbb{R}$ be given by $f(x) = x^2$. Then for any $c \in I$,

$$\begin{aligned}
f'(c) &= \lim_{x \to c} \frac{f(x) - f(c)}{x - c} \\
&= \lim_{x \to c} \frac{x^2 - c^2}{x - c} \\
&= \lim_{x \to c} \frac{(x - c)(x + c)}{x - c} \qquad (\text{☎}) \\
&= \lim_{x \to c} (x + c)
\end{aligned}$$

Recall that since x is a variable and c is a constant, the '$x+c$' function is a continuous function and hence we were allowed to just plug in c for x.

$$\begin{aligned}
&= c + c \\
&= 2c.
\end{aligned}$$

That's the derivative at every fixed c, so we can instead rewrite this using the variable x: $f'(x) = 2x$. Or, writing it out as the derivative of a function:

$$\frac{d}{dx} x^2 = 2x.$$

$\square$

We can actually prove a much more general formula, that for any $n \in \mathbb{N}$ we have $\frac{d}{dx} x^n = n x^{n-1}$. We show that now.

Example 7.5. Let I be an interval and $f : I \to \mathbb{R}$ be given by $f(x) = x^n$ where $n \in \mathbb{N}$. Then for any $c \in I$,

$$\frac{d}{dx} x^n = \lim_{x \to c} \frac{f(x) - f(c)}{x - c} = \lim_{x \to c} \frac{x^n - c^n}{x - c}$$

Remember how we factored $x^2 - c^2$ in (☎)? We can factor $x^n - c^n$ in a similar way.[6] Note that

$$x^n - c^n = (x - c)(x^{n-1} + x^{n-2}c + x^{n-3}c^2 + \cdots + xc^{n-2} + c^{n-1}),$$

because

$$(x - c)(x^{n-1} + x^{n-2}c + x^{n-3}c^2 + \cdots + xc^{n-2} + c^{n-1})$$
$$= x^n + x^{n-1}c + x^{n-2}c^2 + \cdots + x^2c^{n-2} + xc^{n-1}$$
$$\quad - x^{n-1}c - x^{n-2}c^2 - \cdots - x^2c^{n-2} - xc^{n-1} - c^n$$
$$= x^n - c^n.$$

With this factorization, we can now compute the derivative at an arbitrary point c.

$$f'(c) = \lim_{x \to c} \frac{x^n - c^n}{x - c}$$
$$= \lim_{x \to c} \frac{(x - c)(x^{n-1} + x^{n-2}c + x^{n-3}c^2 + \cdots + xc^{n-2} + c^{n-1})}{x - c}$$
$$= \lim_{x \to c}(x^{n-1} + x^{n-2}c + x^{n-3}c^2 + \cdots + xc^{n-2} + c^{n-1})$$
$$= (c^{n-1} + c^{n-2}c + c^{n-3}c^2 + \cdots + c \cdot c^{n-2} + c^{n-1})$$
$$= n \cdot c^{n-1}.$$

Since c was arbitrary, $\frac{d}{dx}x^n = nx^{n-1}$ for any natural number n.[7] $\quad\square$

Speaking of important derivatives, in Exercise 7.34 you will prove that $\frac{d}{dx}\sin(x) = \cos(x)$, $\frac{d}{dx}\cos(x) = -\sin(x)$, and $\frac{d}{dx}\tan(x) = \sec^2(x)$.

Another good example is Example 7.7, which is coming soon. But it will be helpful to think about that example in the context of the following theorem, so we start there.

7.3   Continuity and Differentiability

There are some important questions that we should now ask. We have spent much time recently studying properties of functions, the biggest being continuity. As we continue our study of differentiability, it is important to keep in mind that both continuity and differentiability are *local* conditions—a function is either *continuous at a point c*, or it's not; a function is either *differentiable at a point c*, or it's not. It is of course possible for a function to be continuous at *every* point, in which case we simply call it *continuous*, but the definitions are both point-centric. It's therefore quite natural to ask how continuity and differentiability relate.

Do you think a function is continuous if and only if it is differentiable? If not, which of those implications fails? Is every continuous function differentiable, but not

---

[6]Indeed, one way to see that this should be possible is to note that if you plug c into $x^n - c^n$ you get out zero: $c^n - c^n = 0$. So c is a root of the polynomial $x^n - c^n$, and hence $x - c$ should be a factor. (In general, this would hold for any *algebraic* function.)

[7]In fact, this holds for any real number n, but that's harder to prove.

every differentiable function continuous? Or vice versa? Or maybe neither implication holds?

If f is continuous, is it the derivative of some function g? And if so, is g continuous? What about if f is instead differentiable? And what about with uniform continuity? Extreme values? Boundedness? Intermediate value property? We have all sorts of notions that we have looked at, and so it would be interesting to know how this new notion of a derivative fits in with our theory thus far.

We begin with the first question: Does differentiability imply continuity and vice versa? It turns out that differentiability is the strictly stronger criterion. Thus our goal is to prove that every differentiable function is continuous, and then to find an example of a continuous function that is not differentiable.

Theorem.

Theorem 7.6 (*Differentiable $\Rightarrow$ continuous*). Suppose I is an interval and $f : I \to \mathbb{R}$ is differentiable at $c \in I$. Then f is continuous at $c \in I$.

Proof. By assumption, f's derivative exists at c. That is, there is some $L \in \mathbb{R}$ for which

$$\lim_{x \to c} \frac{f(x) - f(c)}{x - c} = L. \tag{$\maltese$}$$

We wish to show that f is continuous at c. By Theorem 6.17 part (v), it suffices to show that $\lim_{x \to c} f(x) = f(c)$; i.e., that $\lim_{x \to c} [f(x) - f(c)] = 0$. Since the limit ($\maltese$) exists by assumption, and the limit $\lim_{x \to c} (x - c)$ clearly exists as it equals 0, this follows quickly from the func-y limit laws (Theorem 6.14):

$$\begin{aligned}
\lim_{x \to c} [f(x) - f(c)] &= \lim_{x \to c} \left[\frac{f(x) - f(c)}{x - c} \cdot (x - c) \right] \\
&= \left(\lim_{x \to c} \frac{f(x) - f(c)}{x - c} \right) \cdot \left(\lim_{x \to c} (x - c) \right) \quad \text{(Limit laws)} \\
&= L \cdot 0 \\
&= 0.
\end{aligned}$$

$\square$

When you have a single-direction implication, it is worthwhile to consider the contrapositive, which is logically equivalent. The contrapositive of Theorem 7.6 is: If f is discontinuous at some $c \in I$, then f is also non-differentiable at $c \in I$.

The converse, though, is not true. That is, we can find a function which is continuous at some point but not differentiable at that point. (And if you tried to use the above proof in reverse, it would fail because we can not break up the limit like that if we do not already know that f is differentiable at c.)

Example 7.7. Let $f(x) = |x|$ and $c = 0$.

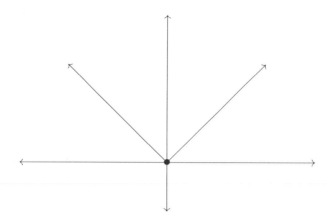

Then f is continuous at 0 since clearly $\lim_{x \to 0} f(x) = f(0)$.[8] However, f is not differentiable at 0. To see this note that

$$\lim_{x \to c} \frac{f(x) - f(c)}{x - c} = \lim_{x \to 0} \frac{f(x) - f(0)}{x - 0}$$
$$= \lim_{x \to 0} \frac{f(x)}{x}$$
$$= \lim_{x \to 0} \frac{|x|}{x},$$

and this limit does not exist since if $x > 0$ this quotient is 1, but if $x < 0$ this quotient equals -1. Thus, this limit does not exist by Exercise 6.13 (or by an application of Note 6.13, with, say $a_n = 1/n$ and $b_n = -1/n$).

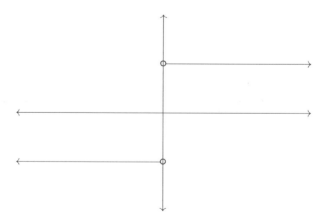

$\square$

---

[8]To prove it, let $\varepsilon > 0$ and $\delta = \varepsilon$, and apply the definition of a limit.

7.4 Differentiability Rules

First we recall some notation that you have probably seen in previous courses.

> **Notation.**
>
> **Notation 7.8.** Suppose f and g are functions and $k \in \mathbb{R}$. Then define
>
> - $(f + g)(x) = f(x) + g(x)$
> - $(f - g)(x) = f(x) - g(x)$
> - $(fg)(x) = f(x)g(x)$
> - $\left(\dfrac{f}{g}\right)(x) = \dfrac{f(x)}{g(x)}$
> - $(kf)(x) = k \cdot f(x)$

In 1684, while Isaac Newton was working out the calculus of dynamics in his masterpiece *Principia*, the first result laws of calculus were published (without proof) by Gottfried Wilhelm Leibniz. This section covers those rules, and we begin with the fact that derivatives satisfy linearity conditions.

> **Proposition.**
>
> **Proposition 7.9** (*Linearity of the derivative*). Let I be an interval, let $f, g : I \to \mathbb{R}$ be differentiable at $c \in I$, and let k be a constant. Then
>
> (i) $(f + g)'(c) = f'(c) + g'(c)$.
>
> (ii) $(f - g)'(c) = f'(c) - g'(c)$.
>
> (iii) $(kf)'(c) = k \cdot f'(c)$.

Proof. The proofs of (i) and (ii) are exactly the same, so we will prove them together. Note that we use a func-y limit law (Theorem 6.14) to move from the second line to the third line, which was allowed because by assumption both of those limits existed.

$$
\begin{aligned}
\lim_{x \to c} \frac{(f \pm g)(x) - (f \pm g)(c)}{x - c} &= \lim_{x \to c} \frac{\big[f(x) \pm g(x)\big] - \big[f(c) \pm g(c)\big]}{x - c} \\
&= \lim_{x \to c} \frac{\big[f(x) - f(c)\big] \pm \big[g(x) - g(c)\big]}{x - c} \\
&= \lim_{x \to c} \frac{f(x) - f(c)}{x - c} \pm \lim_{x \to c} \frac{g(x) - g(c)}{x - c} \\
&= f'(c) \pm g'(c).
\end{aligned}
$$

Now we prove (iii), which again just amounts to writing out the definition of the derivative, moving a couple things around, and applying a limit law.

$$\lim_{x \to c} \frac{(kf)(x) - (kf)(c)}{x - c} = \lim_{x \to c} \frac{k \cdot f(x) - k \cdot f(c)}{x - c}$$

$$= k \cdot \lim_{x \to c} \frac{f(x) - f(c)}{x - c}$$

$$= k \cdot f'(c). \qquad \square$$

Example 7.10. We have shown that $\frac{d}{dx} x^{12} = 12x^{11}$ and $\frac{d}{dx} x^5 = 5x^4$. And so, by Proposition 7.9 part (i), we now know that

- $\dfrac{d}{dx}(x^{12} + x^5) = 12x^{11} + 5x^4$, and

- $\dfrac{d}{dx}(7 \cdot x^5) = 7 \cdot \dfrac{d}{dx} x^5 = 7 \cdot 5x^4 = 35x^4$.

> **Theorem.**
>
> **Theorem 7.11** (*The product rule*). Let I be an interval and let $f, g : I \to \mathbb{R}$ be differentiable at $c \in I$. Then
>
> $$(fg)'(c) = f'(c)g(c) + f(c)g'(c).$$

Proof Idea. At the moment we haven't proved much about derivatives that would be helpful here, so any proof of this will likely be via the definition of the derivative. When solving problems, it is always a good idea to keep in mind where you are trying to reach, and how you might reach that point. Indeed, let's write out where we will begin and start working down, and let's write where we will end and start working up, and let's see how far we can move inward before needing to be smart.

$$(fg)'(c) = \lim_{x \to c} \frac{(fg)(x) - (fg)(c)}{x - c}$$

$$= \lim_{x \to c} \frac{f(x)g(x) - f(c)g(c)}{x - c}$$

. . . then a miracle occurs . . .

$$= \lim_{x \to c} \frac{f(x) \cdot g(c) - f(c) \cdot g(c)}{x - c} + \lim_{x \to c} \frac{f(c) \cdot g(x) - f(c) \cdot g(c)}{x - c}$$

$$= \lim_{x \to c} \frac{f(x) - f(c)}{x - c} \cdot g(c) + f(c) \cdot \lim_{x \to c} \frac{g(x) - g(c)}{x - c}$$

$$= f'(c)g(c) + f(c)g'(c).$$

Now, as the famous comic says, we should probably be more explicit in the miracle step. Note, though, that things *almost* look good. After the miracle we have an $f(x)g(x)$ in the first numerator and a $-f(c)g(c)$ in the second—both of these are in the numerator before the miracle occurred. Moreover, the other post-miracle terms are *almost* canceling each other out. Only problem is that in one term there is a $g(x)$ and the other there is a $g(c)$. How do we turn a $g(x)$ into a $g(c)$? Another limit! Check it out:

Proof. Note that by the func-y limit laws (Corollary 6.14),

$$\begin{aligned}
(fg)'(c) &= \lim_{x \to c} \frac{(fg)(x) - (fg)(c)}{x - c} \\
&= \lim_{x \to c} \frac{f(x)g(x) - f(c)g(c)}{x - c} \\
&= \lim_{x \to c} \frac{f(x)g(x) - f(c)g(x) + f(c)g(x) - f(c)g(c)}{x - c} \\
&= \lim_{x \to c} \frac{f(x)g(x) - f(c)g(x)}{x - c} + \lim_{x \to c} \frac{f(c)g(x) - f(c)g(c)}{x - c} \\
&= \lim_{x \to c} \frac{f(x) - f(c)}{x - c} \cdot g(x) + \lim_{x \to c} f(c) \cdot \frac{g(x) - g(c)}{x - c}
\end{aligned}$$

To move to the next two lines we use that g is assumed to be differentiable and hence by Theorem 7.6, g is continuous. Now by applying Theorem 6.17 (v) we know that $\lim_{x \to c} g(x) = g(c)$:

$$\begin{aligned}
&= \lim_{x \to c} \frac{f(x) - f(c)}{x - c} \cdot \lim_{x \to c} g(x) + f(c) \cdot \lim_{x \to c} \frac{g(x) - g(c)}{x - c} \\
&= f'(c)g(c) + f(c)g'(c).
\end{aligned}$$

$\square$

Theorem.

Theorem 7.12 (*The quotient rule*). Let I be an interval and let $f, g : I \to \mathbb{R}$ be differentiable at $c \in I$. Then

$$\left(\frac{f}{g}\right)'(c) = \frac{f'(c)g(c) - f(c)g'(c)}{[g(c)]^2},$$

provided that $g(c) \neq 0$.

Proof. The proof is similar to the previous proof, and is Exercise 7.5. $\square$

> ### Theorem.
>
> **Theorem 7.13** (*The chain rule*). Let I_1 and I_2 be intervals and $g : I_1 \to I_2$ and $f : I_2 \to \mathbb{R}$. If g is differentiable at $c \in I_1$ and f is differentiable at $g(c) \in I_2$, then
> $$(f \circ g)'(c) = f'(g(c)) \cdot g'(c).$$

<u>Wrong Proof.</u> Below is <u>NOT</u> a correct proof of the chain rule. That said, it is close. And in fact, if you dust off your old calculus textbook and find where they state the chain rule, if they provide a "proof" (as they usually do), then they probably will give the following false proof.

That all said, here is the false proof:

$$
\begin{aligned}
(f \circ g)'(c) &= \lim_{x \to c} \frac{f(g(x)) - f(g(c))}{x - c} \\
&= \lim_{x \to c} \frac{f(g(x)) - f(g(c))}{x - c} \cdot \frac{g(x) - g(c)}{g(x) - g(c)} \\
&= \lim_{x \to c} \frac{f(g(x)) - f(g(c))}{g(x) - g(c)} \cdot \frac{g(x) - g(c)}{x - c} \\
&= \lim_{x \to c} \frac{f(g(x)) - f(g(c))}{g(x) - g(c)} \cdot \lim_{x \to c} \frac{g(x) - g(c)}{x - c} \\
&= f'(g(x)) \cdot g'(x).
\end{aligned}
$$

(Take a minute to see if you can find the mistake on your own...)

As $x \to c$, note that since g is continuous at c we do have $g(x) \to g(c)$. So the problem is *not* that in the first line above we have a $g(x)$ instead of an x.

(If you don't see it yet, take another moment to make your best guess...)

Ok, here is the problem: When you take, say,

$$\lim_{x \to c} \frac{g(x) - g(c)}{x - c},$$

we are never concerned about dividing by zero, because in the definition of a functional limit (Definition 6.8) we never have $x = c$. Likewise, when we take

$$\lim_{x \to c} \frac{f(g(x)) - f(g(c))}{g(x) - g(c)},$$

we again guarantee that $x \neq c$, but it's important to note that this does *not* guarantee that $g(x) \neq g(c)$. For all we know, we could have $g(x) = g(c)$ for x values getting arbitrarily close to c. And in that case, we have a significant divide-by-zero problem.

To fix this, we define a new function which is equal to this quotient, provided the denominator is non-zero—but whenever we do have a $g(x) = g(c)$ issue, our function just redefines this quotient to be what the derivative ought to be. Doing this right will ensure that we do not change the limit in the meantime. Below is this.

Proof. Consider the function

$$d(y) = \begin{cases} \frac{f(y) - f(g(c))}{y - g(c)} & \text{if } y \neq g(c) \\ f'(g(c)) & \text{if } y = g(c) \end{cases}$$

Note that d is continuous at $g(c)$; i.e., $\lim_{y \to g(c)} d(y) = f'(g(c))$, because f is differentiable at $g(c)$. Now we show that

$$\frac{f(g(x)) - f(g(c))}{x - c} = d(g(x)) \cdot \frac{g(x) - g(c)}{x - c}. \tag{☯}$$

If $g(x) \neq g(c)$, then (☯) is just

$$\frac{f(g(x)) - f(g(c))}{x - c} = \frac{f(g(x)) - f(g(c))}{g(x) - g(c)} \cdot \frac{g(x) - g(c)}{x - c},$$

which are the same just by multiplying top-and-bottom by $g(x) - g(c)$, and moving a few things around. If $g(x) = g(c)$, then applying the definition of d to (☯) gives

$$\frac{f(g(x)) - f(g(c))}{x - c} = f'(g(c)) \cdot \frac{g(x) - g(c)}{x - c},$$

and then applying $g(x) = g(c)$,

$$\frac{f(g(c)) - f(g(c))}{x - c} = f'(g(c)) \cdot \frac{g(c) - g(c)}{x - c},$$

which is simply $0 = 0$, which is true.[9] So in either case, (☯) is true. Therefore,

$$\begin{aligned}
(f \circ g)'(c) &= \lim_{x \to c} \frac{f(g(x)) - f(g(c))}{x - c} \\
&= \lim_{x \to c} d(g(x)) \cdot \frac{g(x) - g(c)}{x - c} \\
&= \lim_{x \to c} d(g(x)) \cdot \lim_{x \to c} \frac{g(x) - g(c)}{x - c} \\
&= f'(g(c)) \cdot g'(c).
\end{aligned}$$

$\square$

---

[9]There it is, folks. The most obvious statement in this text.

7.5   Topologist's Sine Curve Examples

Below are some important examples, which all deal with the following general form of a function:

$$g_n(x) = \begin{cases} x^n \sin(1/x) & \text{if } x \neq 0, \\ 0 & \text{if } x = 0. \end{cases}$$

We will focus on the three functions g_0, g_1 and g_2. For each of these we ask: is g_n continuous at 0? Is it differentiable at 0?

Example 7.14. We begin with

$$g_0(x) = \begin{cases} \sin(1/x) & \text{if } x \neq 0, \\ 0 & \text{if } x = 0. \end{cases}$$

Its graph:

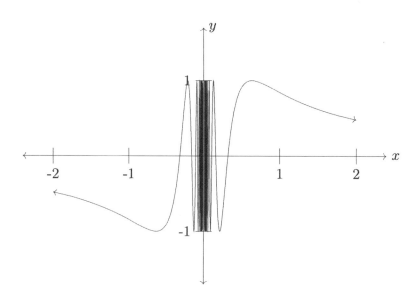

As you can see, g_0 is not continuous at 0, because $\lim\limits_{x \to 0} \sin(1/x) =$ does not exist. Indeed, by looking at the graph and taking $t = \frac{1}{x}$, you can see that the one-sided limits of this limit are equivalent to

$$\lim_{t \to \infty} \sin(t) = \text{D.N.E.} \qquad \text{and} \qquad \lim_{t \to -\infty} \sin(t) = \text{D.N.E.}$$

And then by (the contrapositive to) Theorem 7.6, g_0 being not continuous at 0 implies that it is also not differentiable at 0.   □

Example 7.15. Next, consider the function

$$g_1(x) = \begin{cases} x\sin(1/x) & \text{if } x \neq 0, \\ 0 & \text{if } x = 0. \end{cases}$$

This function has the graph

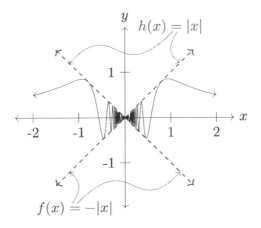

To see that g_1 is continuous at 0, first note that by combining $-1 \leq \sin(1/x) \leq 1$ and $-|x| \leq x \leq |x|$, we have that

$$-|x| \leq x\sin(1/x) \leq |x|.$$

Also,

$$\lim_{x \to 0} -|x| = 0 \qquad \text{and} \qquad \lim_{x \to 0} |x| = 0.$$

So by the func-y squeeze theorem (Corollary 6.15),

$$\lim_{x \to 0} x\sin(1/x) = 0.$$

I.e.,

$$\lim_{x \to 0} g_1(x) = g_1(0),$$

which, by Theorem 6.17 part (v), means that g_1 is continuous at 0.

However, g_1 is *not* differentiable at 0. To see this, we can simply use the definition of the derivative (Definition 7.2):

$$\begin{aligned} g'(0) &= \lim_{x \to 0} \frac{g_1(x) - g_1(0)}{x - 0} \\ &= \lim_{x \to 0} \frac{x\sin(1/x) - 0}{x - 0} \\ &= \lim_{x \to 0} \sin(1/x) \\ &= \text{Does not exist.} \end{aligned}$$

Note, the fact that the final limit does not exist is from Example 7.14. $\qquad \square$

Example 7.16. Lastly, consider the function

$$g_2(x) = \begin{cases} x^2 \sin(1/x) & \text{if } x \neq 0, \\ 0 & \text{if } x = 0. \end{cases}$$

This function has the graph

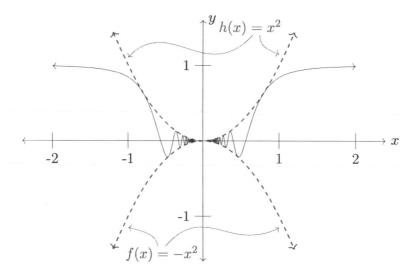

Note that g_2 is continuous at 0 by the func-y squeeze theorem (by essentially the exact same calculation as in Example 7.15). And g_2 is also differentiable at 0:

$$\begin{aligned} g_2'(0) &= \lim_{x \to 0} \frac{g_2(x) - g_2(0)}{x - 0} \\ &= \lim_{x \to 0} \frac{x^2 \sin(1/x) - 0}{x - 0} \\ &= \lim_{x \to 0} x \sin(1/x) \\ &= 0. \end{aligned}$$

Note that we had already found this final limit (via the func-y squeeze theorem) in Example 7.15. □

To summarize the conclusions of the last three examples,[10]

	Continuous at 0?	Differentiable at 0?
g_0	no	no
g_1	yes	no
g_2	yes	yes

[10]Since differentiable implies continuous, by the contrapositive being discontinuous implies being non-differentiable. Therefore a 'yes' in a cell in the right column implies a 'yes' in a cell in the adjacent cell in the left column, and a 'no' in a cell in the left column implies a 'no' in the cell in the adjacent right column. Likewise, by analyzing the limits involved, the answer in a cell in the left column will be the same as the answer in the cell in the right column and in the following row. Using just this, one can show that for all $n > 2$, g_n will have a 'yes' in both columns.

7.6 Local Minimums and Maximums

Derivatives are not just interesting to study, they are darn-right useful too, as you saw when you took calculus. If you recall, all those important optimization problems you did came down to finding critical points of functions, which was nearly the same as finding local maxes and mins.[11]

> ### Definition.
>
> **Definition 7.17.** Let $f : A \to \mathbb{R}$. Then f has a *local maximum* at $c \in A$ if there exists some $\delta > 0$ such that for all $x \in A$ for which $|x - c| < \delta$, we have
>
> $$f(x) \leq f(c).$$
>
> Likewise, f has a *local minimum* at $c \in A$ if there exists some $\delta > 0$ such that for all $x \in A$ for which $|x - c| < \delta$, we have
>
> $$f(x) \geq f(c).$$

An important idea from calculus is that a differentiable function's local mins and maxes only happen at places where its derivative equals zero. Indeed, back in the day you did lots of examples where you took the derivative of a function, set that derivative equal to zero, and solved.[12] In a moment we will prove this result, but first recall the following fact about limits (Exercise 3.22 with $a_n = 0$ for all n).

> ### Recall.
>
> **Recall 7.18.** If (s_n) is a sequence, $s_n \geq 0$ for all n, and $s_n \to L$, then $L \geq 0$.

The following is sometimes called *Fermat's theorem*, but since this chapter is already chock-full with theorems, I'm going to call it a proposition. Fermat predated Calculus, but at times he came tantalizing close to discovering it himself.

> ### Proposition.
>
> **Proposition 7.19** (*Local maxes/mins happen where $f'(c) = 0$*). Let I be an open interval and suppose $f : I \to \mathbb{R}$ is differentiable at $c \in I$. If f has a local max or min at c, then $f'(c) = 0$.

---

[11] But not entirely (cf. saddle points).

[12] Typically you solved it by factoring, using the quadratic formula, or using some other algebraic properties, like those of exponentials or logs.

Proof Idea. Since f is differentiable at c,

$$\lim_{x \to c} \frac{f(x) - f(c)}{x - c} = f'(c).$$

And notice that since c is a local max, $f(x) \leq f(c)$ for all x within some δ-neighborhood of c. And for all these x, notice that

$$\frac{f(x) - f(c)}{x - c} \quad \text{is} \quad \begin{cases} \frac{\text{neg.}}{\text{pos.}} \leq 0 & \text{if } x > c; \\ \frac{\text{neg.}}{\text{neg.}} \geq 0 & \text{if } x < c. \end{cases}$$

The above two observations, combined with Recall 7.18, will imply that both $f'(c) \leq 0$ is both $f'(c) \geq 0$, which can only imply that $f'(c) = 0$.

Proof. Since f is differentiable at c, we know that

$$\lim_{x \to c} \frac{f(x) - f(c)}{x - c} = f'(c).$$

Suppose $c \in I$ is a local max. Then by Definition 7.17 there exists some $\delta > 0$ where $x \in I$ and $|x - c| < \delta$ implies $f(x) \leq f(c)$ (and hence, $f(x) - f(c) \leq 0$). Consequently, for all such x,

$$\frac{f(x) - f(c)}{x - c} \quad \text{is} \quad \begin{cases} \leq 0 & \text{if } x > c; \\ \geq 0 & \text{if } x < c. \end{cases} \tag{$\heartsuit$}$$

By Theorem 6.12, we know that given any sequence (a_n) from I, if $a_n \to c$ then

$$\frac{f(a_n) - f(c)}{a_n - c} \to f'(c).$$

We will now pick one sequence approaching from the left and one approaching from the right; both will be within δ of c, and such sequences exist since I is an open interval. That is, first pick a sequence (a_n) from I for which $c - \delta < a_n < c$ for all n, and also $a_n \to c$. Then by $(\heartsuit)$ we know that

$$\frac{f(a_n) - f(c)}{a_n - c} \geq 0 \quad \text{for all } n, \qquad \text{and} \qquad \frac{f(a_n) - f(c)}{a_n - c} \to f'(c).$$

Therefore, by Recall 7.18, $f'(c) \geq 0$.

Similarly, if we pick a sequence (b_n) from I for which $c < b_n < c + \delta$ for all n, and also $b_n \to c$, then by $(\heartsuit)$ we know that

$$\frac{f(b_n) - f(c)}{b_n - c} \leq 0 \quad \text{for all } n, \qquad \text{and} \qquad \frac{f(b_n) - f(c)}{b_n - c} \to f'(c).$$

Therefore, by Recall 7.18, $f'(c) \leq 0$. And of course, the only way that $f'(c) \geq 0$ and also $f'(c) \leq 0$ is if $f'(c) = 0$. $\qquad \square$

Mathematicians have known of this result for about as long as derivatives themselves have been defined. Fermat, for instance, made use of this to solve optimization problems. It took 200 years, though, for someone to notice an important corollary to this result.

Recall that continuous functions have the intermediate value property, but not every function with the intermediate value property (IVP) is in fact continuous. Also, we know that every differentiable function is continuous, but not every continuous function is differentiable. That is,

$$f \text{ is differentiable } \Rightarrow f \text{ is continuous } \Rightarrow f \text{ has the IVP,}$$

but none of those implications go the other way.

So now, here's where the fun begins. If f is differentiable, it is certainly *not* the case that f' will also have to be differentiable. Ok, let's go weaker: will f' have to be continuous? Again, the answer is no. But the intermediate value property... that we *can* guarantee! This is known as Darboux's theorem.

Theorem.

Theorem 7.20 (*Darboux's theorem*). Suppose $f : [a, b] \to \mathbb{R}$ is differentiable. If α is between $f'(a)$ and $f'(b)$, then there exists $c \in (a, b)$ where $f'(c) = \alpha$.

First we note that "α between $f'(a)$ and $f'(b)$" means that either $f'(a) < \alpha < f'(b)$, or $f'(b) < \alpha < f'(a)$.

Proof. Without loss of generality, assume $f'(b) < \alpha < f'(a)$. Define $g(x) = f(x) - \alpha x$. Then g is differentiable on $[a, b]$, and $g'(x) = f'(x) - \alpha$. Furthermore,

$$g'(a) = f'(a) - \alpha > 0 \qquad \text{and} \qquad g'(b) = f'(b) - \alpha < 0.$$

If we can find a local maximum in (a, b), then we can apply Proposition 7.19 to show that $g'(c) = 0$ for some $c \in (a, b)$, which will prove that $f'(c) = \alpha$. So that is our goal.

Since $[a, b]$ is compact, by the extreme value theorem (Theorem 6.32) g attains a maximum, which of course is also a local maximum.

Claim: g's maximum is not at a.

Assume for a contradiction that the maximum were at a, meaning that $g(x) \leq g(a)$ for all $x \in [a, b]$. This implies that $\frac{g(x) - g(a)}{x - a} = \frac{\text{negative or } 0}{\text{positive}} \leq 0$ for all $x \in (a, b]$, and hence

$$g'(a) = \lim_{x \to a} \frac{g(x) - g(a)}{x - a} \leq 0.$$

But this contradicts our earlier deduction that $g'(a) > 0$. $\square$(claim)

<u>Claim: g's maximum is not at b.</u>

Assume for a contradiction that the maximum were at b, meaning that $g(x) \leq g(b)$ for all $x \in [a, b]$. This implies that $\frac{g(x)-g(b)}{x-b} = \frac{\text{negative or } 0}{\text{negative}} \geq 0$ for all $x \in [a, b)$, and hence

$$g'(b) = \lim_{x \to b} \frac{g(x) - g(b)}{x - b} \geq 0.$$

But this contradicts our earlier deduction that $g'(b) < 0$. $\square$(claim)

<u>Therefore the max is at some $c \in (a, b)$.</u>

Then (by taking $\delta = \min\{b - c, c - a\}$), we see that g has a local maximum at c, which by Proposition 7.19 means that $g'(c) = 0$; i.e., $f'(c) - \alpha = 0$; i.e., $f'(c) = \alpha$. This completes the proof. $\square$

7.7   The Mean Value Theorems

Speed cameras measure how fast a car is moving, and hence detect whether someone is breaking the speed limit at that very moment; if they are, then the camera snaps a picture of the perpetrator and their license plate. If a driver knows where the cameras are, though, then they can simply slow down while they pass each camera, and then speed up again afterwards. How can law enforcement rebut this workaround? They can use mathematics!

Since 2004, the United Kingdom has used so-called *average speed cameras*. On a stretch of road with no on-ramps or off-ramps, a pair of cameras snap a picture of the license plate of each passing car. If the cameras are d feet apart[13] and it took t hours for a car to travel from one camera to the next, then the car's average speed during this time is d/t kilometers per hour. And if a car's average speed exceeds the road's speed limit, then a speeding ticket is issued.

Notice, though, that such a car may have been moving under the speed limit at both moments that their license plate's picture was taken — implying that the camera did not catch them in the act of breaking the law, as the law simply says that it is illegal to exceed the speed limit. However, one consequence of the mean value theorem — which we will prove in this section — is that if your *average speed* exceeds the speed limit, then there was indeed a moment when your *instantaneous speed* exceeded the speed limit; that is, it proves that at some intermediate point you were indeed speeding, and hence demonstrates the soundness of average speed cameras.

That is our goal, but as with the intermediate value theorem and Darboux's theorem, first we prove a "zero version" of the result, which we will then use to prove the main theorem.

---

[13]Typically, d is somewhere between 0.1 kilometers and 10 kilometers.

Theorem.

Theorem 7.21 (*Rolle's theorem*). Let $f : [a, b] \to \mathbb{R}$ be continuous on $[a, b]$ and differentiable on (a, b). If $f(a) = f(b)$, then there exists $c \in (a, b)$ where $f'(c) = 0$.

Proof. By the extreme value theorem, f attains a maximum at some $c_1 \in [a, b]$, and f attains a minimum at some $c_2 \in [a, b]$. Note that if either $c_1 \in (a, b)$ or $c_2 \in (a, b)$, then by Proposition 7.19 such a point c would have $f'(c) = 0$, as desired.

And so we may assume that c_1 and c_2 are the end points; without loss of generality, assume that $c_1 = a$ and $c_2 = b$. Then for all $x \in [a, b]$,

$$f(a) \geq f(x) \geq f(b).$$

But also by assumption, we have $f(a) = f(b)$. Therefore

$$f(a) = f(b) = f(x)$$

for all $x \in [a, b]$. So f is constant, and hence $f'(x) = 0$ for all x, so for any $c \in (a, b)$ we have $f'(c) = 0$, completing the proof. $\qquad\square$

Note that if we are told that f is a continuous function on $[a, b]$, then the most direct route between the points $(a, f(a))$ and $(b, f(b))$, whatever those points may be, is a straight line between them. This line would have slope

$$\frac{f(b) - f(a)}{b - a}.$$

In fact, this line's derivative would be the constant function $f'(x) = \dfrac{f(b) - f(a)}{b - a}$.

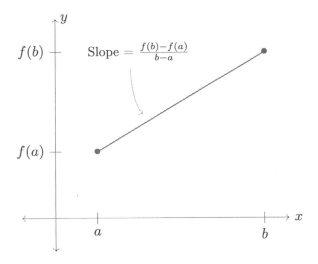

(If the y-axis is a car's distance traveled (in miles) and the x-axis is time (in hours), then this is the graph of a car moving at a constant speed — of $\frac{f(b)-f(a)}{b-a}$ mph.)

Now, if f is still continuous and differentiable but is not a line, like below, there is still a point c for which the derivative at that point (i.e., the slope of the tangent line at that point) is equal to the "average slope," which is $\frac{f(b)-f(a)}{b-a}$:

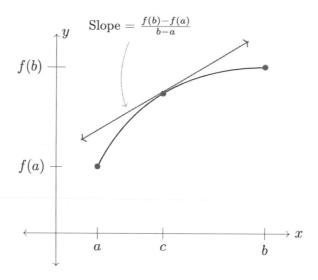

Of course, the function could be even more complicated, and in those situations typically there is more than one such value of c.

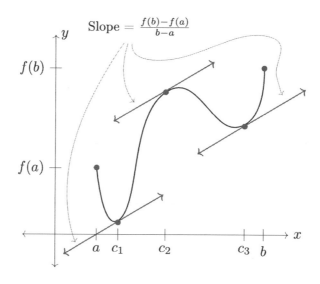

Assume $f : [a, b] \to \mathbb{R}$ is differentiable. If we are hoping there to be some number L for which we are guaranteed that $f'(c) = L$ at some c (no matter f is, such a c exists), then $\frac{f(b)-f(a)}{b-a}$ is our only hope. The first of the three examples graphed above had a derivative that *only* took the value $\frac{f(b)-f(a)}{b-a}$, so we can't hope for anything else. However, as the next two graphs suggested, and as the next theorem states, this value we *can* always guarantee.

> **Theorem.**

Theorem 7.22 (*The (derivative) mean value theorem*). Let $f : [a, b] \to \mathbb{R}$ be continuous on $[a, b]$ and differentiable on (a, b). Then there exists some $c \in (a, b)$ where
$$f'(c) = \frac{f(b) - f(a)}{b - a}.$$

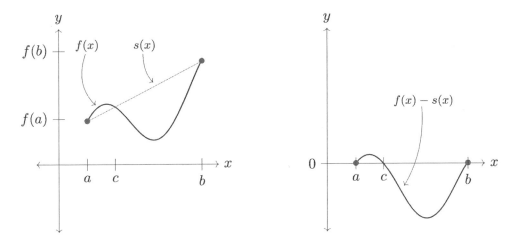

Our plan is to reduce this theorem to Rolle's theorem. If you start with $f(x)$, and then subtract off the secant line from $(a, f(a))$ to $(b, f(b))$ (call this secant line $s(x)$, pictured above), what you get is a function with $f(a) = f(b) = 0$, which the theorem can handle. The theorem will then give us what we seek. Let's Rolle.

Proof. Note that the equation of the secant line from $f(a)$ to $f(b)$ is
$$s(x) = \left(\frac{f(b) - f(a)}{b - a} \right)(x - a) + f(a).$$

Let $g(x) = f(x) - s(x)$. That is, $g(x)$ is what you get when you subtract off this line from $f(x)$:
$$g(x) = f(x) - \left[\left(\frac{f(b) - f(a)}{b - a} \right)(x - a) + f(a) \right].$$

Note that g is continuous on $[a, b]$ and differentiable on (a, b), since f and s are. Also note that $g(a) = f(a) - s(a) = 0$, and $g(b) = f(b) - s(b) = 0$. So we may apply Rolle's theorem to g to say that there exists some $c \in (a, b)$ where $g'(c) = 0$. I.e., $f'(c) - s'(c) = 0$, or equivalently that $f'(c) = s'(c)$. And since
$$s'(x) = \frac{f(b) - f(a)}{b - a},$$

we have indeed found a $c \in (a, b)$ such that $f'(c) = \dfrac{f(b) - f(a)}{b - a}.$ $\qquad\square$

You might be surprised by how important this result is. As Edward Mills Purcell put it, "The mean value theorem is the midwife of calculus — not very important or glamorous by itself, but often helping to deliver other theorems that are of major significance."

Indeed, this theorem will be a crucial piece of many major theorems to come. The first of these important results are a trio of direct corollaries.

Corollary.

Corollary 7.23 ($f'(x) = 0 \Rightarrow f(x) = C$). Let I be an interval and $f : I \to \mathbb{R}$ be differentiable. If $f'(x) = 0$ for all $x \in I$, then f is constant on I.

Proof. Pick $x_1, x_2 \in I$ with $x_1 < x_2$. We aim to show that $f(x_1) = f(x_2)$.

Since f is differentiable on I, it is in particular differentiable on $[x_1, x_2]$, and so by the mean value theorem there exists some $c \in (x_1, x_2)$ such that

$$f'(c) = \frac{f(x_2) - f(x_1)}{x_2 - x_1}.$$

But since by assumption $f'(x) = 0$ for all $x \in I$, we have

$$0 = \frac{f(x_2) - f(x_1)}{x_2 - x_1}.$$

And hence, $0 = f(x_2) - f(x_1)$, which gives $f(x_1) = f(x_2)$, completing the proof. □

The above can also be proved nicely by contradiction. But moving on, do you remember how your calculus teacher would constantly remind not to forget your "$+C$" when taking an antiderivative?[14] Well, the corollary below is why.

Corollary.

Corollary 7.24 (*"Don't forget your $+C$!"*). Let I be an interval and $f, g : I \to \mathbb{R}$ be differentiable. If $f'(x) = g'(x)$ for all $x \in I$, then $f(x) = g(x) + C$ for some $C \in \mathbb{R}$.

Proof. Let $h(x) = f(x) - g(x)$. Note that $h'(x) = f'(x) - g'(x) = 0$ for all $x \in I$. So by Corollary 7.23, $f(x) - g(x) = C$ for some $C \in \mathbb{R}$. And hence $f(x) = g(x) + C$, as desired. □

---

[14]Confession: I was terrible at remembering my "$+C$" when I took calculus. And I got burned by it so often that eventually I would go into an exam chanting to myself "$+C, +C, +C, \ldots$" And then as soon as I got my exam I would write a huge "$+C!$" on the top of every page, so I wouldn't forget. It mostly worked.

Finally, it was important in calculus that a derivative is positive precisely when the differentiable function is increasing.[15]

> ### Corollary.
>
> **Corollary 7.25** ($f' \geq 0 \Leftrightarrow f$ *is increasing*). Let I be an interval and $f : I \to \mathbb{R}$ be differentiable.
>
> (i) f is monotone increasing if and only if $f'(x) \geq 0$ for all $x \in I$.
>
> (ii) f is monotone decreasing if and only if $f'(x) \leq 0$ for all $x \in I$.

Proof. We will prove (i), and simply note that (ii) is extremely similar. For (i), first assume that f is increasing, meaning that for all $x_1, x_2 \in I$ where $x_1 < x_2$, we have $f(x_1) \leq f(x_2)$. Note that this implies that for any $x, c \in I$,

$$f(x) - f(c) \qquad \text{and} \qquad x - c$$

are either both positive (or 0) or are both negative (or 0). And therefore, for any $x \neq c$,

$$\frac{f(x) - f(c)}{x - c} \geq 0$$

for all $x, c \in I$. Thus,

$$f'(c) = \lim_{x \to c} \frac{f(x) - f(c)}{x - c} \geq 0$$

for all $c \in I$, as desired.

Now we show the backwards direction of (i). To that end, assume that $f'(x) \geq 0$ for all $x \in I$. Now pick any $x_1, x_2 \in I$ for which $x_1 < x_2$. To conclude that f is increasing, it suffices to show that $f(x_1) \leq f(x_2)$.

By the mean value theorem, there exists some $c \in (x_1, x_2)$ such that

$$f'(c) = \frac{f(x_2) - f(x_1)}{x_2 - x_1}.$$

I.e.,

$$f(x_2) - f(x_1) = f'(c)(x_2 - x_1).$$

But by assumption $f'(c) \geq 0$, and since $x_1 < x_2$ we also have $(x_2 - x_1) > 0$. So their product is also positive: $f(x_2) - f(x_1) \geq 0$, and hence $f(x_2) \geq f(x_1)$, as desired. $\square$

---

[15]The second derivative determines convexity, and higher derivatives also represent precise things. A fun fact about this: On the reelection campaign trail, Richard Nixon said "the rate of increase of inflation is decreasing." This is believed to be the first time a US president invoked the third derivative to help his electability.[16]Another fun fact: Thomas Jefferson revered Issac Newton, and a careful reading of the Declaration of Independence shows how axiomatically he wrote it. After leaving the presidency he wrote to John Adams, "I have given up newspapers in exchange for Tacitus and Thucydides, for Newton and Euclid; and I find myself much happier."

[16]If a candidate ever invokes the *fourth* derivative, they automatically have my vote.

7.8 L'Hôpital's Rule

By the func-y limit laws (Corollary 6.14),

$$\lim_{x \to a} \frac{f(x)}{g(x)} = \frac{\displaystyle\lim_{x \to a} f(x)}{\displaystyle\lim_{x \to a} g(x)},$$

provided both of these limits exist. If the denominator converges to zero and the numerator either converges to something non-zero or diverges to $\pm\infty$, then the limit of the quotient either does not exist or equals $\pm\infty$, and with a little thought about the signage one can determine which it is. Likewise if the numerator diverges to $\pm\infty$ and the denominator is convergent.

What happens if both the numerator and denominator converge to zero, or both diverge to $\pm\infty$? Intuitively, it should depend on which one is converging/diverging "fastest." For example, if both diverge to ∞ but the numerator does so much more quickly, then the denominator will not slow it down enough to prevent the entire quotient from also diverging to ∞. But if the denominator is diverging to ∞ more quickly, then the ratio should converge to 0. For example,

$$\lim_{x \to 0} \frac{1/x^2}{\ln|x|} = -\infty \qquad \text{and} \qquad \lim_{x \to 0} \frac{1/x^2}{e^{x^{-2}}} = 0.$$

This suggests that each function's *rate of change* will determine what happens to the limit of the quotient. And since derivatives measure rates of change, it is reasonable to think that the derivative will play a role. And, indeed, it does.[17]

We end this chapter with l'Hôpital's rule. This theorem continues a long tradition in mathematics of not being named after the person who first proved it, although the story behind this theorem's name is particularly interesting. Johann Bernoulli first proved the theorem, although it was named after Guillaume de l'Hôpital. So why does it bear l'Hôpital's name? L'Hôpital offered to pay Bernoulli 300 Francs per year *for the rights to some of his theorems*. Seriously. Bernoulli agreed to inform l'Hôpital of his major discoveries, and according to the contract that they both signed, l'Hôpital could do with those results as he pleased. One of those results is what is today called l'Hôpital's rule. But the credit goes to Johann Bernoulli![18]

L'Hôpital did have a meaningful contribution, though: He published the rule in his book *Analyse des Infiniment Petits pour l'ntelligence des Lignes Courbes*, which

---

[17]This might almost seem circular, as we used limits to define derivatives, and now derivatives are being used to solve problems of limits? Fortunately, when done with care, no issues arise.

[18]In academia, people talk about their academic family. One's PhD advisor is considered their academic parent, and their own PhD students are considered their academic children. And other relationships work likewise; e.g., a pair are academic siblings if they share an academic parent. Johann Bernoulli's academic family tree is interesting in that it intersects his blood family tree. His academic father was his blood brother, Jacob Bernoulli; his academic brother was his blood nephew, Nicolaus I Bernoulli; and his academic child was his actual son, Daniel Bernoulli. (Oh, and another of his academic children was perhaps the most prolific mathematician in history, Leonhard Euler.) The Bernoulli (blood) family was a mathematical powerhouse, producing eight world-class mathematicians in just three generations.

was the first ever calculus textbook.[19] Its proof will use the following result, which is a generalization of the mean value theorem (Theorem 7.22).

> **Theorem.**
>
> **Theorem 7.26** (*Cauchy mean value theorem*). If f and g are continuous on $[a, b]$ and differentiable on (a, b), then there is a number $c \in (a, b)$ such that
>
> $$[f(b) - f(a)] \cdot g'(c) = [g(b) - g(a)] \cdot f'(c).$$

I have discovered a truly marvelous proof of this, which this margin is too narrow to contain. So I'll leave it as an exercise to the reader (Exercise 7.29)

Now here's the rule.

> **Theorem.**
>
> **Theorem 7.27** (*L'Hôpital's rule*). Suppose I is an open interval containing a point a, and $f : I \to \mathbb{R}$ and $g : I \to \mathbb{R}$ are differentiable on I (except possibly at a). Suppose also $g'(x) \neq 0$ on I. Then, if
>
> $$\lim_{x \to a} f(x) = 0 \qquad \text{and} \qquad \lim_{x \to a} g(x) = 0,$$
>
> or if
>
> $$\lim_{x \to a} f(x) = \pm\infty \qquad \text{and} \qquad \lim_{x \to a} g(x) = \pm\infty,$$
>
> then
>
> $$\lim_{x \to a} \frac{f(x)}{g(x)} = \lim_{x \to a} \frac{f'(x)}{g'(x)},$$
>
> provided the limit $\lim_{x \to a} \dfrac{f'(x)}{g'(x)}$ exists.

Proof Idea. The big idea behind this proof is that since $\lim_{x \to a} \frac{f'(x)}{g'(x)}$ exists, it equals some $L \in \mathbb{R}$, which by the definition of a functional limit (Definition 6.8, although in the actual proof $\varepsilon/2$ will be playing the role of ε), means that there is some δ blah blah blah such that

$$\left| \frac{f'(x)}{g'(x)} - L \right| < \varepsilon.$$

---

[19]The first book on a subject is sooooooo much harder to write than 20 renditions later. So even though he was cheat, as a real analysis author I still have mad respect for l'Hôpital for paving the way. *pours one out*

We wish to prove that $\lim_{x \to a} \frac{f(x)}{g(x)}$ also equals L, which means that we want to show

$$\left| \frac{f(x)}{g(x)} - L \right| < \varepsilon.$$

To do this we need a means to convert information about the derivative into information about the original function. The Cauchy mean value theorem (Theorem 7.26) is perfectly set up to do this, which (after some simple division) says that for any interval $[a, b]$ on which f and g are differentiable, there is some c for which

$$\frac{f'(c)}{g'(c)} = \frac{f(b) - f(a)}{g(b) - g(a)}.$$

There is some work to go to find the interval(s) $[a, b]$ on which we will apply this theorem, and the above isn't immediately perfect since we which to convert an

$$\frac{f'(c)}{g'(c)} \quad \text{into an} \quad \frac{f(a)}{g(a)}, \quad \text{not into an} \quad \frac{f(b) - f(a)}{g(b) - g(a)},$$

but with the assumptions that $\lim_{x \to a} f(x) = 0$ and $\lim_{x \to a} g(x) = 0$, it all works out nicely.

Proof of the $\frac{0}{0}$ case. Here we will only prove the case when $\lim_{x \to a} f(x) = 0$ and $\lim_{x \to a} g(x) = 0$. The case where both limits are $\pm\infty$ is asked of you in Exercise 7.33.

Let $\varepsilon > 0$. Since $\lim_{x \to a} \frac{f'(x)}{g'(x)}$ exists, there is some $L \in \mathbb{R}$ such that $\lim_{x \to a} \frac{f'(x)}{g'(x)} = L$, which by the definition of a functional limit (Definition 6.8) means that there exists a $\delta > 0$ such that for all $x \in I$ for which $x \in (a - \delta, a + \delta)$,

$$\left| \frac{f'(x)}{g'(x)} - L \right| < \frac{\varepsilon}{2}.$$

By undoing the absolute value and adding L throughout, the above gives that, for all $x \in (a - \delta, a + \delta)$,

$$L - \frac{\varepsilon}{2} < \frac{f'(x)}{g'(x)} < L + \frac{\varepsilon}{2}. \tag{$\mathbf{\Psi}$}$$

Pick any $x_1, x_2 \in (a - \delta, a + \delta)$ where $x_1, x_2 \in I$ and $x_1 < a < x_2$. Notice that f and g being differentiable on I implies that they are differentiable on $[x_1, x_2]$, which by Theorem 7.6 implies that f and g are continuous on $[x_1, x_2]$. We can therefore apply the Cauchy mean value theorem (Theorem 7.26), which produces some $c \in [x_1, x_2] \subseteq (a - \delta, a + \delta)$ such that,

$$\frac{f'(c)}{g'(c)} = \frac{f(x_2) - f(x_1)}{g(x_2) - g(x_1)}. \tag{$\mathbf{f}$}$$

Since $c \in (a - \delta, a + \delta)$, the inequalities in ($\mathbf{\Psi}$) imply $L - \frac{\varepsilon}{2} < \frac{f'(c)}{g'(c)} < L + \frac{\varepsilon}{2}$, which when combined with (**f**) gives

$$L - \frac{\varepsilon}{2} < \frac{f(x_2) - f(x_1)}{g(x_2) - g(x_1)} < L + \frac{\varepsilon}{2}. \tag{$\mathbf{t}$}$$

Since this holds for any such x_1 and x_2, and this quotient is a continuous function in x_2, by taking a limit (as $x_2 \to a$) on all sides we achieve

$$\lim_{x_2 \to a} \left(L - \frac{\varepsilon}{2}\right) \leq \lim_{x_2 \to a} \frac{f(x_2) - f(x_1)}{g(x_2) - g(x_1)} \leq \left(\lim_{x_2 \to a} L + \frac{\varepsilon}{2}\right).$$

By assumption, $\lim_{x_2 \to a} f(x_2) = 0$ and $\lim_{x_2 \to a} g(x_2) = 0$, so by applying the limit to the above we see that

$$L - \frac{\varepsilon}{2} \leq \frac{0 - f(x_1)}{0 - g(x_1)} \leq L + \frac{\varepsilon}{2},$$

which in particular implies

$$L - \varepsilon < \frac{f(x_1)}{g(x_1)} < L + \varepsilon.$$

In summary, we arbitrarily chose $x_1, x_2 \in (a - \delta, a + \delta) \cap I$ where $x_1 < a < x_2$, and we showed that x_1 has the property that

$$L - \varepsilon < \frac{f(x_1)}{g(x_1)} < L + \varepsilon \tag{$\blacktriangleright$}$$

Likewise, if from (**t**) we had taken the limit as $x_1 \to a$ (instead of $x_2 \to a$), then the same sequence of steps would have deduced that x_2 has the property that

$$L - \varepsilon < \frac{f(x_2)}{g(x_2)} < L + \varepsilon. \tag{$\text{\faTwitch}$}$$

Since the inequalities of ($\blacktriangleright$) holds for all $x_1 \in (a - \delta, a)$, and the inequalities of ($\text{\faTwitch}$) holds for all $x_2 \in (a, a+\delta)$, we have what we want: For all x such that $0 < |x - a| < \delta$, we have shown that

$$\left|\frac{f(x)}{g(x)} - L\right| < \varepsilon,$$

which by definition means

$$\lim_{x \to a} \frac{f(x)}{g(x)} = L.$$

Since both equal L, we proven that

$$\lim_{x \to a} \frac{f(x)}{g(x)} = \lim_{x \to a} \frac{f'(x)}{g'(x)},$$

as desired. $\qquad\qquad\qquad\qquad\qquad\qquad\qquad\qquad\qquad\qquad\qquad\square$

The Universal Language

In the opening pages of this book, when we were discussing the building blocks for $\mathbb{R}$, I mentioned a famous quote from Leopold Kronecker: "God made the integers, all else is the work of man." Regardless of whether you believe in a God, it opens up an interesting philosophical discussion as to how much of mathematics lives in nature, and how much lives solely in our collective brains. Kronecker seems to be far to one side of this debate.

Now that we have developed derivatives, if we wanted we could turn to the study of differential equations — equations which include the derivatives of functions, and which do a remarkable just at modeling just about everything in the real world, from moving objects to incoming hurricanes to heat transfer to the stock market. Richard Feynman — one of the greatest modern physicists and an enormously influential science communicator — took quite a different stance than Kronecker, swayed strongly by the universe's deeply mathematical nature. "Have you studied calculus?" he once asked Novelist Herman Wouk after an interview. "No," Wouk admitted. "You had better learn it," Feynman replied. "It's the language God talks."

— Notable Exercises —

- In Exercise 7.5 you are asked to prove the quotient rule. This is one of the principle rules for actually computing derivatives, which is why it plays a prominent role in calculus. Its proof is similar to the proof of the product rule.

- In Exercise 7.7 you are asked to prove a special case of the power rule, another highly useful rule for computing actual derivatives, and one which played a starring role in your calculus class.

- A function is continuous if it is continuous at each point in its domain. Being continuous at each point c means that, for every $\varepsilon > 0$, there is some $\delta > 0$ such that... yada yada yada.... Intuitively, if your function is really steep at c, then you need a really small δ, whereas if your function is not so steep, then although a small δ also works, a larger δ will work too. But note that for a fixed $\varepsilon > 0$, each point c can have its own δ. The idea behind *uniform continuity* is that a single δ works for *all* the points c; essentially, the function never gets "too steep." Exercise 7.13 puts meat on the bones of this idea. It says this if a derivative is bounded (which is another way to say the function never gets "too steep"), then the function will be uniformly continuous.

- Exercise 7.15 deals with Lipschitz functions, which is another way to make precise this important notion of a function never getting "too steep."

- Exercise 7.29 asks you to prove a generalization of the mean value theorem (Theorem 7.22) called the *Cauchy mean value theorem* (Theorem 7.26), which was used to prove the $\frac{0}{0}$ case of l'Hôpital's rule.

- In Exercise 7.33 you are asked to prove the $\frac{\pm\infty}{\pm\infty}$ case of l'Hôpital's rule (Theorem 7.27). Admittedly, this case it significantly harder to prove than the $\frac{0}{0}$ case that was proven a couple pages back.

- In Exercise 7.34 you will find the derivatives of $\sin(x)$, $\cos(x)$ and $\tan(x)$.

— Exercises —

Exercise 7.1. Use the definition of the derivative to find the derivatives of the following functions. For each, let the domain be the positive real numbers.

(a) $f(x) = \sqrt{x} + 3$ (b) $g(x) = \dfrac{1}{x}$ (c) $h(x) = 4x + \dfrac{1}{x^2}$

Exercise 7.2. Assume f and g are functions defined on (a, b), both of which are differentiable at a point $c \in (a, b)$. Also, let $\alpha, \beta \in \mathbb{R}$. Apply the definition of the derivative to the function $(\alpha f + \beta g)$ to prove that $(\alpha f + \beta g)'(c) = \alpha f'(c) + \beta g'(c)$.

Exercise 7.3. Let $f : \mathbb{R} \to \mathbb{R}$ be defined by $f(x) = \lfloor x \rfloor$, and let C be the collection of points on which f is differentiable. Determine C and determine the function f' (which is a function from C to $\mathbb{R}$).

Exercise 7.4. Prove that there do not exist differentiable function f and g for which $f(0) = g(0) = 0$ and $x = f(x)g(x)$.

Exercise 7.5. Prove the *quotient rule* (Theorem 7.12).

Exercise 7.6. Is

$$f(x) = \begin{cases} \frac{1}{2}x & \text{if } x \in \mathbb{Q}, \\ x & \text{if } x \notin \mathbb{Q} \end{cases}$$

differentiable at 0?

Exercise 7.7. Use induction to prove that $\frac{d}{dx}x^n = nx^{n-1}$ for all $n \in \mathbb{N}$, which is the natural number case of the *power rule*.

Exercise 7.8. Let

$$f_a(x) = \begin{cases} x^a & \text{if } x \geq 0 \\ 0 & \text{if } x < 0. \end{cases}$$

(a) For which values of $a \in \mathbb{R}$ is f_a continuous at 0?

(b) For which values of $a \in \mathbb{R}$ is f_a differentiable at 0?

Exercise 7.9. Assume $f : \mathbb{R} \to \mathbb{R}$.

(a) Let $t \in (1, \infty)$. Prove that if $|f(x)| \leq |x|^t$ for all x, then f is differentiable at 0.

(b) Let $t \in (0, 1)$. Prove that if $|f(x)| \geq |x|^t$ for all x, and $f(0) = 0$, then f is not differentiable at 0.

(c) Give a pair of examples showing that if $|f(x)| = |x|$ for all x, then either conclusion is possible.

Exercise 7.10. Recall that a function is *even* if $f(-x) = f(x)$ for all x, and is *odd* if $f(-x) = -f(x)$ for all x. The below two properties are true. Give two proofs of each — one using the definition of the derivative, and one using a result from this chapter — and also draw a picture of each to model the property.

(a) If $f : \mathbb{R} \to \mathbb{R}$ is even and differentiable, then $f'(-x) = -f'(x)$.

(b) If $f : \mathbb{R} \to \mathbb{R}$ is odd and differentiable, then $f'(-x) = f'(x)$.

Exercise 7.11.

(a) Prove that if f is differentiable at some c for which $f(c) \neq 0$, then $|f|$ is also differentiable at c.

(b) Give an example showing that part (a) no longer holds if $f(c) = 0$.

Exercise 7.12. Suppose f is a polynomial of degree n, and $f \geq 0$ (that is, $f(x) \geq 0$ for all x). Prove that

$$f + f' + f'' + \cdots + f^{(n)} \geq 0.$$

You may use, without proof, that $f \geq 0$ implies n is even, and you may use standard properties of even-degree, positive polynomials (e.g. f has a minimum value and $\lim_{x \to \infty} f(x) = \lim_{x \to -\infty} f(x) = \infty$).

Exercise 7.13. Assume that I is an interval and $f : I \to \mathbb{R}$ is differentiable. Prove that if f' is bounded, then f is uniformly continuous.

Exercise 7.14. Give an example of an interval I and a differentiable function $f : I \to \mathbb{R}$ which is uniformly continuous and for which f' is unbounded.

Definition 7.28. A function f is called *Lipschitz continuous* if there exists some $C \in \mathbb{R}$ where

$$|f(x) - f(y)| \leq C \cdot |x - y|$$

for all x and y.

Exercise 7.15. Let $f : [a, b] \to \mathbb{R}$ be differentiable, and assume f' is continuous on $[a, b]$. Prove that f is Lipschitz continuous on $[a, b]$.

Exercise 7.16. Let I be an interval and $f : I \to \mathbb{R}$ be differentiable. Show that f is Lipschitz on I if and only if f' is bounded on I.

Exercise 7.17. Which is greater, e^π or π^e?

Make sure to give a *proof* that your answer is correct—don't just quote your calculator. Feel free to use facts you learned in calculus about $\ln(x)$... and perhaps consider the function $f(x) = \frac{\ln(x)}{x}$.

Exercise 7.18. In this exercise you will explore why each of the three main assumptions in Rolle's theorem is necessary. To that end, give an example of a function f which satisfies of each of the following conditions, and yet $f'(c) \neq 0$ for all $c \in (a, b)$.

(a) A function f which is continuous on $[a, b]$ and differentiable on (a, b) (but we don't assume $f(a) = f(b)$).

(b) A function f which is continuous on $[a, b]$ and $f(a) = f(b)$ (but we don't assume f is differentiable on (a, b)).

(c) A function f which is differentiable on (a, b) and $f(a) = f(b)$ (but we don't assume f is continuous on $[a, b]$).

Exercise 7.19. Use the derivative to find all values of a for which $|x-a| = (x-2)^2$.

Exercise 7.20. Suppose f and g are differentiable functions with $f(a) = g(a)$ and $f'(x) < g'(x)$ for all $x > a$. Prove that $f(b) < g(b)$ for any $b > a$.

Exercise 7.21. A *fixed point* of a function f is a value c for which $f(c) = c$. Prove that if $f : \mathbb{R} \to \mathbb{R}$ is differentiable for all x, and $f'(x) \neq 1$ for all $x \in [0, 1]$, then f can have at most one fixed point in $[0, 1]$.

Exercise 7.22. Prove that if f is differentiable and $f'(x) \geq M$ for all $x \in [a, b]$, then $f(b) \geq f(a) + M(b - a)$.

Exercise 7.23.

(a) Suppose that $g : [0, 5] \to \mathbb{R}$ is differentiable, $g(0) = 0$, and $|g'(x)| \leq M$. Show for all $x \in [0, 5]$ that $|g(x)| \leq Mx$.

(b) Suppose that $h : [0, 5] \to \mathbb{R}$ is twice-differentiable (meaning that it's differentiable, and its derivative is differentiable), that $h'(0) = h(0) = 0$ and $|h''(x)| \leq M$. Show for all $x \in [0, 5]$ that $|h(x)| \leq Mx^2/2$.

(c) Can you give a geometric interpretation of the previous two parts?

Exercise 7.24. Assume that $f(0) = 0$ and $f'(x)$ is an increasing function. Prove that $g(x) = f(x)/x$ is an increasing function on $(0, \infty)$.

Exercise 7.25. Let f be a differentiable function on the interval $[0, 3]$ satisfying that $f(0) = 1$, $f(1) = 2$ and $f(3) = 2$.

(a) Show that there is some point $c \in (0, 3)$ such that $f(c) = c$.

(b) Show that there is some point $d \in [0, 3]$ with $f'(d) = \frac{1}{3}$.

(c) Show that there is some point $e \in [0, 3]$ with $f'(e) = \frac{1}{4}$.

Exercise 7.26. Assume $f : (a, b) \to \mathbb{R}$ is differentiable at some point $c \in (a, b)$. Prove that if $f'(c) \neq 0$ there exists some $\delta > 0$ such that $f(x) \neq f(c)$ for all $x \in (c - \delta, c + \delta)$.

Exercise 7.27. Prove that among all rectangles with a perimeter of p, the square has the greatest area.

Exercise 7.28. Show that the sum of a positive number and its reciprocal is at least 2.

Exercise 7.29. Prove the *Cauchy mean value theorem* (Theorem 7.26), which says this:

If f and g are continuous on $[a, b]$ and differentiable on (a, b), then there is a number $c \in (a, b)$ such that

$$[f(b) - f(a)] \cdot g'(c) = [g(b) - g(a)] \cdot f'(c).$$

(Note that the mean value theorem (Theorem 7.22) is a special case of this, in which $g(x) = x$.)

Exercise 7.30. Suppose f is a strictly monotone continuous function (guaranteeing f's inverse function, f^{-1}, exists) on (a, b). Prove that if $f'(x_0)$ exists and is non-zero, then f^{-1} is differentiable at $f(x_0)$ and

$$(f^{-1})'(f(x_0)) = \frac{1}{f'(x_0)}.$$

Use this to produce the familiar formula for the derivative of the square-root function.

Exercise 7.31. Prove that Thomae's function is not differentiable at any point.

Exercise 7.32. Prove *Leibniz's rule*. That is, prove that if f and g have n^{th} order derivatives on (a, b) and $h = f \cdot g$, then for any $c \in (a, b)$,

$$h^{(n)}(c) = \sum_{k=0}^{n} \binom{n}{k} f^{(k)}(c) g^{(n-k)}(c).$$

Exercise 7.33.   Prove the following "∞/∞" case of l'Hôpital's rule. That is, prove that if $f : (a, b) \to \mathbb{R}$ and $g : (a, b) \to \mathbb{R}$ are differentiable functions and $\lim\limits_{x \to a} f(x) = \lim\limits_{x \to a} g(x) = \infty$, then

$$\lim_{x \to a} \frac{f'(x)}{g'(x)} = L \quad \text{implies} \quad \lim_{x \to a} \frac{f(x)}{g(x)} = L.$$

Exercise 7.34.   In this exercise you will derive the formulas $\frac{d}{dx} \cos(x)$, $\frac{d}{dx} \sin(x)$ and $\frac{d}{dx} \tan(x)$.

First, do you know where $\tan(x)$ got its name? Consider a circle of radius 1, a vertical tangent line, and a ray emanating from the origin at an angle of θ from the positive x-axis.

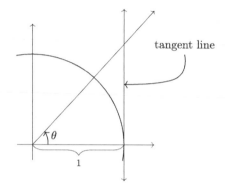

We will focus on just a part of this tangent, and ask for the length of the line segment. Call this length y, and decorate this picture (using the definition of a sine and cosine).

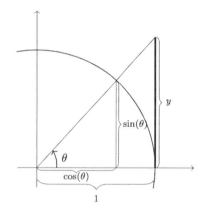

Do you see the similar triangles? The ratio of corresponding sides are thus equal:

$$\frac{\sin(\theta)}{\cos(\theta)} = \frac{y}{1}.$$

But $\tan(\theta) = \dfrac{\sin(\theta)}{\cos(\theta)}$. So we have deduced that

$$y = \tan(\theta).$$

Remember where y came from: it was a tangent line. So now you see why the $\tan(\theta)$ function is call "tangent"—it's the length of a tangent! You will use this diagram to in this exercise:

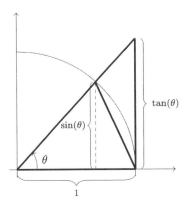

(a) Notice that there is a triangle in this diagram whose base has length 1, and whose height has length $\sin(\theta)$. Notice that this triangle's area is less than the area of the portion of the circle in the θ angle, which is in turn less than the largest triangle's area. Use this reasoning to show that

$$\frac{1}{2}\sin(\theta) < \frac{1}{2}\theta < \frac{1}{2}\tan(\theta).$$

(b) Do some algebra to the above to show that

$$1 \geq \frac{\sin(\theta)}{\theta} \geq \cos(\theta).$$

(c) Use the above and the sequence squeeze theorem to prove that

$$\lim_{x \to 0} \frac{\sin(x)}{x} = 1.$$

(d) Prove that
$$\lim_{x \to 0} \frac{1 - \cos(x)}{x} = 0$$

by multiplying the numerator and denominator of this limit by $1 + \cos(x)$ and, after some quick algebra, applying part (c).

Alternative definition of the derivative. The *derivative* of f is

$$\frac{d}{dx}f(x) = \lim_{h \to 0} \frac{f(x+h) - f(x)}{h}.$$

(e) You will now use the above equivalent definition of the derivative to show that $\frac{d}{dx}\sin(x) = \cos(x)$. Here are the small steps:

  (i) Write $\frac{d}{dx}\sin(x)$ by applying the equivalent definition.

  (ii) Rewrite what you have by using the algebraic identity $\sin(a+b) = \sin(a)\cos(b) + \sin(b)\cos(a)$.

(iii) Do some algebra so that you can apply the limits from parts (c) and (d).

(iv) Conclude that $\frac{d}{dx}\sin(x) = \cos(x)$.

(f) Now that you know $\frac{d}{dx}\sin(x) = \cos(x)$, use the algebraic identity $\cos(x) = \sin(\pi/2 - x)$ to show that $\frac{d}{dx}\cos(x) = -\sin(x)$.

(g) Now that you know $\frac{d}{dx}\sin(x) = \cos(x)$ and $\frac{d}{dx}\cos(x) = -\sin(x)$, show that $\frac{d}{dx}\tan(x) = \sec^2(x)$.

— Open Question —

Question. Notice that the formula for the area of a circle is πr^2. Differentiating this with respect for r gives $\frac{d}{dr}\pi r^2 = 2\pi r$, the formula for perimeter of the circle. The formula for the area of a square with side length $2r$ (so that its "radius" is r) is $(2r)(2r) = 4r^2$. Differentiating this with respect to r gives $\frac{d}{dr}4r^2 = 8r$, which is the perimeter of the square.

Classify all shapes where this property holds (including the higher-dimensional analogues, where you differentiate the volume formula to get the surface area formula). What other geometric interpretations of "radius" are there that give this property?

Chapter 8: Integration

Brief Review

There are a handful of ideas from earlier chapters that will be particularly important in this chapter (and I anticipate this chapter being near the start of the second semester of an analysis course, so a review seems fitting). For those reasons, let's briefly review. Let $A \subseteq \mathbb{R}$.

- U is an *upper bound* of a set A if $x \leq U$ for all $x \in A$. (Definition 1.17)

- L is a *lower bound* of A if $x \geq L$ for all $x \in A$. (Definition 1.17)

- A is *bounded* if there exists an upper bound and a lower bound of A. Or, equivalently, if there exists some M such that $|x| \leq M$ for all $x \in A$. (Definition 1.17)

- If $x \leq y$ for all $x \in A$ and $y \in B$, then $\sup(A) \leq \inf(B)$. (Exercise 8.3)

- A function $f : A \to \mathbb{R}$ is *bounded* if $f(A)$ is bounded. Or, equivalently, if there exists some M such that $|f(x)| \leq M$ for all $x \in A$.

- If $f : [a, b] \to \mathbb{R}$ is continuous, then f is bounded. (Corollary 6.31)

- $f : [a, b] \to \mathbb{R}$ is continuous if and only if f is uniformly continuous. (Proposition 6.40)

- Dirichlet's function
$$f(x) = \begin{cases} 1 & \text{if } x \in \mathbb{Q} \\ 0 & \text{if } x \notin \mathbb{Q} \end{cases}$$
is continuous nowhere. (Example 6.3)

- The modified Dirichlet function
$$g(x) = \begin{cases} x & \text{if } x \in \mathbb{Q} \\ 0 & \text{if } x \notin \mathbb{Q} \end{cases}$$
is continuous only at $x = 0$. (Example 6.5)

- Thomae's function (the function pictured on the cover of this book)

$$h(x) = \begin{cases} 1/n & \text{if } x \neq 0 \text{ and } x = m/n \in \mathbb{Q} \text{ in lowest terms with } n > 0 \\ 0 & \text{if } x \notin \mathbb{Q} \\ 1 & \text{if } x = 0 \end{cases}$$

is continuous on $\mathbb{R} \setminus \mathbb{Q}$, and discontinuous on $\mathbb{Q}$. (Example 6.7)

- Consider

$$g_n(x) = \begin{cases} x^n \sin(1/x) & \text{if } x \neq 0 \\ 0 & \text{if } x = 0. \end{cases}$$

Then (Examples 7.14, 7.15 and 7.16),

	Continuous at 0?	Differentiable at 0?
g_0	no	no
g_1	yes	no
g_2	yes	yes

- If f is differentiable, then f is continuous. However, f' need not be continuous. (Theorem 7.6 and Example 7.16)

- If f is differentiable, then f' has the intermediate value property. (Theorem 7.20)

8.1   The Area of a Circle

Early geometers — from Babylon to Egypt to China to India to Greece to Japan — all deduced the area and volume formulas of the standard straight-lined and flat-surfaced shapes. But for centuries, roundedness left them all bewildered. Even determining the area of a simple circle was too much. Such curiosities inspired some ancient mathematicians-at-heart, and would eventually transform the world.

This transformation began with the realization that cannonballs in flight, planets in orbit and light through a convex prism all rely on the understanding of fundamental curved shapes, and that the geometry developed was critical. But even more transformational than this, was the problem's solution. The key to solving the circle problem was picked up two millennia later by Issac Newton to develop calculus. And without calculus we wouldn't have radio communication, econometrics or ultrasound. But it all began with a circle — and with one particularly eccentric nerd.[1]

---

[1] I've heard Archimedes be referred to as history's first true math nerd. He *loved* math. He reportedly would become so engrossed in his geometry that he would forget to eat or bathe — and when he did bathe he was still focused enough on his work that after once solving a problem mid-bath (a problem tasked to him by the king), he sprang from his tub and ran down the street naked yelling "Eureka!" Whether or not his last words were actually "Do not disturb my circles" is highly debatable, but we do know that he personally requested that his tomb be engraved with his favorite mathematical discovery.

A couple hundred years after Zeno's confounding paradoxes of the infinite, Archimedes — building on the ideas of others — had some revolutionary brainwaves. What is the area of a circle? With his approach and not too much work, he could see that it was a little bigger than $3r^2$; the constant looked suspiciously like the value π (which is defined to be the ratio of the circumference of a circle to that circle's diameter — which is the same for all circles;[2] equivalently, π is defined so that $2\pi r$ is the circumference[3] of the circle of radius r).

His clever idea is known today as the *method of exhaustion*, and was a long precursor to integral calculus. He applied his idea to many different problems, and some such applications remain magnificently brilliant even with two thousand years mathematical hindsight. But for now, our focus here will be on the circle.

What is the area of the circle with radius r? We know that the area is smaller than $4r^2$, since it is smaller than the $2r \times 2r$ square that *circumscribes* it.

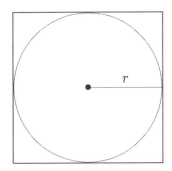

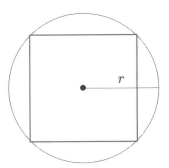

Likewise, you could use a little geometry to figure out the area of the square that *inscribes* a circle. Indeed, if you realize that the square's upper-right vertex is at 45°, and the rest are 90° rotations from there, then you can quickly deduce (by the Pythagorean theorem) that the side length of such a square is $\sqrt{2}r$, and hence its area is $2r^2$.

So we just proved that the area of a circle is between $2r^2$ and $4r^2$! How can we do better? Well, instead of using squares, let's use polygons with more sides, as those more closely resemble a circle.

---

[2]The fact that π is a constant, and does not depend on which circle you pick when you take the circumference divided by the denominator, was not proven in Euclid's *Elements*. It was a century after Euclid that Archimedes finally proved the result (although Greeks thought of it via ratios, not as a number *per se*). Indeed, the famous axioms of Euclid's geometry are insufficient to prove that π is constant — Archimedes had to introduce two more axioms to get a handle on curved arc lengths. For more, see the excellent paper *Circular reasoning: Who first proved C/d is a constant?* by David Richeson here: `arxiv.org/abs/1303.0904`.

[3]Indeed, its symbol 'π' as an abbreviation of the Greek word for *periphery*.

We first focus on the inscribed polygons. A sequence of them looks like so:

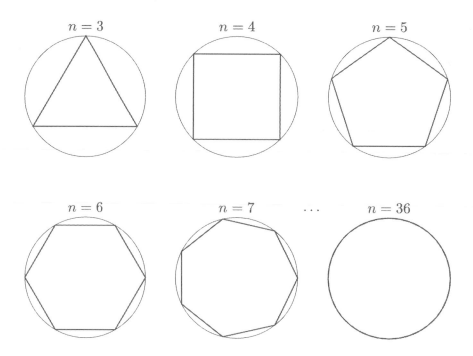

For $n = 36$, you have to zoom in quite a bit just to see any bit of the circle, or to see the edges of the 36-gon![4]

So what is the area of each of these n-gons? Each serves as a lower bound on the actual area of the circle, and presumably they are getting closer and closer to the true value. Let's calculate these values and see what they're doing.

Consider a regular n-gon with side length s inscribed in a circle of radius r. Let $\theta = 2\pi/n$ radians be the measure of a central angle, as shown below (in the picture on the next page, $n = 6$):

---

[4] Archimedes himself moved from 6-gons to 12-gons to 24-gons to 48-gons and eventually as far as a 96-gon, which he used to prove $3 + \frac{10}{71} < \pi < 3 + \frac{10}{70}$, a remarkable feat. He was bested 700 years later by Chinese geometer Zu Chongzhi who continued this doubling *six more times*, and with an astonishing 24,576-gon proved that $3.1415926 < \pi < 3.1415927$. A few hundred years after that, Arabian mathematician al-Khwārizmī estimated it to be 3.1416 in his book. This didn't set a world record, but here's a fun fact: The word *algorithm* is derived from his name, and the word *algebra* is derived from *al-jabr*, meaning restoration, and part of the title of his book. Al-Khwārizmī's work on estimating π is dwarfed by his contributions to algebra. The marriage between geometry and algebra is what makes calculus possible, and although many mathematicians contributed — particularly in China, India and the Middle East — al-Khwārizmī's work was especially important.

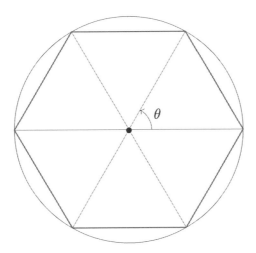

And of course, since the n-gon is regular, every interior angle has this same angle, and all n triangles that are produced are identical. So the area of the n-gon is equal to n times the area of one triangle. Thus, we turn our attention to determining the area of a single triangle. One such triangle looks like this:

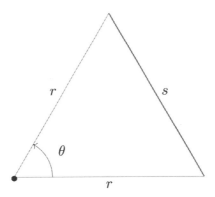

Dropping an altitude from the lower-left vertex to the length-s side gives

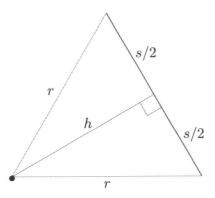

The area of the entire triangle is half of the base times the height: $\frac{1}{2}sh$. Therefore,

$$\text{Area of inscribed } n\text{-gon} = \frac{1}{2}shn.$$

Now here's the awesome observation: sn is equal to the perimeter of the n-gon, and so

$$\text{Area of inscribed } n\text{-gon} = \frac{1}{2} \cdot (\text{perimeter of } n\text{-gon}) \cdot h.$$

The number π is defined to be the circumference divided by diameter of any circle. Equivalently, it is the value such that a circle of radius r has circumference $2\pi r$. But why does a π also appear in the formula for the *area* of a circle? We are just one limit from seeing why!

Indeed, since the area of the inscribed n-gon is $\frac{1}{2} \cdot (\text{perimeter of } n\text{-gon}) \cdot h$, taking the limit as the number of sides, n, gets larger and larger, notice that the perimeter of the n-gon approaches the perimeter of the circle, which as we said is $2\pi r$ by the definition of π. And also in the limit, as n grows, notice that h approaches r. Thus, by taking the limit, the area of the inscribed n-gon approaches

$$\frac{1}{2} \cdot (2\pi r) \cdot r = \pi r^2,$$

which is indeed the area of a circle. And a similar calculation using *circumscribed* n-gons would also approach πr^2.

Now, we could be more precise by using trig functions to describe the area, and using the theory of limits of trig functions we could find what they are tending toward, but you have to be quite careful when doing so to avoid accidentally using something about π that was not already developed.

Also, it should be noted that while Archimedes did use this general idea, he never used a limit. Archimedes used a proof by contradiction: If the area of the circle were assumed to be some A_1 smaller than πr^2, then he found an inscribed n-gon with area larger than A_1, a contradiction. Conversely, if the area of the circle were some A_2 larger than πr^2, then he found a circumscribed n-gon with area smaller than A_2, another contradiction. This proves the answer while never invoking any notion of the infinite.[5] This was called the *method of exhaustion*.

Archimedes was ahead of his time—*way* ahead of his time. It took about eighteen hundred years for Galileo Galilei, Johannes Kepler and others to push further the mathematical study of nature. While Archimedes had focused on static systems, these men studied moving objects—whether rolling down an incline, swinging from a pendulum, or hurtling through space. But time and again their potential scientific advancements were diminished by a lack of understanding of the same curved geometry that had stumped mathematicians for millennia. Science needed savior, and on Christmas Day, a baby boy was born. And his mother called him Issac.

---

[5] Archimedes used similar ideas to prove formulas for the volume and surface area of a sphere. The formula for the volume of a cone had been done a century earlier by another Greek mathematician, Eudoxus of Cnidus, who some regard as the second best mathematician in all of antiquity — second, of course, to Archimedes.

#IntegralGoals

What do we want out of an integral? Here are three important goals:

- Integrals find antiderivatives — at least the indefinite integrals should.[6]

- Integrals relate to area under a curve — at least the definite integrals should.[7]

- We want them to be fairly powerful — we want to be able to integrate as many functions as possible.

The history of integrals is complicated (and the vocabulary has evolved which makes it additionally challenging), but one initial goal in their development was simply to find antiderivatives. This is a perfectly fine goal, but developing them in this way significantly limits their power and utility. For instance, consider the simple *step function*, $s : [0, 2] \to \mathbb{R}$, given by

$$s(x) = \begin{cases} 1 & \text{if } x \in [0, 1] \\ 2 & \text{if } x \in (1, 2] \end{cases}$$

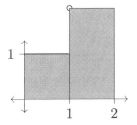

Finding the "area under the curve" of this function is something you could have done when you were 9, but this function is *not* a derivative. Recall that Darboux's theorem (Theorem 7.20) states that every derivative has the intermediate value property — however, $s(x)$ clearly does not satisfy the IVP (for example, nothing maps to 1.5). So if we demand that integrals be antiderivatives, then $\int_0^2 s(x)\, dx$ would not exist, even though we already know in our hearts that the answer should exist and should equal 3.

So in order to also use our area intuition, and in order to have it be fairly powerful (at least powerful enough to integrate $s(x)$), we should develop it from the area perspective — provided the antiderivative angle also works out. Newton noted this area interpretation, but it was Cauchy and Riemann who realized that the best approach was to define it *only* using area, and later prove that antiderivatives pop out from a (fundamental) theorem.

More on this later. For now, let's jump right into familiar pictures from calculus, which most calculus texts call *Riemann sums* (but which we will soon call *Darboux sums*).

---

[6]To see the limitations of this goal, look up Volterra's function.

[7]To see the limitations of this goal, stay tuned.

8.2   Simplistic Approach

Our approach will be similar to the method of exhaustion in which we approached
the area of a circle by finding upper bounds using circumscribed n-gons, and lower
bounds using inscribed n-gons, and then by taking a limit; if the upper bounds and
lower bounds are converging to the same thing, then we have an answer. Likewise,
we will find a sequence of upper bound areas and a sequence of lower bound areas,
and then take a limit, and if they are converging to the same thing then we have an
answer.

– Pictures from Calculus –

Back in calculus you were likely shown some pictures like those below.   For
example, below is the function $f(x) = x^2 + 1$.

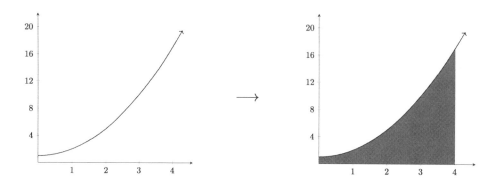

To find this shaded area for this simple increasing function, you learned that the
first thing to do is to find a simple rectangular upper bound, like so:

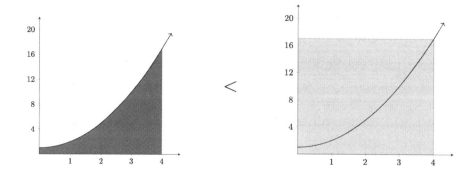

A lower bound was then found by putting a single (dark) rectangle inside the region under the curve, so that it lies entirely below the curve:

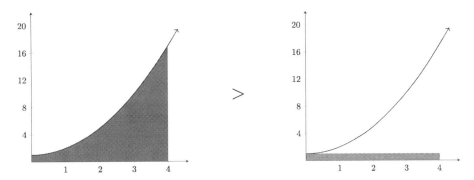

Then you continue with more rectangles. For example, here's a sequence of pictures where the dark and light boxes are overlaid.

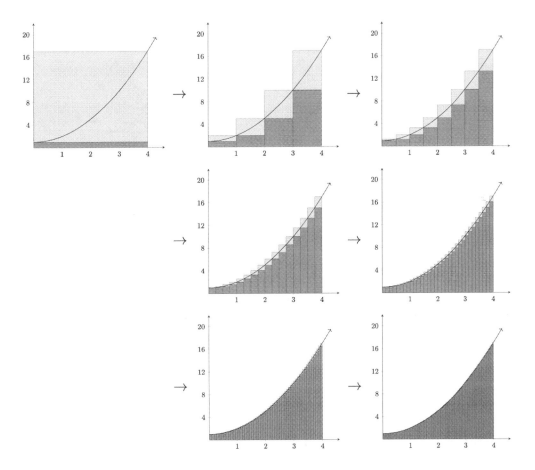

By using more and more rectangles you get a sequence of increasingly better upper and lower bounds. And as you discussed while taking calculus, it certainly is reasonable to wave your hands and say that for a function as simple as this, clearly the sequence of upper bounds and lower bounds are converging to a common number — the area under the curve that we are trying to find. Of course, to do this there is a limit taking place behind the scenes.

In a moment we will generalize this idea, but here is a quick example showing how the area of some curved regions were found by Fermat and others in the 1630s — decades before Newton and Leibniz's monumental papers.

Example 8.1. Recall a fact that you may have been asked to prove when you were first introduced to the principle of mathematical induction:

$$1^2 + 2^2 + 3^2 + \cdots + N^2 = \frac{1}{6}N(N+1)(2N+1). \qquad (\textbf{☻})$$

With this, we can find the area under the curve of $f(x) = x^2$ between $x = 0$ and $x = 1$. Indeed, since the function is increasing we clearly have

$$\text{Area}\left(\vcenter{\hbox{}} \right) \leq \text{Area}\left(\vcenter{\hbox{}} \right) \leq \text{Area}\left(\vcenter{\hbox{}} \right).$$

If the number of rectangles is N, and each rectangle has width $\frac{1}{N}$, then

- For the left area, the ith rectangle has area $\frac{1}{N} \cdot \left(\frac{i-1}{N}\right)^2 = \frac{(i-1)^3}{N^3}$.

- For the right area, the ith rectangle has area $\frac{1}{N} \cdot \left(\frac{i}{N}\right)^2 = \frac{i^3}{N^3}$.

And so, the above is equivalent to

$$\frac{1}{N^3} \cdot \sum_{i=1}^{N}(i-1)^3 \leq \text{Area}\left(\vcenter{\hbox{}} \right) \leq \frac{1}{N^3} \cdot \sum_{i=1}^{N} i^3.$$

And so by using (**☻**),

$$\frac{1}{N^3} \cdot \frac{1}{6}(N-1)N(2N-1). \leq \text{Area}\left(\vcenter{\hbox{}} \right) \leq \frac{1}{N^3} \cdot \frac{1}{6}N(N+1)(2N+1).$$

Since the (middle) area is constant and the above holds for any N, it in fact holds for the limit! And if you take the limit as $N \to \infty$, you will see that

$$\frac{1}{3} \leq \text{Area}\left(\vcenter{\hbox{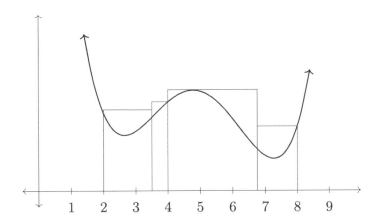}}\right) \leq \frac{1}{3}$$

So this area must indeed be 1/3, completing the argument.[8]   □

Let's now discuss the deficiencies of this simplistic approach, and generalizations thereof.

– More General Approach –

The above works particularly well when the function is continuous and increasing. We will return to the continuity issue in depth later, but two quick changes are:

- We no longer make all the rectangles the same width——we consider all possible widths.

- When the function is increasing, the right endpoint gives the largest function value and the left endpoint gives the smallest function value. If the function is not increasing, then we have to instead find the maximum (or supremum) point on our own, and the minimum (or infimum) point on our own.

For example, here is an upper bound (soon called an "upper sum") where the rectangles have different widths and one of the rectangles is touching the curve at a non-endpoint:

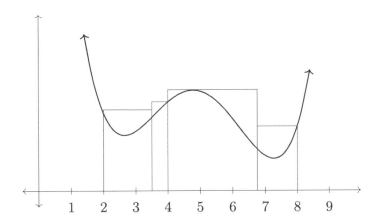

---

[8]I believe this is a startling result. The area under a parabola is something that the standard area formulas from geometry seem completely useless to help find. One could quickly deduce that the area should be less than 1/2, but with that weird curved upper side I don't see any reason that we should expect the answer to be nice at all, let alone one of the simplest possible (rational!) numbers less than 1/2. Math's pretty neat.

Likewise, here is a lower bound (soon called a "lower sum"):

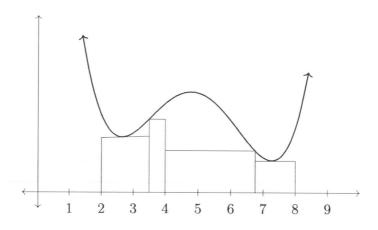

Types of Integrals

It's worthwhile to take a moment to point out that there are many ways to define an integral. Here are versions of the integral that mathematicians have defined:

- Riemann integral

- Generalized Riemann integral

- Riemann-Stieltjes integral

- Lebesgue integral

- Lebesgue-Stieltjes integral

- Daniell integral

- Haar integral

- Ito integral

- Statonovich integral

- Young integral

   ...and *all of these are different!*

We will study the most classic and fundamental integral, which is the *Darboux integral*. Note: This integral is often mistakenly called the *Riemann integral*.[9] Darboux and Riemann each defined an integral and one can (painstakingly) prove that their two integrals are in fact equivalent. Nevertheless, their fundamental definitions are different and we will follow Darboux's approach, which is far simpler to use (and likely the approach you were taught in calculus).

---

[9] In fact, I did so in this text's first edition!

8.3   The Darboux Integral

First we collect the fundamental definitions to allow us to talk about Darboux sums.

– Three Preliminary Definitions/Notation –

First up, the rectangles can have arbitrary widths. To define these widths, we need only define the set of points which are the left/right endpoints of the rectangles. (Note that this is much easier than trying to define the specific widths of each rectangle while also ensuring those widths add up to $b - a$.)

Definition.

Definition 8.2.   A *partition* of $[a, b]$ is a finite set

$$P = \{x_0, x_1, x_2, \ldots, x_n\}$$

such that $a = x_0$, $b = x_n$, and $x_0 < x_1 < x_2 < \cdots < x_n$.

For example, one partition of $[2, 5]$ is $\{2, 2.7, 3.1, \pi, 4.2, 5\}$. There are, of course, infinitely many others.

Next, we need to define the heights of the rectangles. We want the rectangles to be either big enough to completely encompass the region under the curve, or are small enough to be completely encompassed by the region. These two possibilities lead to two different notations that we need to define.

Notation.

Notation 8.3.   Given a partition $\{x_0, \ldots, x_n\}$ of $[a, b]$, consider a subinterval $[x_{i-1}, x_i]$. Denote

- $m_i := \inf\{f(x) : x \in [x_{i-1}, x_i]\}$;

- $M_i := \sup\{f(x) : x \in [x_{i-1}, x_i]\}$.

For example, the following diagram is of a partition of $[x_0, x_4]$ where we labeled $M_3 = \sup\{f(x) : x \in [x_2, x_3]\}$, the height of the third (upper) rectangle.

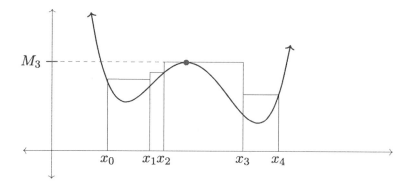

And here is m_3, the height of the third (lower) rectangle:

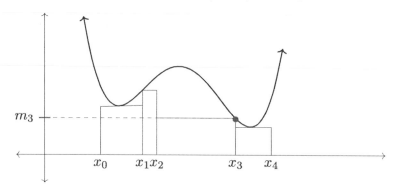

Now that we have a way to talk about the widths and heights of both the big rectangles and the small ones, we can write down what we are really interested in: a formula for the upper bound area and the lower bound area. These are called "upper sums" and "lower sums," respectively.

Definition.

Definition 8.4. Consider a function $f : [a, b] \to \mathbb{R}$, and consider a partition $P = \{x_0, x_1, \ldots, x_n\}$ of $[a, b]$. Define

- the *upper sum* as

$$U(f, P) := \sum_{i=1}^{n} M_i \cdot (x_i - x_{i-1}), \text{ and}$$

- the *lower sum* as

$$L(f, P) := \sum_{i=1}^{n} m_i \cdot (x_i - x_{i-1}),$$

where M_i and m_i were defined in Notation 8.3.

– Refinements –

Suppose you have a partition of $[a, b]$ that contains 10 points. This would give 9 rectangles, and the upper and lower sums may be ok estimates of the integral. But if instead you had 100 points — giving 99 rectangles — that would presumably give a much better estimate. This isn't necessarily the case, though, since maybe all 10 points in the first partition are nicely spread out, while in the second partition 99 of the 100 points are clumped together on one end. Those 99 points would give a fantastic estimate of the area on that one end, but then there's just one huge rectangle to estimate the rest of the area, which will likely have a huge error.

So you see, having more points is not everything, we also want them to be distributed nicely. We begin with an intermediate step, called a *refinement*. The idea here is that if you want to ensure that 100 points are better than 10, construct your 100 points by first starting with the 10 points from the first partition and then adding 90 *more* points. If your original 10 points are still there, then you guarantee that you aren't making things any worse, and are usually making the estimates better and better.

Observe that if P is a partition and Q is another partition with all of P's points plus others, then $P \subseteq Q$. Thus we can use subset notation to define a refinement.

Definition.

Definition 8.5. Consider a partition $P = \{x_0, \ldots, x_n\}$ of $[a, b]$. A partition Q of $[a, b]$ is called a *refinement* of P if $P \subseteq Q$.

We begin with an intuitive result — a refinement really does refine; that is, a refinement makes upper sums smaller and lower sums bigger, or at worst leaves them unchanged.

Proposition.

Proposition 8.6 (*Refinements refine*). Consider a function $f : [a, b] \to \mathbb{R}$ and a partition $P = \{x_0, \ldots, x_n\}$ of $[a, b]$. If Q is a refinement of P, then

$$U(f, P) \geq U(f, Q) \quad \text{and} \quad L(f, P) \leq L(f, Q).$$

Proof Idea. The idea is fairly intuitive when you consider the following picture (of an upper sum), showing an interval from the partition P being broken into two intervals by the refinement Q.

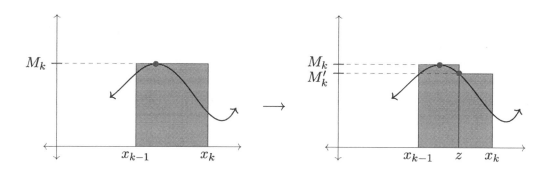

From the picture you can see that adding an additional point did indeed decrease the area of the rectangle(s).

Proof. Consider the upper sum $U(f, P)$ and pick any $z \in Q \setminus P$; note that $z \in [x_{k-1}, x_k]$ for precisely one k. Observe next that in the summation for the upper sum $U(f, P \cup \{z\})$, the only thing that changes is that the term $M_k \cdot (x_k - x_{k-1})$ is replaced by two terms, representing the sum of the areas of the two next rectangles.

The function had a supremum of M_k on $[x_{k-1}, x_k]$, and so f will now have a supremum of M_k on at least one of the intervals $[x_{k-1}, z]$ and $[z, x_k]$, and on the other interval will have a supremum of some M_k'. Suppose, without loss of generality, that the first rectangle is the one of height M_k and the second is the one of height M_k' (like pictured in the Proof Idea). Observe that since M_k is the supremum of f on $[x_{k-1}, x_k]$, and M_k' is the supremum of f on $[z, x_k] \subseteq [x_{k-1}, x_k]$, by Exercise 8.2 this means that $M_k' \leq M_k$.

The two new terms in the upper sum $U(f, P \cup \{z\})$ are then

$$M_k \cdot (z - x_{k-1}) + M_k' \cdot (x_k - z).$$

Thus,

$$
\begin{aligned}
M_k \cdot (x_k - x_{k-1}) &= M_k \cdot (x_k - z + z - x_{k-1}) \\
&= M_k \cdot (x_k - z) + M_k \cdot (z - x_{k-1}) \\
&\geq M_k' \cdot (x_k - z) + M_k \cdot (z - x_{k-1}) \\
&= M_k \cdot (z - x_{k-1}) + M_k' \cdot (x_k - z).
\end{aligned}
$$

And hence refining the partition by adding the point z did indeed decrease the upper sum. Since partitions are finite sets, Q has finitely many more points than P. Doing the above $|Q \setminus P|$ times (or by applying induction to the above) shows that $U(f, P) \geq U(f, Q)$.

The lower sum result is very similar and is left as an exercise. $\square$

Next we have a two-part proposition that ensures that another basic, intuitive result holds: Any upper sum is larger than any lower sum.

Proposition.

Proposition 8.7 (*Lower sums ≤ Upper sums*). Let $f : [a, b] \to \mathbb{R}$.

(a) If P is a partition of $[a, b]$, then $L(f, P) \leq U(f, P)$.

(b) If P_1 and P_2 are any partitions of $[a, b]$, then

$$L(f, P_1) \leq U(f, P_2).$$

Part (a) is of course a special case of part (b), but (a) will be particularly useful so it is helpful to state it explicitly. Also, we will use (a) to prove (b).

Proof. Part (a) is immediate: Clearly $m_k \leq M_k$ for each k since the former is the infimum of f over $[x_{k-1}, x_k]$ and the latter is the supremum of f over the exact same interval. And so clearly

$$L(f, P) = \sum_{i=1}^{n} m_i \cdot (x_i - x_{i-1}) \leq \sum_{i=1}^{n} M_i \cdot (x_i - x_{i-1}) = U(f, P).$$

Now we show (b). The basic idea is this: To show that one thing is less than another, find something that is in-between them. The true integral would be ideal, but we don't know yet what an integral even is. Instead, let's use a sufficiently refined partition—if it is finer than both P_1 and P_2, then it should be closer to the value of the "integral" than either one is, which would be good enough. In particular, if we can find a partition Q that refines *both* P_1 and P_2, then Q's upper and lower sums will be squished between $L(f, P_1)$ and $U(f, P_2)$.

Let $Q = P_1 \cup P_2$. Since P_1 and P_2 are partitions, Q is as well. Moreover, Q is a refinement of both P_1 and of P_2! Therefore by Proposition 8.6,

$$L(f, P_1) \leq L(f, Q) \quad \text{and} \quad U(f, Q) \leq U(f, P_2).$$

Moreover, by (a) we have that

$$L(f, Q) \leq U(f, Q).$$

Combining these,

$$L(f, P_1) \leq L(f, Q) \leq U(f, Q) \leq U(f, P_2),$$

giving $L(f, P_1) \leq U(f, P_2)$, as desired. $\qquad \square$

8.4   Integrability

As you probably already anticipated, the idea is to take better and better partitions to get smaller and smaller upper sums, and larger and larger lower sums. If the upper sums and the lower sums are converging to the same thing, then this common value is what the *integral* is equal to.

But which partitions will make things better and better? If we insist on having equally-spaced rectangles, we will see later that issues can arise. Refinements are good, but if we simply insist on having a sequence of partitions where each is a refinement of the previous, how do you know you're refining everywhere, rather than refining one region better and better while ignoring another?

The solution is simple: Don't try to be clever; instead, consider *all* partitions. Indeed, that's what the following definition does.

Definition.

Definition 8.8. Let $f : [a, b] \to \mathbb{R}$ be a bounded function and let $\mathcal{P}$ be the collection of *all* partitions of $[a, b]$. The *upper integral* of f is defined to be

$$U(f) = \inf\{U(f, P) : P \in \mathcal{P}\}.$$

The *lower integral* of f is defined to be

$$L(f) = \sup\{L(f, P) : P \in \mathcal{P}\}.$$

Here's a quick result:

Lemma.

Lemma 8.9 $(m(b - a) \le L(f) \le U(f) \le M(b - a))$. Let $f : [a, b] \to \mathbb{R}$ be a bounded function; say, $m \le f(x) \le M$ for all $x \in [a, b]$. Then we have

$$m(b - a) \le L(f) \le U(f) \le M(b - a).$$

Proof Sketch. Let's assume for now that the M and m are optimal, meaning that M is the least upper bound on f over this interval, and m is the greatest lower bound on f over this interval.

Remember that $U(f)$ is the infimum of upper sums over *all* partitions of $[a, b]$. So, if you pick any particular partition of $[a, b]$, call it P_0, then certainly $U(f) \le U(f, P_0)$. Likewise, since $L(f)$ is the supremum of lower sums over *all* partitions of $[a, b]$, we also have $L(f, P_0) \le L(f)$. And since $L(f) \le U(f)$, you can already start to see a string of inequalities that looks a little like the one we are trying to prove:

$$L(f, P_0) \leq L(f) \leq U(f) \leq U(f, P_0).$$

What's left? We must find a P_0 that gives the correct bounds.

What is the simplest partition P_0 of $[a, b]$ that we could apply? Remember that a partition of $[a, b]$ is a set of the form $\{x_0, x_1, \ldots, x_n\}$ where $x_0 = a$ and $x_n = b$. Since a and b are the only numbers that *must* be in the partition, the simplest partition would be $P_0 = \{a, b\}$! This is the partition that corresponds to a single rectangle, and since upper and lower sums correspond to area,

$$L(f, P_0) \leq L(f) \leq U(f) \leq U(f, P_0).$$

turns into

$$\text{Area} \left(\begin{array}{c} \includegraphics \end{array} \right) \leq L(f) \leq U(f) \leq \text{Area} \left(\begin{array}{c} \includegraphics \end{array} \right)$$

which gives

$$m(b - a) \leq L(f) \leq U(f) \leq M(b - a),$$

Proof. The inequality $L(f) \leq U(f)$ follows immediately from Proposition 8.7 part (b), and Exercise 8.3. Now we show the bounds on $L(f)$ and $U(f)$, which are a sup/inf over all partitions. The simplest partition possible is the 2-point partition $P_0 = \{a, b\}$ containing only the endpoints (that is, $n = 1$, $x_0 = a$, and $x_1 = b$); this in turn gives a single rectangle in the upper or lower Darboux sum. Thus,

$$L(f) = \sup\{L(f, P) : P \in \mathcal{P}\} \qquad U(f) = \inf\{U(f, P) : P \in \mathcal{P}\}$$
$$\geq L(f, P_0) \qquad\qquad\qquad \leq U(f, P_0)$$
$$= \sum_{i=1}^{n} m_i \cdot (x_i - x_{i-1}) \qquad = \sum_{i=1}^{n} M_i \cdot (x_i - x_{i-1})$$
$$= \sum_{i=1}^{1} m_i \cdot (b - a) \qquad\quad = \sum_{i=1}^{1} M_i \cdot (b - a)$$
$$\geq m \cdot (b - a). \qquad\qquad\quad \leq M \cdot (b - a).$$

Collectively,

$$m(b - a) \leq L(f) \leq U(f) \leq M(b - a).$$

$\square$

We are now ready for a biggie: the definition of the (Darboux) integral.

> **Definition.**
>
> **Definition 8.10.** A bounded function $f : [a, b] \to \mathbb{R}$ is *integrable* if $L(f) = U(f)$. When this happens, we denote $\int_a^b f$ and $\int_a^b f(x)\,dx$ to be this common value. That is,
> $$\int_a^b f(x)\,dx = L(f) = U(f).$$

Note: Officially this is the definition of *Darboux integrable* (which, once again, is equivalent to *Riemann integrable*), but in this text we will just say "integrable." (The formal distinction is due to the fact that there are approximately a bajillion other integral definitions that mathematicians have invented.)

Let's start figuring out which functions are integrable and which ones are not. Let's start as simple as we can: a constant function. We would sure hope that if our "integrable" definition is worth anything at all that this would be integrable.

Example 8.11. Suppose $f : [a, b] \to \mathbb{R}$ is a constant function; say, $f(x) = c$. Then f is integrable and $\int_a^b f(x)\,dx = c(b - a)$.

Proof. To see this, note that if we let $m = c$ and $M = c$, then we have $m \leq f(x) \leq M$ for all $x \in [a, b]$. So by Lemma 8.9,

$$c(b - a) \leq L(f) \leq U(f) \leq c(b - a).$$

Therefore

$$L(f) = U(f) = c(b - a).$$

By definition, we have shown that f is integrable, and moreover that $\int_a^b f(x)\,dx = c(b - a)$, as desired. $\square$

Good, we can integrate a constant function, which is smooth and continuous *everywhere*. What about a function on the other end of the spectrum? Dirichlet's function is continuous *nowhere*, so let's give that one a go.

Example 8.12. Let $f : [0, 1] \to \mathbb{R}$ be the Dirichlet function

$$f(x) = \begin{cases} 1 & \text{if } x \in \mathbb{Q} \\ 0 & \text{if } x \notin \mathbb{Q}. \end{cases}$$

Then f is *not* integrable.

Proof Idea. Given any partition of P of $[0, 1]$, all the upper sum rectangles will have height 1 (because every subinterval $[x_{i-1}, x_i]$ contains a rational) and all lower sum rectangles will have height 0 (because every subinterval $[x_{i-1}, x_i]$ contains an itrational). Thus $U(f, P)$ is the area of a 1×1 square, while $L(f, P)$ is the area of a 1×0 "rectangle."

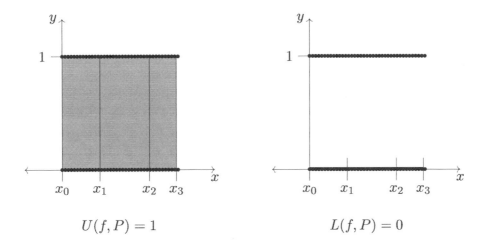

$$U(f, P) = 1 \qquad\qquad\qquad L(f, P) = 0$$

This will imply $U(f) = 1 \neq 0 = L(f)$, and so f is not integrable.

Proof. To see this, let P be an arbitrary partition of $[0, 1]$. By the density of both the rationals $\mathbb{Q}$ and of the irrationals $\mathbb{R} \setminus \mathbb{Q}$, any subinterval $[x_{i-1}, x_i]$ contains both a rational and an irrational. Therefore, $m_i = 0$ and $M_i = 1$ for all i. Thus,

$$L(f, P) = \sum_{i=1}^{n} m_i \cdot (x_i - x_{i-1}) = \sum_{i=1}^{n} 0 \cdot (x_i - x_{i-1}) = 0$$

and

$$U(f, P) = \sum_{i=1}^{n} M_i \cdot (x_i - x_{i-1})$$
$$= \sum_{i=1}^{n} 1 \cdot (x_i - x_{i-1})$$
$$= (x_1 - x_0) + (x_2 - x_1) + (x_3 - x_2) + \cdots + (x_n - x_{n-1})$$

which, by canceling,

$$= -x_0 + x_n$$

and since $\{x_0, x_1, \ldots, x_n\}$ is a partition of $[0, 1]$, $x_0 = 0$ and $x_n = 1$, the above

$$= -0 + 1$$
$$= 1.$$

Since $L(f, P) = 0$ and $U(f, P) = 1$ for *every* partition P,

$$L(f) = \sup\{L(f, P) : P \in \mathcal{P}\} = \sup\{0 : P \in \mathcal{P}\} = 0$$

and

$$U(f) = \inf\{U(f, P) : P \in \mathcal{P}\} = \inf\{1 : P \in \mathcal{P}\} = 1.$$

We've shown that $L(f) \neq U(f)$, implying that f is not integrable.   □

So we have on one end of the spectrum the constant function, which is integrable, and on the other end we have Dirichlet's function, which is not integrable. And there is a *lot* of ground between these two. One major goal of the next couple sections is to investigate how far from the constant function we can travel and still have an integrable function. Is every polynomial integrable? How about an arbitrary continuous functions? Functions which are discontinuous at one point? At a finite number of points? On a dense set like with Thomae's function? Venturing much beyond Thomae's and we will have reached the other end of the spectrum, where we just showed that Dirichlet's function is *not* integrable. So where is the tipping point?

Our goal is to figure that out. We won't want to keep using the clunky definition of integrability to prove these things, though, and so we need a tool—a theorem which will help us check for integrability. That's coming up soon, but before we end this section, below is one last quick proposition and a discussion on boundedness.

Proposition.

Proposition 8.13 ($f \geq 0 \Rightarrow \int f \geq 0$). Assume that a bounded function $f : [a, b] \to \mathbb{R}$ is integrable. If $f(x) \geq 0$ for all $x \in [a, b]$, then

$$\int_a^b f(x)\, dx \geq 0.$$

Proof. By assumption, $0 \leq f(x)$ for all $x \in [a, b]$. Therefore by (the lower bound portion of) Proposition 8.9, $0 \cdot (b - a) \leq L(f)$. I.e.,

$$0 \leq L(f). \qquad (\text{➤})$$

Finally, note that since f was assumed to be integrable, by Definition 8.10 we have that that $L(f) = \int_a^b f(x)\, dx$. Plugging this into (➤),

$$0 \leq \int_a^b f(x)\, dx,$$

as desired.   □

Unbounded Functions

Before proving some theorems about integrability I want to quickly address one subtle point in the definition of *integrable*. In the definition we immediately declared that the function f be bounded. Why is this necessary? If we didn't require boundedness, could an unbounded function be integrable? As you probably remember from calculus, "integrable" essentially means "the integral exists and is finite." This is the same goal in analysis.

So can an unbounded function satisfy this definition? The answer is no. Remember, to be integrable means that $L(f) = U(f)$. Consider for a moment the case where $f : [a, b] \to \mathbb{R}$ is not bounded above. Each upper sum $U(f, P)$ is a finite sum of the form

$$U(f, P) = \sum_{i=1}^{n} M_i \cdot (x_i - x_{i-1}),$$

and each of these M_i is the supremum of f over an interval $[x_{i-1}, x_i]$. But there are only finitely many of these intervals and collectively they form all of $[a, b]$, so if f is unbounded on $[a, b]$, then it is also unbounded on at least one of these intervals. And if it is unbounded on $[x_{k-1}, x_k]$, then $M_k = \infty$. This would in turn mean that

$$U(f, P) = \sum_{i=1}^{n} M_i \cdot (x_i - x_{i-1})$$

has at least one ∞ in this sum (and certainly no $-\infty$'s), so this sum will be ∞.

Now, given such an f, $L(f)$ may or may not be infinite. But either way, we get a "not integrable" result. If $L(f) < \infty$, then $L(f) \neq U(f)$ and so it's not integrable by our old definition. If $L(f) = U(f) = \infty$, then the only reasonable answer would be to say the integral equals infinity, which again is undesirable.

So the standard operating procedure is to avoid it at the outset by demanding in the definition of "integrable" that f be bounded.

8.5 Integrability Criteria

First, we recall an important — and highly useful — result from Chapter 1.

Recall.

Theorem 1.24 (*Suprema analytically*). Let $A \subseteq \mathbb{R}$. Then $\sup(A) = \alpha$ if and only if

    (i) α is an upper bound of A, and

    (ii) Given any $\varepsilon > 0$, $\alpha - \varepsilon$ is *not* an upper bound of A. That is, there is some $x \in A$ for which $x > \alpha - \varepsilon$.

The above theorem also included a similar statement about infimums, which here we omit.

This theorem was quite useful in that it could turn a problem involving an infinite set and infinitely many upper bounds into one in which, for $\varepsilon > 0$, all you have to do is find a single element of A that is within ε of α. We would like to find a similar property about integrability.

By definition, f is integrable when $L(f) = U(f)$. I.e.,

$$\sup\{L(f, P) : P \in \mathcal{P}\} = \inf\{U(f, P) : P \in \mathcal{P}\}.$$

And for a supremum of one set to equal an infimum of another set, elements from the first set must be arbitrarily close to elements of the second set. Thus, it makes sense that an analytic condition for this is that f is integrable if and only if for every $\varepsilon > 0$, there exists some partition P such that $U(f, P) - L(f, P) < \varepsilon$. Indeed, this is the content of the following theorem.

> ### Theorem.
>
> **Theorem 8.14** (*Integrals analytically*). Let $f : [a, b] \to \mathbb{R}$ be bounded. Then f is integrable if and only if, for all $\varepsilon > 0$ there exists a partition P_ε of $[a, b]$ where
>
> $$U(f, P_\varepsilon) - L(f, P_\varepsilon) < \varepsilon.$$

Proof. First we prove the backwards direction. Let $\varepsilon > 0$ and choose P_ε so that $U(f, P_\varepsilon) - L(f, P_\varepsilon) < \varepsilon$. Recall that $L(f, P_\varepsilon) \leq L(f)$, since $L(f)$ is a supremum of a set containing $L(f, P_\varepsilon)$; and likewise that $U(f) \leq U(f, P_\varepsilon)$. Therefore,

$$|U(f) - L(f)| = U(f) - L(f) \leq U(f, P_\varepsilon) - L(f, P_\varepsilon) < \varepsilon.$$

But since ε was arbitrary, this means that $U(f) - L(f) = 0$, and hence $L(f) = U(f) = \int_a^b f(x)\, dx$, meaning that f is integrable.

Now we prove the forward direction. Assume that f is integrable; that is, assume $L(f) = U(f)$. Let $\varepsilon > 0$. We wish to find some P_ε such that $U(f, P_\varepsilon) - L(f, P_\varepsilon) < \varepsilon$.

Since $L(f) = \int_a^b f(x)\, dx$, and $L(f) = \sup\{L(f, P) : P \in \mathcal{P}\}$, by the suprema analytically theorem (Theorem 1.24) there must be some P_1 for which $L(f, P_1) > \int_a^b f(x)\, dx - \frac{\varepsilon}{2}$. Likewise, since $U(f) = \int_a^b f(x)\, dx$, and $U(f) = \inf\{U(f, P) : P \in \mathcal{P}\}$, there must be some P_2 for which $U(f, P_2) < \int_a^b f(x)\, dx + \frac{\varepsilon}{2}$. Since each is within $\varepsilon/2$ of $\int_a^b f(x)\, dx$, they must be within ε of each other:

$$U(f, P_2) - L(f, P_1) < \varepsilon.$$

We are almost there, but our goal is to find a *single* partition P_ε for which the difference is less than ε — a refinement should do the trick! Let $P_\varepsilon = P_1 \cup P_2$. Then,

$$L(f, P_\varepsilon) \geq L(f, P_1) > \int_a^b f(x)\,dx - \frac{\varepsilon}{2},$$

and

$$U(f, P_\varepsilon) \leq U(f, P_2) < \int_a^b f(x)\,dx + \frac{\varepsilon}{2},$$

Thus the upper and lower sums with this new partition are still both within $\varepsilon/2$ of the integral, and hence are within ε of each other:

$$U(f, P_\varepsilon) - L(f, P_\varepsilon) < \varepsilon.$$

$\square$

Corollary.

Corollary 8.15 (*f integrable* $\Rightarrow \lim[U(f, P_n) - L(f, P_n)] = 0$). If $f : [a, b] \to \mathbb{R}$ is integrable, then there exists a sequence P_n of partitions of $[a, b]$ for which

$$\lim_{n \to \infty} [U(f, P_n) - L(f, P_n)] = 0.$$

Proof. This is Exercise 8.8. Hint: Remember that corollaries follow quickly from theorems, so you should probably try to use the previous theorem to prove this... $\square$

8.6 Integrability of Continuous Functions

Is every continuous function integrable? As we have seen, being continuous at a point, or even on a dense collection of points, can still give counterintuitive results. If you know a function is continuous *everywhere*, though, then the function is much more tame. By the integrals analytically theorem (Theorem 8.14), a function f being integrable on $[a, b]$ means that for each $\varepsilon > 0$ there exists a partition $P_\varepsilon = \{x_0, x_1, \ldots, x_n\}$ of $[a, b]$ for which

$$U(f, P_\varepsilon) - L(f, P_\varepsilon) < \varepsilon.$$

Substituting the definitions of $U(f, P_\varepsilon)$ and $L(f, P_\varepsilon)$ into the above gives

$$\sum_{i=1}^n M_i \cdot (x_i - x_{i-1}) - \sum_{i=1}^n m_i \cdot (x_i - x_{i-1}) < \varepsilon$$

$$\sum_{i=1}^n (M_i - m_i) \cdot (x_i - x_{i-1}) < \varepsilon.$$

We want this to hold — so what sorts of functions give this conclusion? Note that all that matters is controlling $M_i - m_i$, since if we can, say, somehow force

$$M_i - m_i < \frac{\varepsilon}{b-a}$$

for all i, then

$$\sum_{i=1}^{n}(M_i - m_i) \cdot (x_i - x_{i-1}) < \sum_{i=1}^{n} \frac{\varepsilon}{b-a} \cdot (x_i - x_{i-1})$$

$$= \frac{\varepsilon}{b-a} \cdot \sum_{i=1}^{n}(x_i - x_{i-1})$$

$$= \frac{\varepsilon}{b-a} \cdot \Big((x_1 - x_0) + (x_2 - x_1) + (x_3 - x_2) + \cdots + (x_n - x_{n-1})\Big)$$

Note that this sum contains an x_1 and a $-x_1$, so these cancel. It also contains an x_2 and a $-x_2$, and so on, up to the x_{n-1}'s. All that's left is a $-x_0$ and an x_n:

$$= \frac{\varepsilon}{b-a} \cdot (-x_0 + x_n)$$

And by definition of a partition, $x_0 = a$ and $x_n = b$. So the above

$$= \frac{\varepsilon}{b-a} \cdot (-a + b)$$

$$= \frac{\varepsilon}{b-a} \cdot (b - a)$$

$$= \varepsilon.$$

So if we can find a partition for which we can guarantee $M_i - m_i$ is small, we can deduce integrability. As it turns out, continuity is a near-perfect condition to guarantee this.

Why? Consider the partition P_n where each rectangle has width $(b-a)/n$. If n is large, each $x_i - x_{i-1}$ is small. So what we want to say is that whenever x and y are close, that $f(x)$ and $f(y)$ are also close — this is precisely the definition of uniform continuity! And remember that if $f : [a,b] \to \mathbb{R}$ is continuous, then in particular it is a continuous function on a compact set and hence (by Proposition 6.40) is also uniformly continuous.

This is the proof idea behind the following theorem.

> **Theorem.**

Theorem 8.16 (*Continuous $\Rightarrow$ integrable*). If $f : [a, b] \to \mathbb{R}$ is continuous, then f is integrable.

Proof. Let $\varepsilon > 0$ and note that, by the extreme value theorem (Theorem 6.32), f is bounded. Thus by the integrals analytically theorem (Theorem 8.14) it suffices to find a partition P_ε where $U(f, P_\varepsilon) - L(f, P_\varepsilon) < \varepsilon$. This is our goal.

Since f is continuous on a compact set, f is uniformly continuous (by Proposition 6.40). By this and the fact that

$$\frac{\varepsilon}{b - a} > 0,$$

for this "ε" there exists some $\delta > 0$ for which $|x - y| < \delta$ implies that

$$|f(x) - f(y)| < \frac{\varepsilon}{b - a}.$$

Choose $n \in \mathbb{N}$ so that $\frac{b-a}{n} < \delta$, and let $P_\varepsilon = \{x_0, x_1, \ldots, x_n\}$ where $x_i = a + i \cdot \frac{b-a}{n}$. Note that P_ε is a partition of $[a, b]$.

We now claim, for each i, that $M_i - m_i < \frac{\varepsilon}{b-a}$. To see this, recall that since f is continuous on $[x_{i-1}, x_i]$, by the extreme value theorem we know that f attains its supremum M_i and its infimum m_i in this interval. Say, $f(x_{\min}) = m_i$ and $f(x_{\max}) = M_i$. Since $x_{\min}, x_{\max} \in [x_{i-1}, x_i]$, we know that $|x_{\min} - x_{\max}| < \delta$, and hence by the uniform continuity of f we deduce that

$$|f(x_{\max}) - f(x_{\min})| < \frac{\varepsilon}{b - a},$$

completing the claim.

Now, with this partition we have that

$$U(f, P_\varepsilon) - L(f, P_\varepsilon) = \sum_{i=1}^{n} M_i \cdot (x_i - x_{i-1}) - \sum_{i=1}^{n} m_i \cdot (x_i - x_{i-1})$$

$$= \sum_{i=1}^{n} (M_i - m_i) \cdot (x_i - x_{i-1})$$

$$< \sum_{i=1}^{n} \frac{\varepsilon}{b - a} \cdot (x_i - x_{i-1})$$

$$= \frac{\varepsilon}{b - a} \cdot \sum_{i=1}^{n} (x_i - x_{i-1})$$

$$= \frac{\varepsilon}{b - a} \cdot \Big((x_1 - x_0) + (x_2 - x_1) + (x_3 - x_2) + \cdots + (x_n - x_{n-1}) \Big)$$

Note that this sum contains an x_1 and a $-x_1$, so these cancel. It also contains an x_2 and a $-x_2$, and so on, up to the x_{n-1}'s. All that's left is a $-x_0$ and an x_n. And by definition of a partition, $x_0 = a$ and $x_n = b$:

$$
\begin{aligned}
&= \frac{\varepsilon}{b-a} \cdot (-x_0 + x_n) \\
&= \frac{\varepsilon}{b-a} \cdot (-a + b) \\
&= \frac{\varepsilon}{b-a} \cdot (b - a) \\
&= \varepsilon.
\end{aligned}
$$

$\square$

8.7   Integrability of Discontinuous Functions

So continuous functions are integrable. Are any discontinuous functions integrable? We have so far only tested one discontinuous function: In Example 8.12 we showed that the *highly* discontinuous Dirichlet function was *not* integrable. Let's go to the other extreme—let's take an extremely simple continuous function and make it discontinuous at just a single point and see if the result is integrable.

Example 8.17. Let $f_1 : [0, 2] \to \mathbb{R}$ be defined by

$$
f_1(x) = \begin{cases} 1 & \text{if } x \neq 1 \\ 0 & \text{if } x = 1. \end{cases}
$$

Then f_1 is integrable.

Proof. First, here is f_1's graph:

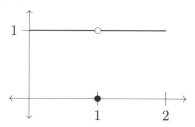

We will not only prove that f_1 is integrable, but also find what its integral equals. If it weren't for that one point, it would be clear that the "area under the curve" should equal 2. As it turns out, changing one point doesn't change this answer.

Indeed, we will show that $\int_0^2 f_1(x)\,dx = 2$, and will do so by using the integrals analytically theorem (Theorem 8.14); to this end, note first that f_1 is clearly bounded. Now let $\varepsilon > 0$. We will find a particular partition P_ε for which $U(f_1, P_\varepsilon) - L(f_1, P_\varepsilon) < \varepsilon$. Here's the idea: The upper sums will always equal 2, and to get the lower sums within $2 - \varepsilon$ we will use 3 rectangles: Two big rectangles will get close to $x = 1$ on each side and will combine to have area $2 - \varepsilon$. Then one really skinny "rectangle" (whose height will be 0), will be around $x = 1$; its height is 0 because for the lower sum the height m_i is the infimum of the function values, and since $f_1(1) = 0$, that infimum will be 0.

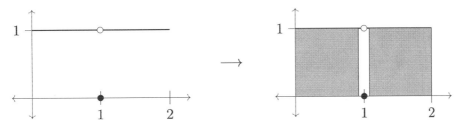

More precisely, consider the partition $P_\varepsilon = \{0, 1 - \varepsilon/4, 1 + \varepsilon/4, 2\}$. Then the sups/infs for our sums are:

	$i = 1$	$i = 2$	$i = 3$
M_i	1	1	1
m_i	1	0	1

So,

$$
\begin{aligned}
U(f_1, P_\varepsilon) &= \sum_{i=1}^{n} M_i \cdot (x_i - x_{i-1}) \\
&= M_1 \cdot (x_1 - x_0) + M_2 \cdot (x_2 - x_1) + M_3 \cdot (x_3 - x_2) \\
&= 1 \cdot [(1 - \varepsilon/4) - 0] + 1 \cdot [(1 + \varepsilon/4) - (1 - \varepsilon/4)] + 1 \cdot [2 - (1 + \varepsilon/4)] \\
&= 1 - \frac{\varepsilon}{4} + 1 + \frac{\varepsilon}{4} - 1 + \frac{\varepsilon}{4} + 2 - 1 - \frac{\varepsilon}{4} \\
&= 2.
\end{aligned}
$$

And,

$$
\begin{aligned}
L(f_1, P_\varepsilon) &= \sum_{i=1}^{n} m_i \cdot (x_i - x_{i-1}) \\
&= m_1 \cdot (x_1 - x_0) + m_2 \cdot (x_2 - x_1) + m_3 \cdot (x_3 - x_2) \\
&= 1 \cdot [(1 - \varepsilon/4) - 0] + 0 \cdot [(1 + \varepsilon/4) - (1 - \varepsilon/4)] + 1 \cdot [2 - (1 + \varepsilon/4)] \\
&= 1 - \frac{\varepsilon}{4} + 2 - 1 - \frac{\varepsilon}{4} \\
&= 2 - \frac{\varepsilon}{2}.
\end{aligned}
$$

Therefore $U(f_1, P_\varepsilon) - L(f_1, P_\varepsilon) = 2 - (2 - \varepsilon/2) = \frac{\varepsilon}{2} < \varepsilon$. And so by the integrals analytically theorem (Theorem 8.14), f_1 is indeed integrable. And moreover, since ε is arbitrary, we can also see that $\int_0^2 f_1(x)\, dx = 2$. $\qquad \square$

A corollary is the direct result of a proposition or a theorem. A *porism* is the direct result of a *proof*.[10] For instance, since the above proof could be done almost

---

[10]Fun fact: Euclid used porisms in *Elements*.

exactly the same regardless of where the point of discontinuity is and regardless of the length of the interval, we have the following porism.

> **Porism.**
>
> **Porism 8.18** (f_1 *integrable for any discontinuity c*). Let $c \in [a, b]$ and suppose $f_1 : [a, b] \to \mathbb{R}$ is given by
>
> $$f_1(x) = \begin{cases} 1 & \text{if } x \neq c \\ 0 & \text{if } x = c \end{cases}.$$
>
> Then $f_1(x)$ is integrable.

Two Discontinuities

What if this function had two points of discontinuity? For example, what about

$$f_2(x) = \begin{cases} 1 & \text{if } x \neq 1 \text{ or } 1.5 \\ 0 & \text{if } x = 1 \text{ or } 1.5 \end{cases},$$

whose graph looks like

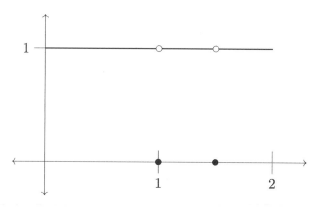

You can probably already imagine that a proof very similar to f_1's would work here. If you used the partition $P = \{0, 1 - \varepsilon/8, 1 + \varepsilon/8, 1.5 - \varepsilon/8, 1.5 + \varepsilon/8, 2\}$, then the upper sum would again equal 2 and the lower sum would equal

$$L(f_2, P) = 1 \cdot \left(1 - \frac{\varepsilon}{8}\right) + 0 + 1 \cdot \left(0.5 - 2 \cdot \frac{\varepsilon}{8}\right) + 0 + 1 \cdot \left(0.5 - \frac{\varepsilon}{8}\right)$$
$$= 2 - \frac{\varepsilon}{2}$$
$$> 2 - \varepsilon.$$

So the upper and lower sums are within ε of each other, showing that this very similar approach does indeed work.

Note: A good way to think about these lower sums is to think about the *missing* area. The area of the entire rectangle would be $2 \times 1 = 2$. So how much of that is missing? It's the shaded area:

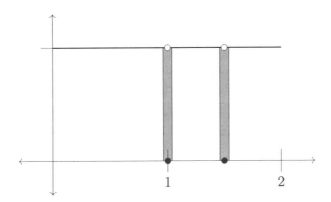

And the rectangle around $x = 1$ has width $\varepsilon/4$ and height 1, and the rectangle around $x = 1.5$ has width $\varepsilon/4$ and height 1. Each has area $\varepsilon/4$, so combined they have area $\varepsilon/2$. Since this is the missing area, the lower sum in total has area $2 - \varepsilon/2$.

But still there is an even better way to handle this, by using the following lemma.

Lemma.

Lemma 8.19 (*f integrable on $[a, c]$ and $[c, b]$ $\Leftrightarrow$ on $[a, b]$*). Assume $f : [a, b] \to \mathbb{R}$ and $a < c < b$. Then f is integrable on $[a, b]$ if and only if f is integrable on both $[a, c]$ and on $[c, b]$

Proof. We will use the integrals analytically theorem (Theorem 8.14), so note that f is bounded on both of $[a, c]$ and $[c, b]$ if and only if f is bounded on $[a, b]$. So the boundedness condition is satisfied. We now prove the rest.

First assume that f is integrable on $[a, c]$ and on $[c, b]$. Since f is integrable on $[a, c]$, by the integrals analytically theorem (Theorem 8.14) there is a partition $P_\varepsilon^{(1)}$ of $[a, c]$ such that

$$U(f, P_\varepsilon^{(1)}) - L(f, P_\varepsilon^{(1)}) < \frac{\varepsilon}{2}.$$

Since f is integrable on $[c, b]$, by the integrals analytically theorem (Theorem 8.14) there is a partition $P_\varepsilon^{(2)}$ of $[c, b]$ such that

$$U(f, P_\varepsilon^{(2)}) - L(f, P_\varepsilon^{(2)}) < \frac{\varepsilon}{2}.$$

Then $Q_\varepsilon := P_\varepsilon^{(1)} \cup P_\varepsilon^{(2)}$ is a partition of $[a, b]$ and

$$\begin{aligned}
U(f, Q_\varepsilon) - L(f, Q_\varepsilon) &= \left[U(f, P_\varepsilon^{(1)}) + U(f, P_\varepsilon^{(2)})\right] - \left[L(f, P_\varepsilon^{(1)}) + L(f, P_\varepsilon^{(2)})\right] \\
&= \left[U(f, P_\varepsilon^{(1)}) - L(f, P_\varepsilon^{(1)})\right] + \left[U(f, P_\varepsilon^{(2)}) - L(f, P_\varepsilon^{(2)})\right] \\
&< \frac{\varepsilon}{2} + \frac{\varepsilon}{2} \\
&= \varepsilon.
\end{aligned}$$

So f is integrable on $[a, b]$.

Now, conversely, assume that f is integrable on $[a, b]$. Then there is a partition P of $[a, b]$ such that

$$U(f, P) - L(f, P) < \varepsilon.$$

We may assume that $c \in P$, since if not we can add it to P and by Proposition 8.6 the resulting refinement would, if anything, only bring the upper and lower sums closer together.

Let $P\big|_{[a,c]} = P \cap [a, c]$; that is, $P\big|_{[a,c]}$ is the partition of $[a, c]$ that is the restriction of P to the first interval $[a, c]$. Then we can write $P = \{x_0, x_1, \ldots, x_T, x_{T+1}, \ldots, x_N\}$, where $P\big|_{[a,c]} = \{x_0, x_1, \ldots, x_T\}$. Intuitively, it's probably clear that the difference in the areas of the upper and lower rectangles from a to c is smaller than that of a to b, and since the latter difference is less than ε, so is the former. Writing this out formally:

$$\begin{aligned}
U\big(f, P\big|_{[a,c]}\big) - L\big(f, P\big|_{[a,c]}\big) &= \sum_{i=1}^{T} M_i(x_i - x_{i-1}) - \sum_{i=1}^{T} m_i(x_i - x_{i-1}) \\
&= \sum_{i=1}^{T} (M_i - m_i)(x_i - x_{i-1})
\end{aligned}$$

And since each $(M_i - m_i) \geq 0$ and each $(x_i - x_{i-1}) \geq 0$, and $T \leq N$,

$$\begin{aligned}
&\leq \sum_{i=1}^{N} (M_i - m_i)(x_i - x_{i-1}) \\
&= \sum_{i=1}^{N} M_i(x_i - x_{i-1}) - \sum_{i=1}^{N} m_i(x_i - x_{i-1}) \\
&= U(f, P) - L(f, P) \\
&< \varepsilon.
\end{aligned}$$

Thus, we have found a partition of $[a, c]$ for which $U\big(f, P\big|_{[a,c]}\big) - L\big(f, P\big|_{[a,c]}\big) < \varepsilon$, which by the integrals analytically theorem (Theorem 8.14) implies that f is integrable on $[a, c]$.

An identical argument shows that f is also integrable on $[c, b]$. $\square$

Using this theorem, you can see another reason why having two discontinuities like with f_2 does not prevent integrability. By Porism 8.18, f_2 is integrable on $[0, 1.25]$ and on $[0, 1.25]$, and so by the lemma f_2 is integrable on $[0, 2]$ as well.

A Finite Number of Discontinuities

You can now imagine that there is nothing special about 1 or 2 discontinuities. Suppose a function f_N has N points of discontinuity, say, $x_1, x_2, \ldots, x_N$. If f_N is equal to 1 everywhere except at these N points of discontinuity, where f_N equals 0, then just create a partition where you include the points $x_i - \frac{\varepsilon}{4N}$ and $x_i + \frac{\varepsilon}{4N}$ in your partition.[11] Indeed, by doing this each of the N missing rectangles has width $\frac{\varepsilon}{2N}$ and height 1; hence each of the N rectangles has area $\frac{\varepsilon}{2N}$ and so collectively the missing area has a total area of $\varepsilon/2$, giving a lower sum of $2 - \varepsilon/2$.

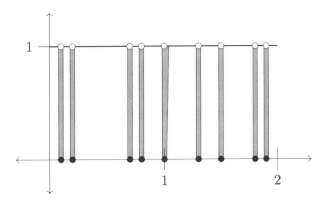

Or, you could apply Lemma 8.19: Break up the domain into $N + 1$ closed intervals so that there is just one discontinuity in each interval. Then f will be integrable in each of these intervals by Porism 8.18, and hence by applying Lemma 8.19 f is integrable on all of $[0, 2]$.

A Countable Number of Discontinuities

As we've seen, the goal is to have the "missing area" add up to less than ε; in the above we always made it add up to $\varepsilon/2$. (The rest of this paragraph is *almost* correct... but does have one error.) If you have countably many discontinuities you can still make it add up to $\varepsilon/2$, and therefore such a function is still integrable! In particular, if your discontinuities are $x_1, x_2, x_3, \ldots$, then just include in your partition the points $x_i - \frac{1}{2^{i+1}}$ and $x_i + \frac{1}{2^{i+1}}$ for each i. Then the rectangle around the first point will have width $\frac{\varepsilon}{2^2}$, around the second will have width $\frac{\varepsilon}{2^3}$, around the third will have width $\frac{\varepsilon}{2^4}$, and so on. Adding these up, the total amount of "missing" area is

$$\sum_{i=1}^{\infty} \frac{\varepsilon}{2^{i+1}} = \frac{\varepsilon}{2} < \varepsilon,$$

as desired.

---

[11] If one of these is outside $[0, 2]$, then you don't include it.

What was the one error in the last paragraph? By Definition 8.2, a partition can not have infinitely many points in it; it must be a *finite* set, and so we can't put each point in its own rectangle. Indeed, some rectangle will include infinitely many points! So we must be more careful. Consider, for instance, the following example.

Example 8.20. The function $f_\infty : [0, 2] \to \mathbb{R}$ where

$$f_\infty(x) = \begin{cases} 1 & \text{if } x \neq \frac{1}{n} \text{ for any } n \in \mathbb{N} \\ 0 & \text{if } x = \frac{1}{n} \text{ for some } n \in \mathbb{N} \end{cases}$$

is integrable.

Proof Idea. Here is the idea behind the following proof. First, as we just mentioned, a partition must be a *finite* set. So we have to find a way to have a finite number of skinny rectangles which contain all the points of discontinuity and their areas collectively add up to less than ε (once again, they will add up to $\varepsilon/2$). And then we will use the integrals analytically theorem (Theorem 8.14).

So how do we get the "missing" areas to add up to $\varepsilon/2$? To do this, we will utilize the fact that the points of discontinuity are converging to 0. So for any $\varepsilon > 0$, only finitely many of them are outside of $[0, \varepsilon/4]$. So if we pick a partition that includes $\varepsilon/4$ (and, as is required, includes 0), then the rectangle

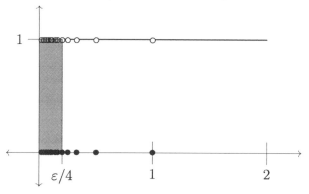

will capture all but finitely many of the points, and do so in a rectangle of area $(\varepsilon/4) \cdot 1 = \varepsilon/4$. Then we just add skinny rectangles around the finitely many remaining points, and make sure they are skinny enough that the sum of their areas equals $\varepsilon/4$.

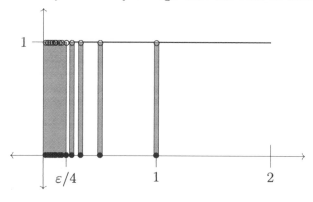

Then collectively the sum of this "missing" area will be $\varepsilon/4 + \varepsilon/4 = \varepsilon/2$, which is less than ε, as required by the integrals analytically theorem (Theorem 8.14). Ok, now here's the proof.

Proof. Let $\varepsilon > 0$. By the Archimedean principle (Lemma 1.26) there exists some (smallest) $N \in \mathbb{N}$ such[12] that $\frac{1}{N+1} \leq \varepsilon/4$. Now consider the partition containing $0, \varepsilon/4, 2$ and, for each point of discontinuity x outside of $[0, \varepsilon/4]$, also including the points $x \pm \frac{\varepsilon}{8N}$. That is,[13]

$$P = \left\{ 0, \frac{\varepsilon}{4}, \frac{1}{N} - \frac{\varepsilon}{8N}, \frac{1}{N} + \frac{\varepsilon}{8N}, \frac{1}{N-1} - \frac{\varepsilon}{8N}, \frac{1}{N-1} + \frac{\varepsilon}{8N}, \ldots, \frac{1}{1} - \frac{\varepsilon}{8N}, \frac{1}{1} + \frac{\varepsilon}{8N}, 2 \right\}.$$

This is clearly a partition of $[0, 2]$. Moreover, clearly all of the points of discontinuity are contained inside the interval $[0, \varepsilon/4]$ or inside one of the N intervals $\left[x - \frac{\varepsilon}{8N}, x + \frac{\varepsilon}{8N} \right]$, where x is one of the N points of discontinuity outside of $[0, \varepsilon/4]$.

We are getting ready to compute $U(f_\infty, P) - L(f_\infty, P)$. Note that when we do, the intervals that do not contain a point of discontinuity can be ignored since the upper and lower sums are equal in those intervals, as $M_i = m_i$ there. Thus, we will focus only on these $N + 1$ intervals containing points of discontinuity; this is the "missing" area.

Furthermore, note that the intervals $\left[x - \frac{\varepsilon}{8N}, x + \frac{\varepsilon}{8N} \right]$ each have length $\frac{\varepsilon}{4N}$; so if these equal x_{i-1} and x_i, then $x_i - x_{i-1} = \frac{\varepsilon}{4N}$. Lastly, on each of these intervals containing points of discontinuity, note that $m_i = 0$ while $M_i = 1$. Putting this all together,

$$U(f_\infty, P) - L(f_\infty, P) = \sum_{i=1}^{N+1} M_i(x_i - x_{i-1}) - \sum_{i=1}^{N+1} m_i(x_i - x_{i-1})$$

$$= \sum_{i=1}^{N+1} (M_i - m_i)(x_i - x_{i-1})$$

---

[12]Note: The reason why we included a "+1" is so there are N points of discontinuity,

$$\frac{1}{1}, \frac{1}{2}, \frac{1}{3}, \ldots, \frac{1}{N},$$

outside of $[0, \varepsilon/4]$, while the rest,

$$\frac{1}{N+1}, \frac{1}{N+2}, \frac{1}{N+3}, \ldots,$$

are inside of $[0, \varepsilon/4]$. This makes the notation easier.

[13]Technically we should assume that $\varepsilon \leq 8$ to ensure that all of these points are in $[0, 2]$, but this is unimportant. There are many easy fixes, such as just removing any points from P that fall outside of $[0, 2]$.

The first term of this sum corresponds to the rectangle of width $\varepsilon/4$. So the above

$$= (1-0)(\varepsilon/4 - 0) + \sum_{i=2}^{N+1} (1-0)\left(\frac{\varepsilon}{4N}\right)$$

$$= \frac{\varepsilon}{4} + N \cdot (1-0)\left(\frac{\varepsilon}{4N}\right)$$

$$= \frac{\varepsilon}{4} + \frac{\varepsilon}{4}$$

$$= \frac{\varepsilon}{2}$$

$$< \varepsilon.$$

And since f_∞ is clearly bounded, the integrals analytically theorem (Theorem 8.14) now says that f_∞ is integrable.   □

Example 8.21. Thomae's function on $[0,2]$ is integrable.

Proof. This is Exercise 8.17 and Exercise 8.23.   □

An Uncountable Number of Discontinuities

Amazingly, an integrable function can even have an uncountable number of discontinuities. It requires a weird uncountable set. It requires an uncountable set that is "small" in the visual sense. It requires... The Cantor Set! We won't go into the details, but here is such an example:

$$g(x) = \begin{cases} 1 & \text{if } x \in \mathcal{C} \\ 0 & \text{if } x \notin \mathcal{C}, \end{cases}$$

where $\mathcal{C}$ is the Cantor set. (See Example B.1 in Appendix B.)

Other Continuous Functions

You can imagine that if you take *any* function f on an interval $[a, b]$ that is discontinuous at just 1 point, a similar argument would show that that function is also integrable; like above, you would hollow out a tiny region around the point of discontinuity, then the rest of the function's upper and lower bounds get as close as you wish, and the error caused by the point of discontinuity you can make as small as you want by making the tiny region around it as tiny as you need. There's no concern about the function getting too big or small on the subinterval because f being continuous implies that f is bounded.

Then, by Lemma 8.19, we can handle any finite number of discontinuities. And, like above, in certain situations we can even handle an infinite number of discontinuities.

But how can we tell in general that we have an integrable function? The answer takes a little care to write down, but it's awesome. It's called the *measure zero criterion*.

8.8   The Measure Zero Integrability Criterion

In Example 8.20 we showed in detail that f_∞ was integrable. Why did it work out? Well, it was highly convenient that the infinitely many points were all bunching up together around 0. This allowed all but a finite number of the points to be contained inside in a tiny interval. And the remaining finite number of points could be handled easily. This suggests one condition that guarantees integrability:

"f is integrable if for all $\varepsilon > 0$ the points of discontinuity can be covered by a finite number of intervals, $I_1, I_2, I_3, \ldots, I_N$, such that sum of the lengths of these intervals is less than ε."

Here, by "length" we mean the following (for more on this, see the Example B.1 in Appendix B).

Definition.

Definition 8.22. The *length* of the interval $[a, b]$ is equal to $b - a$, and is denoted $\mathscr{L}([a, b])$. This is also the length of (a, b), $(a, b]$ and $[a, b)$. Intervals that involve $\pm \infty$ are said to have length ∞.

So above we argued informally that if, for every $\varepsilon > 0$, the points of discontinuity can be covered by a finite set of intervals whose lengths add up to less than ε, then f is integrable. In fact, even more is true—a countable collection of sets works too! This is the idea of a set having "measure zero."

Definition.

Definition 8.23. A set A has *measure zero* if for all $\varepsilon > 0$ there exists a countable collection $I_1, I_2, I_3, \ldots$ of intervals such that

$$A \subseteq \bigcup_{k=1}^{\infty} I_k \qquad \text{and} \qquad \sum_{k=1}^{\infty} \mathscr{L}(I_k) < \varepsilon.$$

This leads to a famous theorem sometimes called *Lebesgue's integrability condition*—an insightful and delightful criterion for integrability.

> **Theorem.**
>
> **Theorem 8.24** (*f integrable $\Leftrightarrow$ discontinuities have measure zero*). Assume $f : [a, b] \to \mathbb{R}$ is a bounded function and let $\mathcal{D}$ be the set of points at which f is discontinuous. Then f is integrable if and only if $\mathcal{D}$ has measure zero.

The proof of this theorem is long, technical, and boring; we will skip it.[14]

One can use this theorem to prove that if $f : [a, b] \to \mathbb{R}$ and $g : [a, b] \to \mathbb{R}$ are integrable and $k \in \mathbb{R}$, then kf, $f + g$ and fg are also integrable. You will be asked to work out the details of this in Exercise 8.24.

8.9   Linearity Properties of the Integral

In this section we will prove many familiar properties from calculus. Here's the first result.

> **Proposition.**
>
> **Proposition 8.25** ($\int_a^b f = \int_a^c f + \int_c^b f$). Assume that $f : [a, b] \to \mathbb{R}$ is integrable. If $a < c < b$, then
>
> $$\int_a^b f(x)\, dx = \int_a^c f(x)\, dx + \int_c^b f(x)\, dx.$$

Proof. Recall from Lemma 8.19 that f is integrable on $[a, b]$ if and only if f is integrable on both $[a, c]$ and on $[c, b]$. Thus we only have to show that the asserted sum holds.

By the integrals analytically theorem (Theorem 8.14), for every $\varepsilon > 0$ we can find a partition P_1 of $[a, c]$ and a partition P_2 of $[c, b]$ for which

$$U(f, P_1) - L(f, P_1) < \frac{\varepsilon}{2} \qquad \text{and} \qquad U(f, P_2) - L(f, P_2) < \frac{\varepsilon}{2}.$$

Letting $P = P_1 \cup P_2$ produces a partition of $[a, b]$ where

$$
\begin{aligned}
U(f, P) - L(f, P) &= [U(f, P_1) + U(f, P_2)] - [L(f, P_1) + L(f, P_2)] \\
&= [U(f, P_1) - L(f, P_1)] + [U(f, P_2) - L(f, P_2)] \\
&< \frac{\varepsilon}{2} + \frac{\varepsilon}{2} \\
&= \varepsilon.
\end{aligned}
$$

---

[14]You're welcome.

Lower sums are always less than the integral, and upper sums are always greater than the integral:

$$L(f,P) \leq \int_a^b f(x)\,dx \leq U(f,P).$$

We saw above that $U(f,P) - L(f,P) < \varepsilon$, which is equivalent to $U(f,P) < L(f,P) + \varepsilon$ and to $U(f,P) - \varepsilon < L(f,P)$. Thus,

$$U(f,P) - \varepsilon < L(f,P) \leq \int_a^b f(x)\,dx \leq U(f,P) < L(f,P) + \varepsilon.$$

Simplifying by removing the extra upper and lower sums,

$$U(f,P) - \varepsilon < \int_a^b f(x)\,dx < L(f,P) + \varepsilon.$$

And since $L(f,P) = L(f,P_1) + L(f,P_2)$ and $U(f,P) = U(f,P_1) + U(f,P_2)$,

$$[U(f,P_1) + U(f,P_2)] - \varepsilon < \int_a^b f(x)\,dx < [L(f,P_1) + L(f,P_2)] + \varepsilon.$$

Once again, upper sums are bigger than their integrals, and lower sums are smaller than their integrals. Thus,

$$\left[\int_a^c f(x)\,dx + \int_c^b f(x)\,dx\right] - \varepsilon < \int_a^b f(x)\,dx < \left[\int_a^c f(x)\,dx + \int_c^b f(x)\,dx\right] + \varepsilon.$$

So we know that $\int_a^c f(x)\,dx + \int_c^b f(x)\,dx$ and $\int_a^b f(x)\,dx$ are within ε of each other. And since ε was arbitrary, this implies that they are in fact equal:

$$\int_a^c f(x)\,dx + \int_c^b f(x)\,dx = \int_a^b f(x)\,dx,$$

as desired. $\square$

Yes, I agree, that proof was fairly boring. And the bad news is that this mostly continues. The theorems about integrals are quite beautiful, but since they are defined in terms of infima and suprema of sums of other infima and suprema over arbitrary partitions... the elegance is often lost in the weeds. Nevertheless, we persevere. Here's a cute  puppy for motivation. Next up is a proposition on linearity properties of integrals.

> **Proposition.**

Proposition 8.26 (*Linearity of the integral*). Assume that $f : [a, b] \to \mathbb{R}$ and $g : [a, b] \to \mathbb{R}$ are integrable. Then,

(i) For any $k \in \mathbb{R}$, kf is also integrable on $[a, b]$, and

$$\int_a^b k \cdot f(x)\, dx = k \cdot \int_a^b f(x)\, dx.$$

(ii) The function $f + g$ is also integrable on $[a, b]$, and

$$\int_a^b (f(x) + g(x))\, dx = \int_a^b f(x)\, dx + \int_a^b g(x)\, dx.$$

Proof. First we prove (i). If $k > 0$, then intuitively all we are doing is making each rectangle in our upper/lower sums k times bigger. Indeed, each M_i in $U(f, P)$ will turn into $k \cdot M_i$ in $U(kf, P)$. And so,

$$U(k \cdot f, P) = \sum_{i=1}^n k \cdot M_i(x_i - x_{i-1}) = k \cdot \sum_{i=1}^n M_i(x_i - x_{i-1}) = k \cdot U(f, P).$$

And likewise for the lower sums we have $L(k \cdot f, P) = k \cdot L(f, P)$.

Since f is integrable on $[a, b]$, by Corollary 8.15 there exists a sequence (P_n) of partitions of $[a, b]$ for which

$$\int_a^b f(x)\, dx = \lim_{n \to \infty} U(f, P_n) = \lim_{n \to \infty} L(f, P_n).$$

But using what we argued above and properties of limits,

$$\lim_{n \to \infty} U(kf, P_n) = \lim_{n \to \infty} k \cdot U(f, P_n) = k \cdot \lim_{n \to \infty} U(f, P_n) = k \cdot \int_a^b f(x)\, dx$$

and

$$\lim_{n \to \infty} L(kf, P_n) = \lim_{n \to \infty} k \cdot L(f, P_n) = k \cdot \lim_{n \to \infty} L(f, P_n) = k \cdot \int_a^b f(x)\, dx.$$

So we have shown that

$$\lim_{n \to \infty} U(kf, P_n) = \lim_{n \to \infty} L(kf, P_n) = k \cdot \int_a^b f(x)\, dx,$$

which by Corollary 8.15 concludes (i) in this case where $k > 0$. The case where $k < 0$ is very similar, the only difference is that the upper sums for f become lower sums for kf, and vice versa. In particular, $U(kf, P) = k \cdot L(f, P)$ and $L(kf, P) = k \cdot U(f, P)$.

The rest is almost identical.[15] Finally, the case where $k = 0$ is immediate.

The proof of (ii) can use the same general approach, but is just messier. It's left as an exercise.[16] □

8.10   More Properties of the Integral

The next result is a corollary of an earlier result.

Recall 8.27. In Proposition 8.13 we proved the following:

Assume that $f : [a, b] \to \mathbb{R}$ is integrable. If $f(x) \geq 0$ for all $x \in [a, b]$, then

$$\int_a^b f(x)\, dx \geq 0.$$

> **Corollary.**
>
> **Corollary 8.28** ($f \leq g \Rightarrow \int f \leq \int g$)**.** Let $f : [a, b] \to \mathbb{R}$ and $g : [a, b] \to \mathbb{R}$ be integrable functions. If $f(x) \leq g(x)$ for all $x \in [a, b]$, then
>
> $$\int_a^b f(x)\, dx \leq \int_a^b g(x)\, dx.$$

Proof. Since $f(x) \leq g(x)$ we have $0 \leq g(x) - f(x)$. By Proposition 8.26 part (i), since f is integrable, $(-1) \cdot f = -f$ is also integrable. And since g is also assumed to be integrable, by Proposition 8.26 part (ii) we have $g + (-f) = g - f$ is integrable.

We have shown that $g - f$ is a non-negative, integrable function and thus by Proposition 8.13 we have that

$$\int_a^b (g - f)(x)\, dx \geq 0.$$

---

[15]Interviewer: "So why did you become a mathematician?" Mathematician: "Because I never liked working with numbers."

[16]Note: In Exercise 8.24 you will be asked to prove that kf and $f + g$ (and fg) are integrable using the measure zero criterion.

Applying Proposition 8.26 one more time,

$$\int_a^b [g(x) - f(x)]\, dx \geq 0$$

$$\int_a^b g(x)\, dx - \int_a^b f(x)\, dx \geq 0$$

$$\int_a^b g(x)\, dx \geq \int_a^b f(x)\, dx.$$

$\square$

The next result can be thought of as an "integral triangle inequality." When adding up discrete numbers you can use the triangle inequality: $|x + y| \leq |x| + |y|$. More generally, $|\sum x_i| \leq \sum |x_i|$. And in some sense, the way we "add up" a continuous range of numbers is using the integral. So this is the continuous version of the triangle inequality.

Corollary.

Corollary 8.29 ($|\int f| \leq \int |f|$). If $f : [a, b] \to \mathbb{R}$ is integrable, then $|f|$ is also integrable, and

$$\left| \int_a^b f(x)\, dx \right| \leq \int_a^b |f(x)|\, dx.$$

Proof. Let $\varepsilon > 0$. Note that since f is integrable, f is bounded and hence $|f|$ is also bounded. Furthermore, since f is integrable there exists a partition $P_\varepsilon = \{x_0, x_1, \ldots, x_n\}$ such that

$$U(f, P_\varepsilon) - L(f, P_\varepsilon) < \varepsilon.$$

Over each interval $[x_{i-1}, x_i]$ we have notation m_i and M_i to represent the infimum and supremum of f over the interval, respectively. Likewise, define

$$\overline{m}_i = \inf\{|f(x)| : x \in [x_{i-1}, x_i]\}$$
$$\overline{M}_i = \sup\{|f(x)| : x \in [x_{i-1}, x_i]\}.$$

Since $|f(x)| \geq f(x)$ for all x, clearly $\overline{m}_i \geq m_i$ and $\overline{M}_i \geq M_i$ for all i. Moreover,

$$M_i - m_i \geq \overline{M}_i - \overline{m}_i$$

since if M_i and m_i are the same sign (both positive or both negative), then these two are equal, while if they are opposite signs, then taking the absolute value only makes them closer to each other.

Then,

$$U(|f|, P_\varepsilon) - L(|f|, P_\varepsilon) = \sum_{i=1}^{n} \left(\overline{M}_i - \overline{m}_i \right)(x_i - x_{i-1})$$

$$\leq \sum_{i=1}^{n} (M_i - m_i)(x_i - x_{i-1})$$

$$= U(f, P_\varepsilon) - L(f, P_\varepsilon)$$

$$< \varepsilon.$$

By the integrals analytically theorem (Theorem 8.14), $|f|$ is then integrable.

To show that the integral inequality holds, first note that using the fact that $-|f(x)| \leq f(x) \leq |f(x)|$ and applying Corollary 8.28 (twice) we have that

$$-\int_a^b |f(x)|\, dx \leq \int_a^b f(x)\, dx \leq \int_a^b |f(x)|\, dx. \qquad (\text{🐾})$$

Finally, simply recall that $-t \leq s \leq t$ if and only if $|s| \leq t$ for any real numbers s and t. Using this, (🐾) is equivalent to

$$\left| \int_a^b f(x)\, dx \right| \leq \int_a^b |f(x)|\, dx. \qquad \square$$

We only defined the integral on intervals $[a, b]$ where $a < b$. Therefore the below have to be stated as definitions, rather than deduced as properties. Note, though, that they are consistent with a natural generalization of our definitions.

Definition.

Definition 8.30.

- For any function f, define

$$\int_a^a f(x)\, dx = 0.$$

- For any integrable $f : [a, b] \to \mathbb{R}$, define

$$\int_a^b f(x)\, dx = -\int_b^a f(x)\, dx.$$

The Mean Value Theorems

Recall the (derivative) mean value theorem.

Recall.

Theorem 7.22 (*The (derivative) mean value theorem*). Let $f : [a, b] \to \mathbb{R}$ be continuous on $[a, b]$ and differentiable on (a, b). Then there exists some $c \in (a, b)$ where
$$f'(c) = \frac{f(b) - f(a)}{b - a}.$$

This theorem will have practical value very soon, but first it motivates the next result, known as the *integral mean value theorem*.

When looked at just right, this new mean value theorem looks a lot like the derivative mean value theorem, provided you recall the fundamental theorem of calculus from when you took calculus. Instead of viewing f' as being the derivative f, you could instead think about f as being an *antiderivative* of f'. We haven't yet formally defined an antiderivative, but using your intuition let's let F be an antiderivative of f. Applying the (derivative) MVT to F,

$$F'(c) = \frac{F(b) - F(a)}{b - a}.$$

As an antiderivative, $F'(c) = f(c)$, and by what will soon be the fundamental theorem of calculus, $F(b) - F(a) = \int_a^b f(x)\,dx$. With this intuition, we reach the *integral mean value theorem*, which we state now.

Proposition.

Proposition 8.31 (*The integral mean value theorem*). If f is a continuous function on $[a, b]$, then there is some $c \in [a, b]$ such that
$$f(c) = \frac{1}{b - a} \int_a^b f(x)\,dx.$$

Proof. Since f is continuous on a closed interval, f is bounded (by the extreme value theorem) and integrable (by Theorem 8.16). The extreme value theorem furthermore implies that f attains a minimum $f(x_{\min}) = m$ and a maximum $f(x_{\max}) = M$ at some points $x_{\min}, x_{\max} \in [a, b]$. Therefore by Lemma 8.9, since $m \le f(x) \le M$ for all $x \in [a, b]$, we have

$$m(b - a) \le \int_a^b f(x)\,dx \le M(b - a).$$

I.e.,

$$m \leq \frac{1}{b-a} \int_a^b f(x)\, dx \leq M.$$

In summary, f is a continuous function and $\frac{1}{b-a} \int_a^b f(x)\, dx$ is some number between $f(x_{\min})$ and $f(x_{\max})$. Therefore, by the intermediate value theorem, there is some c between $x_{\min}$ and $x_{\max}$ for which

$$f(c) = \frac{1}{b-a} \int_a^b f(x)\, dx.$$

$\square$

That proof was quite a tour of our major theorems on functions!

8.11   The Fundamental Theorem of Calculus

The fundamental theorem of calculus (FTC) is remarkable. And it's a tragedy that you all already know the grand reveal. If tonight you watch Star Wars for the first time, you probably won't be surprised to learn that Darth Vader is Luke Skywalker's father; that has been common knowledge for over 40 years now, even among those who haven't seen the movies. But for those who learned it in the theaters for the first time on May 21, 1980, this was an astounding development. How can two disparate forces in this grand story be linked in such an intimate way??[17]

Likewise, due to the evils of calculus classes, you already know what FTC says. But for a moment, pretend you don't. Under this veil of ignorance, we have an amazing plot twist to our story.

In Chapter 7 we developed the theory of derivatives as a limit, motivated by finding slopes of tangent lines, in order to solve physics problems. And we then pushed the theory further and proved lots of nice theorems. During this chapter we did something completely different. We developed integrals as a way to approximate areas under curves, by using infima and suprema of upper and lower sums. This purely-geometric endeavor also produced some nice results with some fairly intuitive properties.

There is absolutely no reason at all to think these two have anything to do with each other. With the veil over our eyes, we should even be confused why these two topics are contained in the same class. They are not even remotely related, right? Right?? Well with four words with implications as startling as "I am your father," we should be amazed at how the "fundamental theorem of calculus" intimately ties together these seemingly-disparate ideas in the most incredible way imaginable.

So either take off that veil or hold onto your hats, because we are about to state the fundamental theorem of calculus.

---

[17]Millennials version: How shocked would you have been if Voldemort was Harry Potter's father? And what if there was an amazing backstory why, and breadcrumb clues had been subtly dropped along the way? What a twist!

Super Cool Theorem.

Theorem 8.32 (*The fundamental theorem of calculus*).

(i) If $f : [a,b] \to \mathbb{R}$ is integrable, and $F : [a,b] \to \mathbb{R}$ satisfies $F'(x) = f(x)$ for all $x \in [a,b]$, then

$$\int_a^b f(x)\,dx = F(b) - F(a).$$

(ii) Let $g : [a,b] \to \mathbb{R}$ be integrable and define $G : [a,b] \to \mathbb{R}$ by

$$G(x) = \int_a^x g(t)\,dt.$$

Then G is continuous. Moreover, if g is continuous, then G is differentiable and $G'(x) = g(x)$. That is, $\dfrac{d}{dx}\displaystyle\int_a^x g(t)\,dt = g(x)$.

The fundamental theorem of calculus was found by two people, one of which was Issac Newton's PhD advisor. Surprisingly, the proof of part (i) just comes down to an application of the derivative mean value theorem[18] and a cute use of telescoping.

Proof of (i). Let $P = \{x_0, \dots, x_n\}$ be an arbitrary partition of $[a,b]$. Consider the behavior of F on some subinterval $[x_{i-1}, x_i]$. Since F is differentiable, by the derivative mean value theorem (stated above) there must exist some $c_i \in (x_{i-1}, x_i)$ where

$$F'(c_i) = \frac{F(x_i) - F(x_{i-1})}{x_i - x_{i-1}}$$

That is,

$$F(x_i) - F(x_{i-1}) = F'(c_i) \cdot (x_i - x_{i-1})$$

Using now the assumption that $F'(x) = f(x)$,

$$F(x_i) - F(x_{i-1}) = f(c_i) \cdot (x_i - x_{i-1}).$$

Consider now the upper sum $U(f, P)$ and the lower sum $L(f, P)$. These sums contain the sups/infs M_i and m_i which satisfy $m_i \leq f(c_i) \leq M_i$, implying that

$$\sum_{i=1}^n m_i(x_i - x_{i-1}) \leq \sum_{i=1}^n f(c_i)(x_i - x_{i-1}) \leq \sum_{i=1}^n M_i(x_i - x_{i-1}).$$

---

[18]The mean value theorem's fingerprints are all over real analysis. If it didn't already have an important name I definitely would have called it "the fundamental lemma of calculus."

The left and the right sums are just the lower and upper sums, and the terms of the middle sum we already noted are just $F(x_i) - F(x_{i-1})$:

$$L(f, P) \leq \sum_{i=1}^{n} \Big(F(x_i) - F(x_{i-1}) \Big) \leq U(f, P).$$

But notice that the middle sum telescopes! Indeed,

$$\sum_{i=1}^{n} \Big(F(x_i) - F(x_{i-1}) \Big)$$
$$= \Big(\cancel{F(x_1)} - F(x_0) \Big) + \Big(\cancel{F(x_2)} - \cancel{F(x_1)} \Big) + \Big(\cancel{F(x_3)} - \cancel{F(x_2)} \Big) + \dots$$
$$\dots + \Big(\cancel{F(x_{n-1})} - \cancel{F(x_{n-2})} \Big) + \Big(F(x_n) - \cancel{F(x_{n-1})} \Big)$$
$$= -F(x_0) + F(x_n)$$
$$= F(x_n) - F(x_0)$$
$$= F(b) - F(a).$$

Using this to simplify the middle term of the previous inequality,

$$L(f, P) \leq F(b) - F(a) \leq U(f, P).$$

But notice that this holds no matter which partition we choose! So we also have that

$$\sup\{L(f, P) : P \in \mathcal{P}\} \leq F(b) - F(a) \leq \inf\{U(f, P) : P \in \mathcal{P}\},$$

where $\mathcal{P}$ is the collection of all partitions of $[a, b]$. That is,

$$L(f) \leq F(b) - F(a) \leq U(f). \qquad (\text{🚗})$$

And this is perfect, since f being integrable means that $L(f) = U(f) = \int_a^b f(x)\,dx$. This and (🚗) imply that

$$\int_a^b f(x)\,dx = F(b) - F(a),$$

completing the proof of part (i). $\qquad \square_{(i)}$

Proof of (ii). First, assume that g is integrable; we will prove that $G(x) = \int_a^x g(t)\,dt$ is continuous, by using the definition of continuity. First, note that if $g(x) = 0$ for all x, then also $G(x) = 0$ on all x, and we are done. Therefore, we may assume that

$$M = \sup\{|g(x)| : x \in [a, b]\}$$

is greater than zero.

Pick any $x_0 \in [a, b]$. To prove that G is continuous at x_0, let $\varepsilon > 0$, and then let $\delta = \frac{\varepsilon}{M}$. Assume now that $|x - x_0| < \delta$; we will prove that $|G(x) - G(x_0)| < \varepsilon$. Note that

$$\begin{aligned}
|G(x) - G(x_0)| &= \left| \int_a^x g(t)\, dt - \int_a^{x_0} g(t)\, dt \right| \\
&= \left| \int_{x_0}^x g(t)\, dt \right| \\
&\leq \int_{x_0}^x |g(t)|\, dt \qquad\qquad\qquad \text{(Corollary 8.29)} \\
&\leq M \cdot |x - x_0| \qquad\qquad\qquad \text{(Lemma 8.9)} \\
&< M \cdot \delta \qquad\qquad\qquad\qquad \text{(Choice of } x) \\
&= M \cdot \frac{\varepsilon}{M} \qquad\qquad\qquad\quad \text{(Choice of } \delta) \\
&= \varepsilon.
\end{aligned}$$

This proves G is continuous at x_0, and since x_0 was arbitrary, it proves that G is continuous on all of $[a, b]$, concluding the first half of (ii).

Next assume that g is also continuous; we will now prove that G is differentiable and that $G'(x) = g(x)$. To prove this, pick an arbitrary $c \in [a, b]$; by the definition of the derivative we must prove

$$\lim_{x \to c} \frac{G(x) - G(c)}{x - c} = g(c).$$

By the sequential definition of limits (Theorem 6.17 (iv)), this is equivalent to proving that every sequence (x_n) from $[a, b] \setminus \{c\}$ such that $x_n \to c$ has the property that

$$\lim_{n \to \infty} \frac{G(x_n) - G(c)}{x_n - c} = g(c).$$

We now prove this. To that end, choose such a sequence (x_n). Note that

$$G(x_n) - G(c) = \int_a^{x_n} g(t)\, dt - \int_a^c g(t)\, dt = \int_c^{x_n} g(t)\, dt. \qquad (\spadesuit)$$

By the integral mean value theorem (Proposition 8.31, but with the "$b - a$" moved to the other side), there is some c_n between x_n and c such that

$$\int_c^{x_n} g(t)\, dt = g(c_n) \cdot (x_n - c).$$

Applying this to ($\spadesuit$), we get $G(x_n) - G(c) = g(c_n) \cdot (x_n - c)$. I.e.,

$$\frac{G(x_n) - G(x_0)}{x_n - c} = g(c_n).$$

Moreover, since c_n was between x_n and c for each n, and $x_n \to c$, we also know that

$c_n \to c$. We are ready for the final computation:

$$G'(c) = \lim_{n\to\infty} \frac{G(x_n) - G(c)}{x_n - c}$$
$$= \lim_{n\to\infty} g(c_n)$$

And since g is continuous, by Theorem 6.17 (iv),

$$= g\left(\lim_{n\to\infty} c_n\right)$$
$$= g(c). \qquad\qquad \square_{(ii)} \quad \square$$

There's also a neat way to prove FTC (i) from FTC (ii) in the case that f is continuous. Suppose $f : [a, b] \to \mathbb{R}$ is continuous (and hence integrable), and $F : [a, b] \to \mathbb{R}$ satisfies $F'(x) = f(x)$ for all x. We'll show

$$\int_a^b f(x)\, dx = F(b) - F(a).$$

To that end, let

$$h(x) = \int_a^x f(t)\, dt. \qquad\qquad (\text{🍴})$$

Then, using that f is continuous, by FTC (ii) we know $h'(x) = f(x)$. To summarize, we now have that $F' = h' = f$ (i.e., both h and F are *antiderivatives* of f), which by the "$+C$ corollary" (Corollary 7.24) means that these two functions can only differ by a constant; that is, $h(x) = F(x) + C$, for some $C \in \mathbb{R}$. Flipping around (🍴) and plugging in $h(x) = F(x) + C$ then gives

$$\int_a^x f(t)\, dt = F(x) + C. \qquad\qquad (\text{🤘})$$

Equation[19] (🤘) holds for any x, so in particular when $x = b$,

$$\int_a^b f(t)\, dt = F(b) + C.$$

We are nearly there! As long as we can show that $C = -F(a)$, we will have it. And why is that true? Well, (🤘) also holds when $x = a$! And plugging that in gives

$$0 = \int_a^a f(t)\, dt = F(a) + C,$$

implying that $C = -F(a)$, completing the proof. $\qquad\qquad \square$

---

[19]The spatula and 'rock on' symbols were designed by my former student, Jessie Loucks. She used the spatula as her end-of-proof symbol when she took my analysis class and designed the 'rock on' for another student to use as his end-of-proof symbol. She graciously let me use them in this book.

It is perhaps overdue to formally record the definition of an *antiderivative*.

> **Definition.**
>
> **Definition 8.33.** Assume F and f are functions on $[a, b]$ for which $F'(x) = f(x)$ for all x. Then F is called an *antiderivative* of f.

8.12 Integration Rules

The fundamental theorem of calculus has a number of nice consequences, all of which will be familiar from calculus.

Rule 1. Integration by Parts — The "Inverse Product Rule"

Here is our first integration rule.

> **Corollary.**
>
> **Corollary 8.34** (*Integration by parts*). If f and g are differentiable functions which have continuous derivatives on $[a, b]$, then fg' and $f'g$ are integrable and
>
> $$\int_a^b fg' = f(b)g(b) - f(a)g(a) - \int_a^b f'g.$$

Note that in calculus this is commonly written $\int_a^b u\,dv = uv\Big|_a^b - \int_a^b v\,du$.

Integration by parts can be thought of as an "inverse product rule." Indeed, the proof essentially just integrates a product rule and applies FTC.

Proof. We assumed that f and g are differentiable, and hence are continuous. We also assumed that f' and g' are continuous. This all implies that both fg' and $f'g$ are continuous, and hence by Theorem 8.16 are integrable.

Now, by the Product Rule,

$$(fg)' = f'g + fg'.$$

Integrating both sides,

$$\int_a^b (fg)' = \int_a^b f'g + \int_a^b fg'. \qquad (\clubsuit)$$

We now use FTC to rewrite the left-hand side. Clearly fg is an antiderivative of $(fg)'$, and so by the fundamental theorem of calculus part (i),

$$\int_a^b (fg)' = (fg)(b) - (fg)(a) = f(b)g(b) - f(a)g(a).$$

Plugging this into (🐾) gives

$$f(b)g(b) - f(a)g(a) = \int_a^b f'g + \int_a^b fg',$$

which is equivalent to what we are trying to prove. $\qquad\square$

Rule 2. w-Substitution — The "Inverse Chain Rule"

Next is what most people call u-substitution, but for various calculus-based pedagogical reasons I call w-substitution.

> ### Corollary.
>
> **Corollary 8.35** (*w-substitution*). Suppose g is a function whose derivative g' is continuous on $[a, b]$, and suppose that f is a function that is continuous on $g([a, b])$. Then
> $$\int_a^b f(g(x))g'(x)\, dx = \int_{g(a)}^{g(b)} f(w)\, dw.$$

First, don't be confused by "$g([a, b])$." Remember, this is a set; it is the set $\{g(x) : x \in [a, b]\}$. Since we are taking $f(g(x))$, the domain of f must contain the range of g, and are assuming f is continuous on this set.

Second, w-substitution can be thought of as an "inverse chain rule"—the substitution essentially collapses down an application of the chain rule. The proof drives this point home by showing that it holds via a combination of the chain rule with FTC. Indeed, if you let $F(x) = \int_{g(a)}^{x} f(t)\, dt$, then the 30,000 foot view of our proof is that

$$\int_a^b f(g(x))g'(x)\, dx = F(g(b)) - F(g(a)) = \int_{g(a)}^{g(b)} f(w)\, dw.$$

Both equalities look like variations on FTC, but to make it all work in the formal proof takes some care.

Proof. Since g is continuous, note that $g([a, b])$ is either a closed interval or a single point.[20] Thus we have two cases to consider.

Case 1: $g([a, b]) = \{L\}$. In this case, $g(x) = L$ for all $x \in [a, b]$. Since the derivative of constant functions are identically zero, $g'(x) = 0$ for all x. Indeed, the equality in the corollary holds because both sides simply equal 0:

$$\int_a^b f(g(x)) \cdot g'(x) \, dx = \int_a^b f(g(x)) \cdot 0 \, dx$$
$$= 0$$
$$= \int_L^L f(w) \, dw$$
$$= \int_{g(a)}^{g(b)} f(w) \, dw,$$

proving the equality holds for Case 1.

Case 2: $g([a, b]) = [c, d]$. Define

$$F(x) = \int_{g(a)}^x f(t) \, dt.$$

Note that $F(g(b))$ is the right-hand side of the equality we are trying to show, so that it suffices to prove that $F(g(b)) = \int_a^b f(g(x)) \cdot g'(x) \, dx$. Furthermore, note that $F(g(a)) = \int_{g(a)}^{g(a)} f(t) \, dt = 0$. Thus it is also sufficient to prove that

$$F(g(b)) - F(g(a)) = \int_a^b f(g(x)) \cdot g'(x) \, dx.$$

This is what we will show, and we will combine the chain rule and FTC to do so.

By assumption, $f(g(x))$ and $g'(x)$ are continuous, implying that $f(g(x)) \cdot g'(x)$ is continuous and hence integrable. Applying the chain rule to $F(g(x))$,

$$\frac{d}{dx} F(g(x)) = F'(g(x)) \cdot g'(x) = f(g(x)) \cdot g'(x).$$

Said differently, $F(g(x))$ is an antiderivative of $f(g(x)) \cdot g'(x)$. And by FTC part (i), this means that

$$\int_a^b f(g(x)) \cdot g'(x) \, dx = F(g(b)) - F(g(a)),$$

as desired.   □

---

[20]If you're interested in a detailed reason for this, here's a sketch for how to prove it: By the extreme value theorem, g attains a max and a min. If the max equals the min, then $g([a, b])$ is a single point. Otherwise, let c be the min and d be the max, which implies that $g([a, b]) \subseteq [c, d]$. But also note that by the intermediate value theorem, since g hits c and d, it also must hit everything in between, meaning that $g([a, b]) = [c, d]$. □

Shoulders of *which* giants?

Issac Newton often gets most of the credit for developing differential and integral calculus. Part of his genius was the ability to combine many different ideas into a comprehensive theory. Newton took the ancient Greek's method of exhaustion, but expressed it using algebra from the Middle East; he adopted the analytic geometry of the French to analyze functions, which he then expressed in an ingenious way via a novel perspective on Indian decimals. In just a couple years in his early 20s, Issac Newton developed the unifying theory of calculus, and in doing so, thousands of problems which at that time only the best mathematicians would have a chance to (inelegantly) solve, were instantly turned into today's high school homework problems.

However, perhaps there is another reason that Newton gets more credit than he deserves. German mathematician Gottfried Wilhelm Leibniz discovered many of the same results that Newton did. And even though he did so nearly 10 years after Newton, it seems likely that much (or maybe all — although Newton's crew suggested otherwise) of it was completely independent of Newton. Moreover, Newton didn't bother to formally publish or widely promote his results, which allowed Leibniz was to publish first.

That said, Leibniz's landmark paper was rather odd. He begins it by immediately listing general rules of differentiation, with little explanation for why any of his claims are true. John Bernoulli — one of the best mathematicians of the day — struggled to understand it, calling it "an enigma rather than an explication." But despite this inadequate introduction, Leibniz did popularize the newly discovered field, giving lectures and inspiring others to do likewise, including Marquis de l'Hôpital, who would go on to write the first calculus textbook.

So as you can see, assigning credit was a bit complicated. And neither man would relinquish their claim — particularly as the accusations turned ugly, and their character was on the line. Indeed, Newton and his supporters accused Leibniz of scooping his work, Leibniz fired back as you'd imagine an irritated mathematician might do, and a controversy for the mathematical ages erupted throughout Europe, with surprisingly consequential backlashes. The mystery continues to this day, but in the end both men likely deserve significant credit.

In closing, even though it is common to focus primarily on Newton when one wants to give a passing reference to some of the history of calculus without falling deeply into controversial weeds, it's some of Leibniz's personal touches that we most explicitly duplicate time and time again: The familiar notation we use repeatedly, including "$\frac{d}{dx} f(x)$" and "$\int f(x)\, dx$," are Leibniz's own. And this contribution should not be overlooked — a great thing about calculus is its usability, and Leibniz's notation is an important part of this usability. His legacy is firmly rooted.

— Notable Exercises —

- Exercise 8.11 asks you to prove some special cases of what is sometimes called the *inverse power rule*, which in particular implies that an antiderivative of x^n is $\frac{x^{n+1}}{n+1}$. A twenty-two-year-old Isaac Newton discovered this result, recording the below table in his college notebook:[21]

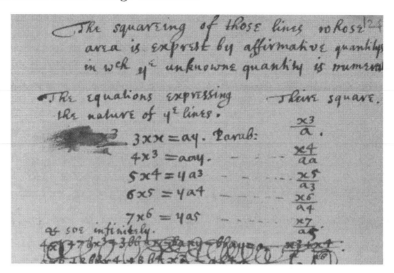

The language and notation has evolved some since these first pen strokes. The left column he labeled "The equations expressing the nature of ye lines." Their antiderivatives give the area under the curve, which used to be called "squaring the curve." So the right column he labels "Theire square." Note also that he viewed each of these expressions as having dimensions, and therefore divided by various powers of a, which acted as an arbitrary unit length.[22]

- In Chapter 6 on continuity, Theorem 6.12 gave a method to convert theorems on sequences into theorems on functions, which quickly produced func-y versions of the sequence limit laws and the sequence squeeze theorem. Exercise 8.8 is in some ways similar: It (halfway) turns the esoteric integrability criterion of Definitions 8.8 and 8.10 into a much more practical form.

- You are asked for two proofs that Thomae's function is integrable, one in Exercise 8.17 and one in Exercise 8.23. Knowing multiple proofs of a theorem is an underrated way to understand a result. A minor reason for proving things in this text is to know the theorem is true. Proofs foremost allow us to understand *why* it is true, so digesting multiple proofs richens our understanding. And in the same spirit, Exercise 8.27 asks for a second proof of the intermediate value theorem (Theorem 6.38).

---

[21] 9/10. Remember your $+C$ next time, Isaac.

[22] The above image was reproduced by kind permission of the Syndics of Cambridge University Library. Their collection is amazing — check it out. The above page of his notebook is here: https://cudl.lib.cam.ac.uk/view/MS-ADD-04000/260. And you can flip through each page!

— Exercises —

Exercise 8.1. Travell, Morgan and Quin are splitting an Uber to get to their homes; they all live on the same straight street, which is also the same street they are getting picked up on. After driving 10 miles on this street they reach Travell's house and he gets out, after another 10 miles they reach Morgan's house and she gets out, after another 10 miles they reach Quin's house and he gets out.

The cost of the Uber is \$1.50 per mile, so in total they have to pay \$45. They discuss how much should each person have to pay for it to be fair. They agree that since their rides were different distances, they should pay different amounts, but Travell and Quin come up with different ways to divide the cost. Travell sends his reasoning, and it looks like this:

$$\begin{cases} \text{Travell} & = \frac{1}{3} \cdot 15 = \$5 \\ \text{Morgan} & = \frac{1}{3} \cdot 15 + \frac{1}{2} \cdot 15 = \$5 + \$7.50 = \$12.50 \\ \text{Quin} & = \frac{1}{3} \cdot 15 + \frac{1}{2} \cdot 15 + 15 = \$5 + \$7.50 + \$15 = \$27.50 \end{cases}$$

But Quin sends his reasoning and it looks like this:

Travell: ▢

Morgan: ▢ ▢

Quin: ▢ ▢ ▢ $\implies$ $\begin{cases} \text{Travell} & = \frac{1}{6} \cdot 45 = \$7.50 \\ \text{Morgan} & = \frac{2}{6} \cdot 45 = \$15 \\ \text{Quin} & = \frac{3}{6} \cdot 45 = \$22.50 \end{cases}$

Explain each person's reasoning. Why do they get different answers? Who do you think is correct? Explain why theirs is better than the other.

Exercise 8.2. Prove that if A and B are nonempty bounded sets and $A \subseteq B$, then $\sup(A) \le \sup(B)$ and $\inf(B) \le \inf(A)$.

Exercise 8.3. Suppose A and B are nonempty sets of real numbers such that for any $x \in A$ and $y \in B$ we have $x \le y$. Prove that $\sup(A) \le \inf(B)$.

Exercise 8.4. Let $f : [0,2] \to \mathbb{R}$ be defined by $f(x) = x^2 - x$, and let $P = \{0, 1, \frac{3}{2}, 2\}$. Compute $U(f, P)$ and $L(f, P)$ with respect to this particular partition P.

Exercise 8.5. Suppose $a \le b < c \le d$ and that f is integrable on $[a, d]$. Prove that f is also integrable on $[b, c]$.

Exercise 8.6. Give an example of a function $f : [0,1] \to \mathbb{R}$ for which f is *not* integrable, but $(f(x))^2$ *is* integrable.

Exercise 8.7. Recall that the *modified Dirichlet function* is defined to be

$$g(x) = \begin{cases} x & \text{if } x \in \mathbb{Q}, \\ 0 & \text{if } x \notin \mathbb{Q}. \end{cases}$$

(a) Let P be a partition of $[0, 4]$. Compute $L(g, P)$.

(b) Find $\inf\{U(g, P) : P \text{ a partition of } [0, 4]\}$.

Exercise 8.8. Prove Corollary 8.15. That is, prove that if $f : [a, b] \to \mathbb{R}$ is integrable, then there exists a sequence P_n of partitions of $[a, b]$ for which

$$\lim_{n \to \infty} [U(f, P_n) - L(f, P_n)] = 0.$$

Exercise 8.9. In this exercise you do <u>not</u> need to prove that your example works. Give an example of a function $f : [0, 2] \to \mathbb{R}$ which has the following two properties:

- When restricted to the domain $[0, 1]$ (so now $f : [0, 1] \to \mathbb{R}$), we have

$$L(f, P) < U(f, P)$$

    for every partition P of $[0, 1]$.

- But when restricted to the domain $[1, 2]$ (so now $f : [1, 2] \to \mathbb{R}$), we have

$$L(f, P) = U(f, P)$$

    for every partition P of $[1, 2]$.

Exercise 8.10. Give an example of numbers a and b, and of integrable functions $f, g : [a, b] \to \mathbb{R}$, where

$$U(f + g, P) < U(f, P) + U(g, P)$$

for every partition P of $[a, b]$. Make sure to prove your answer.

Exercise 8.11.

(a) Prove that $\displaystyle\int_0^b x \, dx = \frac{b^2}{2}$ by considering partitions into n equal subintervals.

(b) Prove that $\displaystyle\int_0^b x^2 \, dx = \frac{b^3}{3}$ by considering partitions into n equal subintervals.

(c) Prove that $\displaystyle\int_0^b x^3 \, dx = \frac{b^4}{4}$ by considering partitions into n equal subintervals.

Exercise 8.12. Consider the function $f : [0,3] \to \mathbb{R}$ given by

$$f(x) = \begin{cases} 1 & \text{if } x \in [0,1) \\ 2 & \text{if } x \in [1,2) \\ 3 & \text{if } x \in [2,3] \end{cases}$$

Using the integrals analytically theorem, prove that f is integrable.

Exercise 8.13. Consider the function $s : [0,2] \to \mathbb{R}$ given by

$$s(x) = \begin{cases} 1 & \text{if } x \in [0,1) \\ 5 & \text{if } x = 1 \\ 2 & \text{if } x \in (1,2] \end{cases}$$

Using the integrals analytically theorem, prove that s is integrable.

Exercise 8.14. Let f be the function graphed below, where the zig-zag pattern continues, and where $f(0) = 0$. Is f integrable on $[0,2]$?

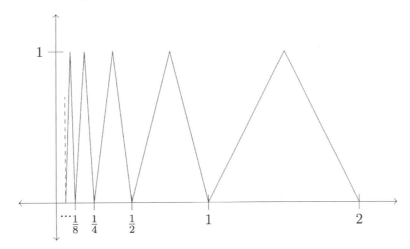

Exercise 8.15. Suppose f and g are integrable on $[a,b]$.

(a) Prove that if there exists some $c \in [a,b]$ such that $f(x) = g(x)$ for all $x \neq c$, then
$$\int_a^b f(x)\,dx = \int_a^b g(x)\,dx.$$

(b) If $f(x) = g(x)$ for all but countably many x values in $[a,b]$, must it be the case that $\int_a^b f(x)\,dx = \int_a^b g(x)\,dx$?

Exercise 8.16. A function f is *strictly increasing* on $[a,b]$ if for any x_1 and x_2 from $[a,b]$ where $x_1 < x_2$, we have that $f(x_1) < f(x_2)$. Prove that if f is strictly increasing on $[a,b]$, then f is integrable.

Exercise 8.17.   Consider Thomae's function h restricted to $[0, 2]$. That is, $h : [0, 2] \to \mathbb{R}$ is given by

$$h(x) = \begin{cases} 1/n & \text{if } x \neq 0 \text{ and } x = m/n \in \mathbb{Q} \text{ in lowest terms with } n > 0 \\ 0 & \text{if } x \notin \mathbb{Q} \\ 1 & \text{if } x = 0 \end{cases}$$

This function is pictured on the cover of this book. In Example 6.7 we discussed how h is continuous at every irrational number and discontinuous at every rational number. In this question you will prove that Thomae's function is integrable. To do this, complete the following steps:

(a) Prove that $L(h, P) = 0$ for any partition P of $[0, 2]$.

(b) Note why h is bounded.

(c) Let $\varepsilon > 0$. Determine whether there are finitely many points x such that $h(x) > \varepsilon/4$, or infinitely many such points. Explain your answer.

(d) Explain how to construct a partition P_ε of $[0, 2]$ where $U(h, P_\varepsilon) < \varepsilon$. And prove that partition works.

(e) State which theorem completes this proof.

Exercise 8.18.

(a) Suppose f is continuous on $[a, b]$ and $\int_a^x f(t)dt = 0$ for all $x \in [a, b]$. Prove that $f(x) = 0$ for all $x \in [a, b]$.

(b) Suppose f is continuous on $[a, b]$ and $\int_a^x f(t)dt = \int_x^b f(t)dt$ for all $x \in [a, b]$. Prove that $f(x) = 0$ for all $x \in [a, b]$.

Exercise 8.19.   For each function f_k below, find a formula for $F_k(x) = \int_{-1}^x f_k(t) \, dt$. Where is F_k continuous? Where is F_k differentiable? Where does $F_k' = f_k$? You do not need to prove your answers.

(a) $f_1 : [-1, 1] \to \mathbb{R}$ given by $f_1(x) = |x|$.

(b) $f_2 : [-1, 1] \to \mathbb{R}$ given by

$$f_2(x) = \begin{cases} 1 & \text{if } x < 0 \\ 2 & \text{if } x \geq 0. \end{cases}$$

(c) $f_3 : [-1, 1] \to \mathbb{R}$ given by

$$f_3(x) = \begin{cases} 1 & \text{if } x \neq 0 \\ 2 & \text{if } x = 0. \end{cases}$$

Exercise 8.20.   Suppose f and g are continuous functions on $[a, b]$, and $g(x) \geq 0$ for all $x \in [a, b]$. Prove that there exists some $c \in [a, b]$ such that

$$\int_a^b f(x)g(x)\,dx = f(c)\int_a^b g(x)\,dx.$$

Exercise 8.21.   Suppose that f is continuous on $[0, 1]$. Prove that

$$\lim_{n \to \infty} \int_0^1 f(x^n)\,dx = f(0).$$

Exercise 8.22.   The *average value* of an integrable function f on an interval $[a, b]$ is defined to be

$$\mathrm{avg}(f) = \frac{1}{b - a}\int_a^b f.$$

(a) Explain in your own words why this is a sensible definition of "average value."

(b) Prove that if f is the derivative of another function F, then $\mathrm{avg}(f)$ is the average rate of change of F over $[a, b]$.

(c) Prove that

$$\int_a^b \mathrm{avg}(f) = \int_a^b f.$$

(d) Suppose that $m \leq f(x) \leq M$ for all $x \in [a, b]$. Prove that $m \leq \mathrm{avg}(f) \leq M$.

(e) Prove that if f is continuous, then f achieves its average value: precisely, there is some $c \in [a, b]$ such that $f(c) = \mathrm{avg}(f)$.

(f) Give an example to show that the previous part cannot be extended to all discontinuous f.

Exercise 8.23.   Use Theorem 8.24 to provide a second proof (easier than the way outlined in Exercise 8.17) that Thomae's function is integrable.

Exercise 8.24.   Assume that $f : [a, b] \to \mathbb{R}$ and $g : [a, b] \to \mathbb{R}$ are integrable and $k \in \mathbb{R}$. In this exercise we will use Theorem 8.24 to prove that kf, $f + g$ and fg are also integrable. To that end, let $\mathcal{D}_f$ be the set of discontinuities of f; likewise for $\mathcal{D}_g$, $\mathcal{D}_{kf}$, $\mathcal{D}_{f+g}$ and $\mathcal{D}_{fg}$.

(a) Prove that if sets A and B have measure zero and $C \subseteq A \cup B$, then C has measure zero.

(b) Prove that $\mathcal{D}_{kf} \subseteq \mathcal{D}_f$, $\mathcal{D}_{f+g} \subseteq \mathcal{D}_f \cup \mathcal{D}_g$, and $\mathcal{D}_{fg} \subseteq \mathcal{D}_f \cup \mathcal{D}_g$.

(c) Explain why this implies that kf, $f + g$ and fg are all integrable.

Exercise 8.25. Give an example of each of the following, or state that no such example exists.

(a) A non-empty compact set of measure zero.

(b) A non-empty open set of measure zero.

Exercise 8.26.

(a) Assume f is integrable on $[a, b]$, and $f(x) \geq 0$ for all x. Moreover, assume that $\int_a^b f(x)\, dx > 0$. Prove that there are infinitely many points x for which $f(x) > 0$.

(b) Assume g is integrable on $[a, b]$ and $g(x) \geq 0$ for all x. If $g(x) > 0$ for an infinite number of values of x, must it be the case that $\int_a^b g(x)\, dx > 0$?

Exercise 8.27. Use the fundamental theorem of calculus (Theorem 8.32) and Darboux's theorem (Theorem 7.20) to give a second proof of the intermediate value theorem (Theorem 6.38).

Exercise 8.28.

(a) Suppose f is integrable on $[a, b]$ and $f(x) \geq 0$ for all x. Prove that $\int_a^b f(x)\, dx \geq 0$.

(b) Suppose f and g are integrable on $[a, b]$ and $f(x) \geq g(x)$ for all x. Prove that
$$\int_a^b f(x)\, dx \geq \int_a^b g(x)\, dx.$$

(c) Suppose f is continuous on $[a, b]$ and $f(x) \geq 0$ for all x. Prove that if $f(x_0) > 0$ for some $x_0 \in [a, b]$, then $\int_a^b f(x)\, dx > 0$.

(d) Give an example of a function f on $[a, b]$ where $f(x) \geq 0$ for all x, $f(x_0) > 0$ for some $x_0 \in [a, b]$, and $\int_a^b f(x)\, dx = 0$.

Exercise 8.29.

(a) Let $F(x) = \int_0^x x f(t)\, dt$. What is $F'(x)$?

(b) Prove that if f is continuous, then
$$\int_0^x f(t)(x - t)\, dt = \int_0^x \left(\int_0^t f(s)\, ds \right) dt.$$

Exercise 8.30. Find a function f such that

$$\int_0^x tg(t)\,dt = x^2 + 2x^3.$$

Exercise 8.31. Prove that

$$\int_1^a \frac{1}{x}\,dx + \int_1^b \frac{1}{x}\,dx = \int_1^{ab} \frac{1}{x}\,dx.$$

Exercise 8.32. Prove that

$$\int_{ca}^{cb} f(x)\,dx = c \cdot \int_a^b f(cx)\,dx.$$

Exercise 8.33. Define a function $L\colon (0, \infty) \to \mathbb{R}$ by

$$L(x) = \int_1^x \frac{1}{t}\,dt.$$

You should intuitively think of this as the natural log function — although we haven't formally defined that yet.

(a) What is $L(1)$? Explain why L is differentiable and compute its derivative.

(b) Prove the identity $L(xy) = L(x) + L(y)$.

(c) Show that $L(x/y) = L(x) - L(y)$.

(d) Define
$$\gamma_n = \left(1 + \tfrac{1}{2} + \tfrac{1}{3} + \cdots + \tfrac{1}{n}\right) - L(n).$$

Prove that the sequence $(\gamma_n)_{n \geq 1}$ converges. The constant $e = \lim \gamma_n$ is called the *Euler–Mascheroni constant*.

(e) By considering the sequence $\gamma_{2n} - \gamma_n$ or otherwise, show that $L(2)$ is the limit of the alternating harmonic series.

(f) Prove that $L(x^y) = yL(x)$ for all $x \in (0, \infty)$, $y \in \mathbb{R}$. Deduce that $L(e^x) = x$ for all $x \in \mathbb{R}$. Deduce also that $\frac{d}{dx}e^x = e^x$.

Exercise 8.34. Suppose that $f : [a, b] \to \mathbb{R}$ is bounded. Suppose further that f is continuous at every point in $[a, b]$ with the exception of a single point $x_0 \in (a, b)$. Prove that f is integrable on $[a, b]$.

Exercise 8.35. Suppose $f : [a, b] \to \mathbb{R}$ is integrable.

(a) Prove that there is some $x_0 \in [a, b]$ such that

$$\int_a^{x_0} f(t)\, dt = \int_{x_0}^b f(t)\, dt.$$

(b) Give an example showing that it's not always the case that $x_0 \in (a, b)$.

Exercise 8.36. Suppose $f : [a, b] \to \mathbb{R}$ is integrable. Follow the following steps to prove that f is continuous at infinitely many points (a dense set, in fact). Parts (a)–(e) will prove that there exists at least one point $c \in (a, b)$ where f is continuous at c.

(a) By a theorem from this chapter, there exists a partition $P = \{x_0, x_1, \ldots, x_n\}$ of $[a, b]$ with $U(f, P) - L(f, P) < b - a$. Which theorem gives this?

(b) Under the partition P, prove that $M_i - m_i < 1$ for some i. Let $I_1 = [x_{i-1}, x_i]$; that is,
$$\sup\left(\{f(x) : x \in I_1\}\right) - \inf\left(\{f(x) : x \in I_1\}\right) < 1.$$

(c) Prove that there exists a subinterval $I_2 \subseteq I_1$ where
$$\sup\left(\{f(x) : x \in I_2\}\right) - \inf\left(\{f(x) : x \in I_2\}\right) < \frac{1}{2}.$$

(d) Prove that, in general, there exists a subinterval $I_{k+1} \subseteq I_k$ where
$$\sup\left(\{f(x) : x \in I_{k+1}\}\right) - \inf\left(\{f(x) : x \in I_{k+1}\}\right) < \frac{1}{k+1}.$$

(e) Apply the *nested intervals theorem* (Exercise 1.34) to prove that there exists a point c at which f is continuous.

(f) Using what you just proved, show why f is continuous on a dense subset of $[a, b]$.

Exercise 8.37. Suppose that $f : [a, b] \to [0, \infty)$ is a continuous function with maximum value 2. Prove that
$$\lim_{n \to \infty} \left(\int_a^b f^n\right)^{1/n} = 2.$$

Exercise 8.38. Suppose that $f : [0, 1] \to \mathbb{R}$ is integrable. Prove that there is a point $c \in (0, 1)$ at which f is continuous.

Exercise 8.39.

(a) Prove that if f is integrable on $[a, b]$ and $m \leq f(x) \leq M$ for all x in $[a, b]$, then

$$\int_a^b f(x)\, dx = (b - a) \cdot \gamma$$

for some γ with $m \leq \gamma \leq M$.

(b) Prove that if f is continuous on $[a, b]$, then

$$\int_a^b f(x)\, dx = (b - a) \cdot f(x_0)$$

for some $x_0 \in [a, b]$.

(c) Show by example that (b) need not hold if f is not continuous.

(d) Prove the following more general result (part (c) is the special case $g(x) = 1$ for all x), known as the *general mean value theorem for integrals*: If f is continuous on $[a, b]$ and g is integrable and nonnegative on $[a, b]$, then

$$\int_a^b f(x)g(x)\, dx = f(x_0) \cdot \int_a^b g(x)\, dx$$

for some $x_0 \in [a, b]$.

Exercise 8.40.   Say that a set $X \subseteq [a, b]$ has *content zero* if for every $\varepsilon > 0$ there are finitely many intervals $[a_1, b_1], [a_2, b_2], \ldots, [a_n, b_n]$ such that $X \subseteq \bigcup_{k=1}^n [a_k, b_k]$ and $\sum_{k=1}^n (b_k - a_k) < \varepsilon$. (The first condition says that the intervals cover X; the second says that their total length is small.)

(a) Show that every finite set has content zero.

(b) Show that the Cantor set C has content zero. (A definition of the Cantor set is in Example B.1 of Appendix B.

(c) Suppose that f is a bounded function on $[a, b]$ and that the set of points $x \in [a, b]$ at which f is discontinuous has content zero. Prove that f is integrable.

(d) Prove that the function

$$h(x) = \begin{cases} 1 & \text{if } x \in C \\ 0 & \text{if } x \notin C \end{cases}$$

is integrable and compute $\int_0^1 h$. (Here C is the Cantor set.)

Exercise 8.41. Prove that the set $\{\int_1^b \frac{1}{x^2}\, dx : b \in [1, \infty)\}$ is bounded above and compute its supremum, which we might sensibly denote $\int_1^\infty \frac{1}{x^2}\, dx$.

Exercise 8.42. A function $s : [a, b] \to \mathbb{R}$ is called a *step function* if there is a partition $P = \{x_0, x_1, \ldots, x_n\}$ of $[a, b]$ such that s is constant on each interval (x_{i-1}, x_i).

(a) Prove that if f is integrable on $[a, b]$, then for any $\varepsilon > 0$ there is a pair of step functions s_1 and s_2 where $s_1(x) \le f(x) \le s_2(x)$ for all x, and

$$\int_a^b [f(x) - s_1(x)]\, dx < \varepsilon \qquad \text{and} \qquad \int_a^b [s_2(x) - f(x)]\, dx < \varepsilon.$$

(b) Suppose that for any $\varepsilon > 0$ there is a pair of step functions s_1 and s_2 where $s_1(x) \le f(x) \le s_2(x)$ for all x, and

$$\int_a^b [f(x) - s_1(x)]\, dx < \varepsilon \qquad \text{and} \qquad \int_a^b [s_2(x) - f(x)]\, dx < \varepsilon.$$

Prove that f is integrable.

(c) Give an example of a function f which is not a step function but for which $\int_a^b f(x)\, dx = L(f, P)$ for some partition P of $[a, b]$. You don't need to prove your answer.

Exercise 8.43. Prove that if $f : [a, b] \to \mathbb{R}$ is integrable, then for any $\varepsilon > 0$ there is a continuous function g where $g(x) \le f(x)$ for all x, and

$$\int_a^b f(x)\, dx - \int_a^b g(x)\, dx < \varepsilon.$$

Exercise 8.44. Prove the *Cauchy-Bunyakovsky-Schqarz inequality*, which says that if f and g are integrable functions on $[a, b]$, then

$$\left(\int_a^b fg \right)^2 \le \left(\int_a^b f \right) \left(\int_a^b g \right).$$

— Open Questions —

Question 1. Under which conditions on f does the inequality

$$\int_a^b f^{\alpha+\beta}(x)\,dx \geq \left(\int_a^b (x-a)^\alpha f^\beta(x)\,dx\right)^\lambda$$

hold for some α, β and λ, and for such an f, what are the most general conditions on α, β and λ for which the above inequality holds?

Question 2. Under which conditions on f does the inequality

$$\frac{\displaystyle\int_a^b f^{\alpha+\beta}(x)\,dx}{\displaystyle\int_a^b f^{\alpha+\gamma}(x)\,dx} \geq \frac{\left(\displaystyle\int_a^b (x-a)^\alpha f^\beta(x)\,dx\right)^\delta}{\left(\displaystyle\int_a^b (x-a)^\alpha f^\gamma(x)\,dx\right)^\lambda}$$

hold for some $\alpha, \beta, \gamma, \delta$ and λ, and for such an f, what are the most general conditions on α, β, γ and λ for which the above inequality holds?

Question 3. Let $h : [0,\infty) \to \infty$ be a non-negative function and let $\alpha > 0$. Prove that if

$$\int_0^1 \frac{h(tx)}{x}(1-x)^{n-1}\,dx \leq t^\alpha$$

for all $t \in [0,\infty)$, then

$$\int_0^\infty \frac{h(t)}{t+t^{2\alpha+1}}\,dt \leq \frac{\pi}{2}\prod_{k=1}^{n-1}\left(1+\frac{\alpha}{k}\right).$$

Chapter 9:   Sequences and Series of Functions

9.1   Introduction to Pointwise Convergence

> **Recall.**
>
> **Recall 9.1.**
>
> - In Chapter 3 we studied sequences of numbers. For example, if $a_k = k^2$, then (a_k) denotes the sequence $a_1, a_2, a_3, a_4, \ldots$. For this particular a_k, this is the sequence $1, 4, 9, 16, \ldots$.
>
> - $a_k \to a$ if for every $\varepsilon > 0$ there is some $N \in \mathbb{N}$ for which $|a_k - a| < \varepsilon$ for all $k \geq N$.

In this chapter we will study sequences of *functions*. Instead of having each term of the sequence being a number, each term is a function. We begin by generalizing the above notation.

> **Notation.**
>
> **Notation 9.2.** If, for each $k \in \mathbb{N}$, we have a function $f_k : A \to \mathbb{R}$, then
>
> $$f_1, f_2, f_3, f_4, \cdots$$
>
> is a *sequence of functions* and is denoted (f_k).

With sequences of numbers we studied convergence, and with sequences of functions we also have a notion of convergence. (Note: This is one of two types of functional convergence we will study.)

> **Definition.**
>
> **Definition 9.3.** Suppose (f_k) is a sequence of functions, each defined on $A \subseteq \mathbb{R}$. The sequence (f_k) of functions *converges pointwise* to a function $f : A \to \mathbb{R}$ if, for each $x_0 \in A$,
>
> $$\lim_{k \to \infty} f_k(x_0) = f(x_0).$$
>
> In this case, we write "$f_k \to f$ pointwise."

It is important to note that since x_0 is fixed, the sequence $f_k(x_0)$ is a sequence of *numbers*, it's *not* a sequence of functions. Indeed, $f_1(x_0) \in \mathbb{R}$, $f_2(x_0) \in \mathbb{R}$, $f_3(x_0) \in \mathbb{R}$, and so on. So the sequence (f_k) of functions *converges pointwise* to a function $f : A \to \mathbb{R}$ if, for each $x_0 \in A$, the sequence of real numbers $(f_k(x_0))$ converges to the real number $f(x_0)$.

The following is an example of this. As you are reading through it, see if you can think of what function this sequence of functions is converging pointwise to. This is challenging, as it may be surprising that this sequence converges to anything at all!

Example 9.4. For each $k \in \mathbb{N}$, let $f_k : \mathbb{R} \to \mathbb{R}$ be given by $f_k(x) = \dfrac{x^2 + kx}{k}$. That is, for each fixed $k \in \mathbb{N}$ we have defined a function of x. Collectively we have defined a *sequence of functions*:

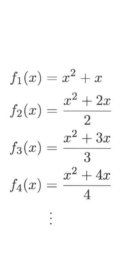

$$f_1(x) = x^2 + x$$
$$f_2(x) = \frac{x^2 + 2x}{2}$$
$$f_3(x) = \frac{x^2 + 3x}{3}$$
$$f_4(x) = \frac{x^2 + 4x}{4}$$
$$\vdots$$

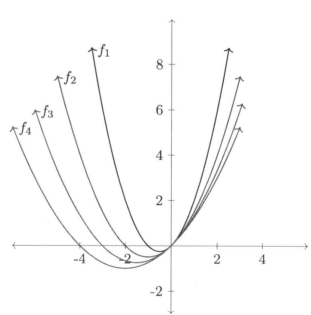

The first four iterations (f_1, f_2, f_3 and f_4) are graphed[1] above.

---

[1] Geogebra and Desmos can demonstrate the whole sequence well, with an enormously satisfying slider bar. Check it out on your own!

These functions look like nice parabolas, but take a look what happens when k gets large. Here's f_{100}:

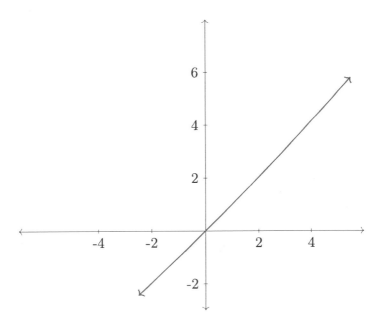

Why does it look more like a line than a parabola? We are zoomed in too much! Zooming out:

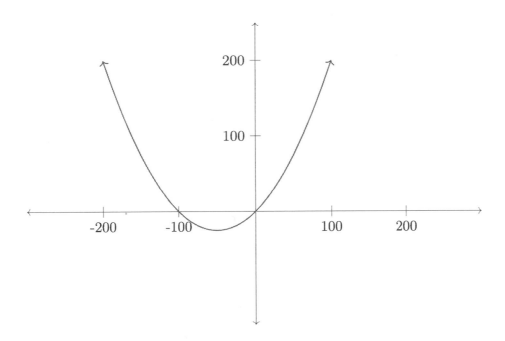

But if we then plot $f_{1,000}$ on these same x-values, it again looks like a line:

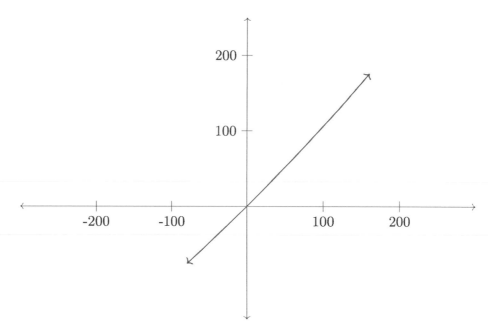

So you see, for any particular range of x-values like this, we can crank up k high enough to make the parabola look like a line (even though if you zoom out far enough it will look like a parabola again).

The way we think about this is that, for any particular $x_0 \in \mathbb{R}$, if you let $k \to \infty$ we have

$$\lim_{k \to \infty} f_k(x_0) = \lim_{k \to \infty} \frac{x_0{}^2 + kx_0}{k} = \lim_{k \to \infty} \left(\frac{x_0{}^2}{k} + x_0 \right) = 0 + x_0 = x_0.$$

That is, for each fixed x we have $\lim_{k \to \infty} f_k(x) = x$, meaning that (f_k) converges pointwise to the function $f(x) = x$. □

9.2 Continuity and Functional Convergence

Suppose we know $f_k \to f$ pointwise. We want to start investigating which properties of f_k will carry over to f. Here's a basic question:

Question 9.5. If $f_k \to f$ pointwise, and each f_k is continuous, must f also be continuous?

After some thought you might realize that the answer is 'no.' What's a counterexample? We need to come up with continuous functions f_k and a discontinuous function f for which $f_k \to f$ pointwise. Let's start with f. One simple way to have a function be discontinuous is for it to have a jump discontinuity at just a single point, as the next example shows.

Example 9.6. Let $f_k : [0,1] \to \mathbb{R}$ where $f_k(x) = x^k$. This gives us a sequence of functions (f_k). The first five iterations ($k = 1,2,3,4,5$) look like this:

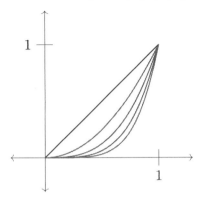

As you can see, for any fixed $x \in [0,1)$, as k gets large we see that x^k is tending to zero. For example, fix $x = 0.8$; the point $f_k(0.8)$ is marked on each of the graphs, and you can tell it's tending towards zero:

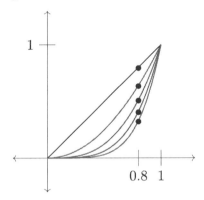

For x-values close to 1 it takes longer, but all will eventually get arbitrarily close to 0. For instance, below is $f_{20}(x)$'s graph (with $f_{20}(0.8)$ again marked):

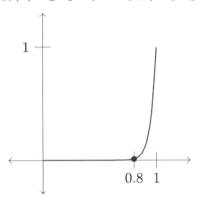

Plugging a few points into the function, we have $f_{20}(0.8) = 0.01$, $f_{20}(0.9) = 0.12$, and $f_{20}(0.99) = 0.82$, so all but the last 10% of the domain is becoming quite close to 0 by $k = 20$. And $f_{500}(0.99) = 0.01$, so by $k = 500$ far less than 1% of interval remains far from 0.

There is of course one point that will never get close to 0: the point $x = 1$. For any k, $f_k(1) = 1^k = 1$. To summarize:

$$\begin{cases} \lim\limits_{k \to \infty} f_k(x) = 0 & \text{if } x \in [0, 1), \\ \lim\limits_{k \to \infty} f_k(x) = 1 & \text{if } x = 1. \end{cases}$$

So we have seen that the sequence of continuous functions $f_k = x^k$ converges pointwise to the discontinuous function

$$f(x) = \begin{cases} 0 & \text{if } x \in [0, 1), \\ 1 & \text{if } x = 1. \end{cases}$$

$\square$

This is kind of a bummer. We might have hoped that pointwise convergence would preserve the continuity property, but sadly it is not strong enough. Now, there is another, stronger notion of a functional limit that *is* strong enough to maintain continuity in the limit. To understand where it comes from, let's try to prove that pointwise convergence preserves continuity, and see where the proof breaks down (and hence how to fix it).

$Almost$-correct proof that pointwise convergence preserves continuity.
Assume that each $f_k : A \to \mathbb{R}$ is continuous, and that $f_k \to f$ pointwise. Fix a point $c \in A$ and an $\varepsilon > 0$. To show that f is continuous at c we would have to show that there is some $\delta > 0$ such that

$$|x - c| < \delta \qquad \Longrightarrow \qquad |f(x) - f(c)| < \varepsilon.$$

Rewriting $|f(x) - f(c)|$ in a tricky way and applying the triangle inequality twice,

$$\begin{aligned} |f(x) - f(c)| &= |f(x) - f_k(x) + f_k(x) - f_k(c) + f_k(c) - f(c)| \\ &\leq |f(x) - f_k(x)| + |f_k(x) - f_k(c)| + |f_k(c) - f(c)| \\ &= |f_k(x) - f(x)| + |f_k(x) - f_k(c)| + |f_k(c) - f(c)|. \end{aligned}$$

Notice how this works for *any* choice of k. If we can make each term less than $\varepsilon/3$, then we would have a complete proof; so what goes wrong? You might think:

- We can make the first term small because $f_k \to f$ pointwise.

- We can make the second term small because each f_k is continuous.

- We can make the third term small because $f_k \to f$ pointwise.

One of these doesn't work though... Can you spot it...? It's actually the first term, but to see why, let's look at the others first, in reverse order:

- We can indeed make the third term small: Since $f_k \to f$ pointwise and c is fixed, we can choose some k_0 such that $|f_{k_0}(c) - f(c)| < \varepsilon/3$.

- We can then make the second term small: We had assumed that each f_k is continuous, and so by f_{k_0}'s continuity there exists some $\delta_0 > 0$ such that $|f_{k_0}(x) - f_{k_0}(c)| < \varepsilon/3$ for all x where $|x - c| < \delta_0$.

- But this is a problem for the first term. For each particular x, we can apply pointwise convergence to produce a big enough k_0 for which $|f_{k_0}(x) - f(x)| < \varepsilon/3$. However, each x will likely have a different "k_0" value, and so we have no guarantee that any particular k_0 is big enough for *all* of the infinitely many x for which $|x - c| < \delta_0$.

  But in the event that our k_0 is not big enough, why don't we just go back and pick a large enough k_0 so that it does work for all these x? Well first, maybe no k_0 is big enough for them all (there are infinitely many x-values and each requires a large k, so maybe there's no upper bound on the needed k's). But more to the point, even if there was a max k-value (call it k_1) that was required and we went back and switched out k_0 for this k_1, then suddenly δ_0 likely changes (to some δ_1), since it is now dependent on the function f_{k_1} and not f_{k_0}. So we found a new k_1 that works for all the x-values for which $|x - c| < \delta_0$, but with our new k_1 we instead need it to work for all the x-values for which $|x - c| < \delta_1$!

  See? Because x depends on δ_0, which in turn depends on k_0, we can't just change k_0 without messing everything up.

So the problem is in controlling $|f_k(x) - f(x)|$ for all x. Indeed, in Example 9.6 each $f_k(x) = x^k$ was continuous, but as k got large, x-values close to 1 required much smaller δ-values than did the x away from 1. This shows why we were unable to control the $|f_k(x) - f(x)|$ term in this sequence.

Pointwise convergence allows us to control each individual x, but we want to control them all at the same time; that is, we want *uniform* control. This directly motivates the following definition of *uniform convergence*.

Uniform Convergence

Definition.

Definition 9.7. Let (f_k) be a sequence of functions defined on a set A. We say (f_k) *converges uniformly on A* to a function f if, for every $\varepsilon > 0$, there exists an $N \in \mathbb{N}$ such that $|f_k(x) - f(x)| < \varepsilon$ for all $k \geq N$ and for all $x \in A$.

When this happens we write "$f_k \to f$ uniformly."

Note that if $f_k \to f$ uniformly, then certainly $f_k \to f$ pointwise, so this definition is strictly stronger than our definition of pointwise convergence.

We constructed this form of convergence so that it will preserve continuity, so of course our first result is that it does indeed do this.

> **Proposition.**
>
> **Proposition 9.8** (f_k *continuous,* $f_k \to f$ *uniformly* $\Rightarrow$ f *continuous*). Assume that each $f_k : A \to \mathbb{R}$ is continuous at some $c \in A$. If (f_k) converges uniformly to f, then f is continuous at c.

Proof. Fix a $c \in A$ and let $\varepsilon > 0$. By the uniformity assumption, there exists some $N \in \mathbb{N}$ such that

$$|f_k(x) - f(x)| < \frac{\varepsilon}{3}$$

for all $k \geq N$ and for all $x \in A$. In particular, this means that $|f_N(c) - f(c)| < \frac{\varepsilon}{3}$. We also assumed that each f_k is continuous at c. In particular, since f_N is continuous at c there must exist a $\delta > 0$ for which

$$|x - c| < \delta \qquad \text{implies} \qquad |f_N(x) - f_N(c)| < \frac{\varepsilon}{3}.$$

And so, by rewriting $|f(x) - f(c)|$ in a tricky way, applying the triangle inequality twice and using the above,

$$\begin{aligned}
|f(x) - f(c)| &= |f(x) - f_N(x) + f_N(x) - f_N(c) + f_N(c) - f(c)| \\
&\leq |f(x) - f_N(x)| + |f_N(x) - f_N(c)| + |f_N(c) - f(c)| \\
&= |f_N(x) - f(x)| + |f_N(x) - f_N(c)| + |f_N(c) - f(c)| \\
&< \frac{\varepsilon}{3} + \frac{\varepsilon}{3} + \frac{\varepsilon}{3} \\
&= \varepsilon
\end{aligned}$$

for all $x \in A$ for which $|x-c| < \delta$. Thus by the definition of continuity, f is continuous at c. $\qquad \square$

Picture of uniform convergence, and of not

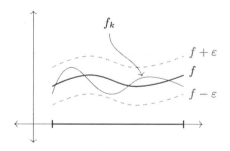

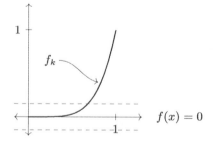

Let $\varepsilon > 0$. Then $f_k \to f$ uniformly if there exists some N for which, for every $f_k(x)$ where $k > N$, $f_k(x)$ is entirely inside the dashed ε region.

Next, $f_k \not\to f$ uniformly when there is an $\varepsilon > 0$ for which, for every N, there is some $k > N$ where f_k is *not* entirely inside the dashed ε region.

9.3 Other Properties with Functional Convergence

We want to investigate which properties are preserved by pointwise and uniform convergence. We aim to complete the following chart.

Assume that each f_k has the below property	If $f_k \to f$ pointwise, must f satisfy property?	If $f_k \to f$ uniformly, must f satisfy property?
Continuous	No, by Example 9.6	Yes, by Proposition 9.8
Bounded	???	???
Unbounded	???	???
Uniformly Continuous	???	???
Differentiable	???	???
Integrable	???	???

A few of these will be left to the exercises, but ticking off the rest will help us better understand the differences between pointwise and uniform convergence. It's also pretty fun to think about finding a proof or counterexample of each of these. Let's start at the top — boundedness.

Boundedness

For each of these I strongly encourage you to think on your own about what you think the answer is. And if you think the answer is no, try to come up with an example demonstrating it.

Assuming you have spent time thinking about whether each f_k being bounded implies f is, you can read on. We will show that for pointwise convergence boundedness might not carry to the limit, but with uniform convergence it does. We begin with an example showing the pointwise claim.

Example 9.9. Assume that each f_k is bounded and $f_k \to f$ pointwise. It need not be the case that f is also bounded. Here is an example of that:

Define $f_k : (0,1] \to \mathbb{R}$ by

$$f_k(x) = \begin{cases} \frac{1}{x} & \text{if } x \in [1/k, 1] \\ 0 & \text{if } x \in (0, 1/k). \end{cases}$$

Clearly $0 \leq f_k(x) \leq k$ for each x, so each f_k is bounded.[2] However, for any $x \in (0,1]$, by the Archimedean principle there is some $N \in \mathbb{N}$ for which $\frac{1}{N} < x$. And thus for all $k \geq N$ we have $f_k(x) = 1/x$. So clearly for each x we have

$$\lim_{k \to \infty} f_k(x) = \frac{1}{x}.$$

---

[2]Recall again that for each particular f_k, the number k is fixed. So, e.g., $0 \leq f_{100}(x) \leq 100$. If we knew that $g : \mathbb{R} \to \mathbb{R}$ had the property $g(x) \leq x$, then this would not mean that g is bounded, because x is variable. But saying $f_k(x) \leq k$ does mean that f_k is bounded above, since k is fixed.

That is, by letting $f(x) = 1/x$ we know that each f_k is bounded, f is unbounded, and $f_k \to f$ pointwise on $(0, 1]$. So we are done. $\square$

But while pointwise convergence does not preserve boundedness, we prove next that uniform convergence does.

> ### Proposition.
>
> **Proposition 9.10** (f_k *bounded,* $f_k \to f$ *uniformly* $\Rightarrow$ f *bounded*)**.** Assume that each $f_k : A \to \mathbb{R}$ is bounded. If (f_k) converges uniformly to f, then f is also bounded.

Proof. Since $f_k \to f$ uniformly, then (by letting $\varepsilon = 1$) we know that there exists some $N \in \mathbb{N}$ such that

$$|f_k(x) - f(x)| < 1$$

for all $k \geq N$ and $x \in A$. In particular,

$$|f_N(x) - f(x)| < 1$$

for all $x \in A$. I.e.,

$$f_N(x) - 1 < f(x) < f_N(x) + 1$$

for all $x \in A$. And since f_N was assumed to be bounded, there exist real numbers L and U such that $L \leq f_N(x) \leq U$ for all $x \in A$. Combining this with above, we see that

$$L - 1 < f(x) < U + 1$$

for all $x \in A$, proving that f is bounded. $\square$

Unboundedness

These will be left for Exercise 9.5.

Uniform continuity

In case you need a reminder, below is the definition of *uniform continuity*.

> ### Recall.
>
> **Definition 6.39.** Let $f : A \to \mathbb{R}$. We say f is *uniformly continuous* if for all $\varepsilon > 0$ there exists some $\delta > 0$ such that, for all $x, y \in A$,
>
> $$|x - y| < \delta \qquad \text{implies} \qquad |f(x) - f(y)| < \varepsilon.$$

This section will be brief. I will give you an example where each f_k is uniformly continuous and $f_k \to f$ pointwise, but yet f is not uniformly continuous. What about if the convergence is uniform? If instead $f_k \to f$ uniformly, then f *will* be uniformly continuous — but the proof of that will be left to the exercises. First up, the example.

Example 9.11. Assume that each f_k is uniformly continuous and $f_k \to f$ pointwise. It need not be the case that f is also uniformly continuous. In fact, our continuity example (Example 9.6) works:

Define $f_k : [0, 1] \to \mathbb{R}$ by $f_k(x) = x^k$. Since each f_k is a continuous function on a closed interval, by Proposition 6.40 we know that each f_k is uniformly continuous. However, as we saw in Example 9.6, $f_k \to f$ pointwise where

$$f(x) = \begin{cases} 0 & \text{if } x \in [0, 1) \\ 1 & \text{if } x = 1 \end{cases}.$$

And note that f is not uniformly continuous since it is not even continuous. □

Proposition.

Proposition 9.12 (f_k *unif. cts.,* $f_k \to f$ *uniformly* $\Rightarrow$ f *unif. cts.*). Assume that each $f_k : A \to \mathbb{R}$ is uniformly continuous. If (f_k) converges uniformly to f, then f is also uniformly continuous.

Proof. This is Exercise 9.6. □

As a quick reminder, here is where we stand in our efforts to complete the table from earlier:

Assume that each f_k has the below property	If $f_k \to f$ pointwise, must f satisfy property?	If $f_k \to f$ uniformly, must f satisfy property?
Continuous	No, by Example 9.6	Yes, by Proposition 9.8
Bounded	No, by Example 9.9	Yes, by Proposition 9.10
Unbounded	Exercise 9.5	Exercise 9.5
Uniformly Continuous	No, by Example 9.11	Yes, by Prop 9.12 and Ex 9.6
Differentiable	???	???
Integrable	???	???

9.4 Convergence of Derivatives and Integrals

Let's keep going. Next up, differentiability and integrability. As you may guess, the weakness of pointwise convergence persists. For these more sophisticated properties, though, it is less obvious right off the bat whether uniform convergence is strong enough to preserve the properties.

Differentiability

Assume each f_k is differentiable at a point c. It turns out that $f_k \to f$ pointwise does not guarantee that f is differentiable. This may not be a surprise since in Example 9.6 we showed that continuity is not preserved by pointwise convergence, and differentiability is strictly stronger than continuity. (That alone does not is not imply the same for differentiability, but it's decent motivation.) For continuity, boundedness and uniform continuity, once we insisted that the convergence is uniform then we could prove that the property carries over to f — for differentiability, though, this is not the case. Indeed, it is possible for each f_k to be differentiable at a point c and $f_k \to f$ uniformly, while f is not differentiable at c. And so for this subsection, we can give just one example that shows that neither convergence is sufficient.[3]

Example 9.13. Assume that each f_k is differentiable and $f_k \to f$ uniformly (and hence also pointwise). It need not be the case that f is also differentiable. Here is an example of that:

We will consider the functions $f_k : [-1, 1] \to \mathbb{R}$, defined by $f_k(x) = x^{1 + \frac{1}{2k-1}}$; note that f_k is indeed defined on $[-1, 1]$, since $2k - 1$ is always odd. For any $k \in \mathbb{N}$ the function f_k is differentiable at $x = 0$. (In fact, each f_k is differentiable on all of $[-1, 1]$ by the power rule.) To understand the limit of this function as $k \to \infty$, note that

$$\lim_{k \to \infty} x^{\frac{1}{2k-1}} = \begin{cases} 1 & \text{if } x \in (0, 1]; \\ -1 & \text{if } x \in [-1, 0); \\ 0 & \text{if } x = 0. \end{cases}$$

And so $f_k(x) = x^{1 + \frac{1}{2k-1}} = x \cdot x^{\frac{1}{2k-1}} \to |x|$ as $k \to \infty$. So $f_k \to f$ pointwise where $f(x) = |x|$, which is not differentiable at $x = 0$, despite each f_k being differentiable at 0. (Exercise 9.8: Show that we also have $f_k \to f$ uniformly.) $\square$

---

[3] Recall that we essentially defined uniform convergence to be just strong enough to guarantee that continuity is preserved. The fact differentiability is strictly stronger than continuity is one reason to believe that things may be different for differentiability.

The first three functions in Example 9.13 are graphed below (note how they already look very much like $f(x) = |x|$).

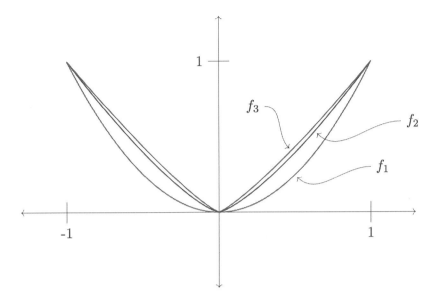

By looking at those first few graphs it seems pretty convincing that the functions do converge uniformly. Showing it requires a little finesse, though.

What's particularly amazing, is that there even exists a function g which has these properties:

1. A sequence (g_k) of functions such that $g_k \to g$ uniformly,

2. Each g_k is differentiable at a point c,

3. g is also differentiable at c,

4. And the limit $\lim\limits_{k \to \infty} g_k'(c)$ exists,

5. But yet $\lim\limits_{k \to \infty} g_k'(c) \neq g'(c)$.

Weird! The idea is that you can be getting really close to a function but yet at a particular point, by using the ε-high wiggle room, you can achieve any slope you want. In the below, for instance, the functions are getting closer and closer to 0 (whose derivative is 0 everywhere). However, right at zero they re-angle themselves so that they go through the origin with a slope of 1, giving $f_k'(0) = 1$ for each k. That's next.

Example 9.14. Let $g_k : [-2, 2] \to \mathbb{R}$ be defined by $g_k(x) = \dfrac{x}{1 + kx^2}$. The first four functions looks like this:

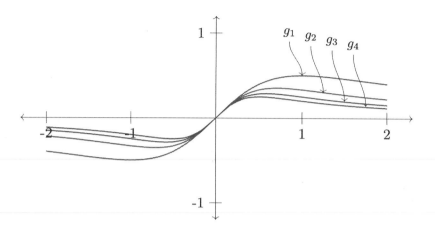

The later functions are closer to the x-axis. For instance, here is g_{20}:

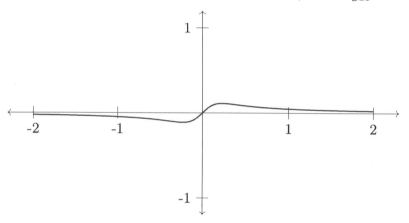

And here is g_{100}:

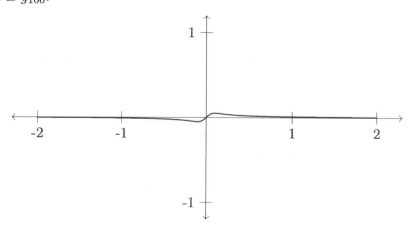

So as you can see, the sequence (g_k) is converging both pointwise and uniformly to the function $g(x) = 0$. However, despite this, each $g_k'(0) = 1$, since right at $x = 0$

each of the functions reorients themselves so that if passes through the origin with a slope of 1. Indeed, by the quotient rule,

$$\frac{d}{dx}\left(\frac{x}{1+kx^2}\right) = \frac{1 \cdot (1+kx^2) - x \cdot (2kx)}{(1+kx^2)^2} = \frac{1-kx^2}{(1+kx^2)^2}.$$

So $g'_k(0) = \dfrac{1-k(0)^2}{(1+k(0)^2)^2} = 1$.

So we have shown that each $g'_k(0) = 1$, but meanwhile the g_k's converge to the zero function $g(x) = 0$ which clearly has $g'(0) = 0$. $\qquad\square$

Integrability

Example 9.15. Assume that each f_k is integrable and $f_k \to f$ pointwise. It need not be the case that f is also integrable. For Exercise 9.9 you will give such an example. $\qquad\square$

What about for uniform limits? We showed in the last section that uniform limits do not preserve differentiability. For integrability, though, this property is preserved.

> **Proposition.**
>
> **Proposition 9.16** (f_k *integrable,* $f_k \to f$ *uniformly* $\Rightarrow$ f *integrable*). Assume that each $f_k : [a, b] \to \mathbb{R}$ is integrable. If (f_k) converges uniformly to f, then f is also integrable. Moreover, $\int_a^b f_k \, dx \to \int_a^b f \, dx$.

Before we begin, recall that if $I = [x_{i-1}, x_i]$ is a subinterval in a Darboux sum,

$$\sup(\{g(x) + h(x) : x \in I\}) \le \sup(\{g(x) : x \in I\}) + \sup(\{h(x) : x \in I\}),$$
$$\inf(\{g(x) + h(x) : x \in I\}) \ge \inf(\{g(x) : x \in I\}) + \inf(\{h(x) : x \in I\}).$$

We will use these in the following proof. As for the second, we will in particular use the equivalent statement

$$-\inf(\{g(x) + h(x) : x \in I\}) \le -\inf(\{g(x) : x \in I\}) - \inf(\{h(x) : x \in I\}).$$

Now we prove the proposition (our last dry, technical integrals proof!).

Proof. We will prove that f is integrable by using the definition of integrability. That is, we will show that f is bounded and $L(f) = U(f)$. For the former, simply note that since each f_k is integrable and $f_k \to f$ uniformly, by Proposition 9.10 we may conclude that f is also bounded.

Now we prove that $L(f) = U(f)$. Let $\varepsilon > 0$. Since $f_k \to f$ uniformly there exists some $N \in \mathbb{N}$ such that

$$|f_k(x) - f(x)| < \frac{\varepsilon}{2(b-a)}$$

for all $k \geq N$ and for all $x \in [a, b]$. Then,

$$U(f) - L(f) = U(f - f_N + f_N) - L(f - f_N + f_N).$$

So by the recall right before this proof,

$$\leq U(f - f_N) + U(f_N) - L(f - f_N) - L(f_N).$$

And since f_N is integrable we have $U(f_N) = L(f_N)$,

$$= U(f - f_N) - L(f - f_N).$$

By our choice of N above, $|f(x) - f_N(x)| < \dfrac{\varepsilon}{2(b - a)}$. That is, $\dfrac{-\varepsilon}{2(b - a)} < f(x) - f_N(x) < \dfrac{\varepsilon}{2(b - a)}$. Thus,

$$\leq U\left(\frac{\varepsilon}{2(b - a)}\right) - L\left(\frac{-\varepsilon}{2(b - a)}\right)$$
$$= \frac{\varepsilon}{2} + \frac{\varepsilon}{2}$$
$$= \varepsilon.$$

So we have that $0 \leq U(f) - L(f) \leq \varepsilon$ for all $\varepsilon > 0$. This implies that $U(f) = L(f)$, and hence that f is integrable.

Next we must prove that the sequence of numbers $(\int_a^b f_k \, dx)$ is converging to the number $\int_a^b f \, dx$. Let $\varepsilon > 0$. Since $f_k \to f$ uniformly and $\frac{\varepsilon}{b-a} > 0$, there exists some N such that $|f_k(x) - f(x)| < \frac{\varepsilon}{b-a}$ for all $k > N$ and for all $x \in [a, b]$. This implies that

$$f_k(x) - \frac{\varepsilon}{b - a} < f(x) < f_k(x) + \frac{\varepsilon}{b - a}$$

for all $k > N$ and $x \in [a, b]$, which by the monotonicity property of the integral (Corollary 8.28) implies that, for $k > N$,

$$\int_a^b \left(f_k(x) - \frac{\varepsilon}{b - a}\right) dx < \int_a^b f(x) \, dx < \int_a^b \left(f_k(x) + \frac{\varepsilon}{b - a}\right) dx.$$

Using the linearity properties of the integral (Proposition 8.26) and the fact that $\int_a^b \frac{\varepsilon}{b-a} \, dx = \frac{\varepsilon}{b-a}(b - a) = \varepsilon$, the above says that

$$\int_a^b f_k(x) \, dx - \varepsilon < \int_a^b f(x) \, dx < \int_a^b f_k(x) \, dx + \varepsilon$$

for all $k > N$. This means that

$$\left| \int_a^b f_k(x) \, dx - \int_a^b f(x) \, dx \right| < \varepsilon$$

for all $k > N$. That is, we have shown that for an arbitrary $\varepsilon > 0$ there exists some N such that $k > N$ implies that

$$\left| \int_a^b f_k(x)\, dx - \int_a^b f(x)\, dx \right| < \varepsilon.$$

By the definition of sequence convergence (Definition 3.7), $\int_a^b f_k(x)\, dx \to \int_a^b f(x)\, dx$. $\square$

Completed Table

Assume that each f_k has the below property	If $f_k \to f$ pointwise, must f satisfy property?	If $f_k \to f$ uniformly, must f satisfy property?
Continuous	No, by Example 9.6	Yes, by Proposition 9.8
Bounded	No, by Example 9.9	Yes, by Proposition 9.10
Unbounded	Exercise 9.5	Exercise 9.5
Uniformly Continuous	No, by Example 9.11	Yes, by Prop 9.12 and Ex 9.6
Differentiable	No, by Example 9.13	No, by Example 9.13
Integrable	Exercise 9.9	Yes, by Proposition 9.16

Table 9.1: Properties of the limit of (f_k)

Three Final Results

I will wrap up our discussion of sequences of functions by mentioning three more results, without proof. For the first, note from the table above there is only one major failing of uniform continuity, and that's that it does not preserve differentiability. I would like to mention that if you additionally know that the *derivatives* converge uniformly (plus one more minor assumption to ensure nothing funny is happening with the "$+C$"s), then you *can* guarantee that the original sequence is converging to a differentiable function. In essence, with these additional assumptions that lone 'no' in the right column of the table can be turned in to a 'yes.' Below is that.

> **Theorem.**
>
> **Proposition 9.17** (*Uniform limit of differentiable functions*). Let $f_k : [a, b] \to \mathbb{R}$ and assume each f_k is differentiable. If (f_k') converges uniformly to some g and there exits some $x_0 \in [a, b]$ for which $(f_k(x_0))$ converges, then (f_k) converges uniformly to some f, this f is differentiable, and $f' = g$.

The second result I'd like to mention will be useful when studying series of functions, and is a generalization of a Chapter 4 result that said a numerical sequence converges if and only if it is Cauchy. Indeed, a sequence of functions (f_k), each defined on a domain A, is said to be *Cauchy* if for every $\varepsilon > 0$ there exists some $N \in \mathbb{N}$ such that $|f_m(x) - f_n(x)| < \varepsilon$ for all $m, n \geq N$ and all $x \in A$.

> ### Proposition.
>
> **Proposition 9.18** (*Cauchy criterion for sequences of functions*). Let $f_k : A \to \mathbb{R}$. The sequence (f_k) converges uniformly if and only if (f_k) is Cauchy.

Proof. This is Exercise 9.11. □

The final result is another generalization from Chapter 3's work on sequences of numbers, in this case of the of the Bolzano-Weierstrass theorem which says this:

> ### Recall.
>
> **Theorem 3.37** (*The Bolzano-Weierstrass theorem*). Every bounded sequence has a convergent subsequence.

There is an analogous theorem for sequences of functions (f_k), where the conclusion is that there exists a subsequence of (f_k) that converges uniformly. It's a famous result known as the *Arzela-Ascoli theorem*. Bolzano-Weierstrass requires that the numerical sequences are bounded (there exists some M such that $|a_k| \leq M$ for all k), and likewise, Arzela-Ascoli requires that the functions be bounded. However note that if $f_k(x) = k$ for all x, then each f_k is bounded, but yet clearly (f_k) is not converging to any particular f. So we need more. It is not enough for each f_k to be bounded—we need the whole sequence of functions to be bounded. Indeed, we need there to exist an M such that $|f_k(x)| \leq M$ for all x and for all k; this is called being *uniformly bounded*.

Unfortunately, being uniformly bounded is alone not enough to get a convergent subsequence (either pointwise or uniform). Convince yourself that every subsequence of $(f_k(x) = x^k)$ on $[0,1]$ converges to the same limit, and no subsequence converges uniformly to that limit. The problem is that the functions are getting steeper and steeper; each is uniformly continuous, but for later and later functions in the sequence, the "ε" in the definition of uniform continuity requires a smaller and smaller "δ."

What we need is that, for every $\varepsilon > 0$, there exists a single $\delta > 0$ that works for *all* functions f_k. Then, for this δ, we have that $|x-y| < \delta$ implies that $|f_k(x) - f_k(y)| < \varepsilon$ for all $x, y \in A$ and for all $k \in \mathbb{N}$. This is called being *equicontinuous*.

> ### Theorem.
>
> **Theorem 9.19** (*The Arzela-Ascoli theorem*). If (f_k) is uniformly bounded and equicontinuous on A, then (f_k) contains a uniformly convergent subsequence.

The proof is a diagonalization-type argument, that we omit.

Brief Review of Series of Numbers

After studying sequences of numbers in Chapter 3, we moved on to studying series of numbers in Chapter 4. Likewise, now that we have studied sequences of functions, we are about to move on to *series of functions*. There are a handful of ideas from Chapter 4 that will be particularly important as we move forward. We will now briefly review those ideas.

- In Chapter 4 we had a definition for what it means for a series to converge, in terms of the sequence of partial sums (Definition 4.1). In the below we write that, plus state a pair of different ways to write that definition, and finally at the end we state the Cauchy Criterion for convergence, which is equivalent to converging.

$$\sum_{k=1}^{\infty} a_k \text{ converges } \Leftrightarrow \text{ The sequence of partial sums converges}$$

$$\Leftrightarrow (S_n) \text{ converges, where } S_n = \sum_{k=1}^{n} a_k$$

$$\Leftrightarrow \lim_{n \to \infty} \sum_{k=1}^{n} a_k \text{ exists}$$

$$\Leftrightarrow \forall \varepsilon > 0 \ \exists N \text{ s.t. } \left| \sum_{j=m}^{n} a_j \right| < \varepsilon \ \forall n \geq m \geq N.$$

- We also proved many tests for convergence, including the kth-term test, geometric series test, comparison test, series p-test and alternating series test (respectively, Propositions 4.5, 4.9, 4.12, 4.16, and 4.17).

- In particular, for $\sum_{k=1}^{\infty} a_k$ to converge, it must be the case that $a_k \to 0$. The converse, though, is not true. For example, the harmonic series $\sum_{k=1}^{\infty} \frac{1}{k}$ diverges to infinity even though $\frac{1}{k} \to 0$ (Proposition 4.15).

9.5   Series of Functions

We will define both pointwise and uniform convergence of series, although almost all of our focus will be on uniform convergence. The reason is, as we have seen, pointwise convergence is so weak that it does almost nothing worthwhile.

Definition.

Definition 9.20. Let (f_k) be a sequence of functions defined on a set A and let $s_n = \sum_{k=1}^{n} f_k$ be the nth partial sum.

(i) The series $\sum_{k=1}^{\infty} f_k$ *converges pointwise* to a function $f : A \to \mathbb{R}$ if the sequence (s_n) converges pointwise to f.

(ii) The series $\sum_{k=1}^{\infty} f_k$ *converges uniformly* to a function $f : A \to \mathbb{R}$ if the sequence (s_n) converges uniformly to f.

Note that if each f_k is continuous, then $s_n = \sum_{k=1}^{n} f_k$ is also continuous. And, likewise, finite sums also preserve boundedness, uniform continuity, differentiability and integrability. Therefore, if you know that each f_k satisfies one of these properties, then s_n will too. This fact, along with Definition 9.20, allows us to recast the results of Table 9.1:

Assume that each f_k has the below property	Must each s_n satisfy property?	If $\sum_{k=1}^{\infty} f_k$ converges uniformly to f, must f satisfy property?
Continuous	Yes	Yes, by Proposition 9.8
Bounded	Yes	Yes, by Proposition 9.10
Uniformly Continuous	Yes	Yes, by Prop 9.12 and Ex 9.6
Differentiable	Yes	No
Integrable	Yes	Yes, by Proposition 9.16

Table 9.2: Properties of $\sum_{k=1}^{\infty} f_k$

Since a series is just a sequence of partial sums, by rewriting each "$\sum_{k=1}^{\infty} f_k$" as "$\lim_{n \to \infty} s_n$" we can apply all of our previous results to series to complete the above table. The one exception to this is with differentiation; we won't go into details here, but to guarantee that the sum is differentiable you need a couple further assumptions.

Next we generalize the Cauchy criterion to series of functions.

Proposition.

Proposition 9.21 (*Cauchy criterion for series of functions*). Let $f_k : A \to \mathbb{R}$. The series $\sum_{k=1}^{\infty} f_k$ converges uniformly on A if and only if for every $\varepsilon > 0$ there exists some $N \in \mathbb{N}$ such that

$$\left| \sum_{k=m}^{n} f_k(x) \right| < \varepsilon$$

for all $n \geq m \geq N$ and all $x \in A$.

Proof. This is Exercise 9.21. □

As with every other use of the Cauchy criterion, the advantage to using Cauchyness is that you can show convergence without knowing what the sequence/series is converging to. That said, this criterion is still usually hard to apply. Its corollary, though, is much easier to apply.

The above gives a condition that guarantees that a series of functions converges uniformly, but oftentimes it is a cumbersome condition to apply. The following corollary will give another such condition. It is weaker and so it does not always work, but when it does it is often much easier to apply. We will build up to it by discussing how "tame" such functions will have to be for their sum to converge uniformly.

- If each $f_k(x) = k$, then the functions are all fairly "tame," since each is a constant. But *collectively* they are not tame. Indeed, certainly $\sum_{k=1}^{\infty} f_k$ does not converge uniformly; it diverges to ∞ everywhere! So having functions which are each individually "nice" is not enough. And certainly they can't go off to infinity.

- But what does "collectively tame" mean? If $f_k(x) = 1$ for all k, then certainly $\sum_{k=1}^{\infty} f_k$ does not converge uniformly; it again diverges to ∞ everywhere! So having functions that don't go off to infinity is not enough, and also having them be close together is not enough, because their sum could still go off to infinity.

- Let $f_k(x) = \frac{x^k}{k}$ on $[0, 1]$. The function $f_1(x) = x$ is already pretty small for most $x \in [0, 1]$, and then for larger k the functions f_k are getting even smaller and smaller. For instance, f_{100} only takes values between 0 and 0.01. Nevertheless, the series $\sum_{k=1}^{\infty} f_k$ is still not converging to any f. The reason is that at $x = 1$ we have $f_k(1) = 1/k$ for each k. And so at this x-value,

$$\sum_{k=1}^{\infty} f_k(1) = \sum_{k=1}^{\infty} \frac{1}{k} = \infty,$$

as this is the harmonic series which in Chapter 4 we saw diverges. So it is *not* true that the sum converges for *all* $x \in [0, 1]$, and hence the sum $\sum_{k=1}^{\infty} f_k$ does *not* converge to an f. Indeed, if it did then we would have to have $f(1) = \infty$, which is not allowed. One way to state this issues is that $|f_k(x)| \leq \frac{1}{k}$ while $\sum_{k=1}^{\infty} \frac{1}{k}$ does not converge.

- So not only do we need each f_k to be bounded, but we also need the sum of all of their bounds to converge. This corollary is called the *Weierstrass M-test*.

Corollary.

Corollary 9.22 (*Weierstrass M-test*). Let $f_k : A \to \mathbb{R}$ and suppose that, for each $k \in \mathbb{N}$, there exists $M_k \in \mathbb{R}$ such that $|f_k(x)| \leq M_k$ for all $x \in A$. If $\sum_{k=1}^{\infty} M_k$ converges, then $\sum_{k=1}^{\infty} f_k$ converges uniformly on A.

Proof Sketch. The general outline of the proof is this:

$$\sum M_k \text{ converges} \Rightarrow M_k \text{ Cauchy}$$
$$\Rightarrow \sum f_k \text{ Cauchy}$$
$$\Rightarrow \sum f_k \text{ converges.}$$

Now we prove it in detail.

Proof. Let $\varepsilon > 0$. Since $\sum_{k=1}^{\infty} M_k$ converges, the partial sums of $\sum_{k=1}^{\infty} M_k$ are Cauchy. Thus, there exists some $N \in \mathbb{N}$ where for all $n \geq m \geq N$ we have

$$\sum_{k=m}^{n} M_k < \varepsilon.$$

Thus, for this same N, by the triangle inequality we have

$$\left| \sum_{k=m}^{n} f_k(x) \right| \leq \sum_{k=m}^{n} |f_k(x)| \leq \sum_{k=m}^{n} M_k < \varepsilon,$$

for all $n \geq m \geq N$. So by the Cauchy criterion for series of functions (Proposition 9.21), $\sum_{k=1}^{\infty} f_k$ converges uniformly on A. $\qquad\square$

Here's a quick example of using the Weierstrass M-test on a specific function.

Example 9.23. Let $f_k(x) = \dfrac{1}{x^4 + 3x^2 k + k^2 + 7}$ on $\mathbb{R}$. We will show that $\sum_{k=1}^{\infty} f_k$ converges uniformly on $\mathbb{R}$.

Solution. To see this, note that every term in the denominator of f_k is non-negative, and hence

$$|f_k(x)| = \frac{1}{x^4 + 3x^2 k + k^2 + 7} \leq \frac{1}{k^2}.$$

Let $M_k = \frac{1}{k^2}$, and note that $\sum_{k=1}^{\infty} M_k = \sum_{k=1}^{\infty} \frac{1}{k^2}$ converges by the series p-test (Proposition 4.16). So by the above we have that $|f_k(x)| \leq M_k$ for all k, and that the sum of the M_k's converges. Therefore by the Weierstrass M-test, $\sum_{k=1}^{\infty} f_k$ converges uniformly on $\mathbb{R}$. $\qquad\square$

9.6   Power Series

We have seen that a series of functions $\sum f_k$ can converge to another function f. But if you are given the f first, are there "simple" functions f_k which sum to f? Working towards this end, we begin with some of the simplest functions: polynomials. We first define *power series*, which essentially are polynomials with an infinite number of terms.

Definition.

Definition 9.24. A *power series* is an expression that can be written as an infinite polynomial. That is,

$$f(x) = \sum_{k=0}^{\infty} a_k x^k = a_0 + a_1 x + a_2 x^2 + a_3 x^3 + \dots$$

where each $a_k \in \mathbb{R}$.

Note that the sum has to start at $k = 0$, since we need a constant term in the sum (which could be zero). Also note that, unlike finite polynomials, for each x-value you plug in, the resulting sum might converge or might not converge. So the first important question to ask is, for a specific sequence (a_k), for which x's does f's sum converge? Of course, for any (a_k), at $x = 0$ we have that $f(0) = a_0$ converges. But where else?

Example 9.25. The geometric series

$$\sum_{k=0}^{\infty} x^k = 1 + x + x^2 + x^3 + \dots$$

is a power series where each $a_k = 1$. And since (in the lingo of geometric series) the common ratio $r = x$, we saw in Proposition 4.9 that this power series converges precisely on the interval $(-1, 1)$. Indeed, Proposition 4.9 also told us that for these x values it converges to $\frac{1}{1-x}$.   □

As it happens, the points for which a power series converges always look somewhat like this. First, they always form an interval. Second, that interval is always centered at zero. (However, note that we do include the extreme cases as an option, like the "interval" that is just the point $x = 0$, as well as all of $\mathbb{R}$, in addition to standard intervals like $(-1, 1)$.)

Intervals of Convergence

> ### Lemma.
>
> **Lemma 9.26** (*P.S. converges on* $(-x_0, x_0) \Rightarrow$ *unif. on closed subintervals*). If a power series $\sum_{k=0}^{\infty} a_k x^k$ converges at some point $x_0 \in \mathbb{R}$ and $0 < c < |x_0|$, then it converges uniformly on $[-c, c]$.
>
> In particular, if a power series converges at some x_0 and $|x| < |x_0|$, then it also converges at x.

The proof is a fun tour through many of the important results we have proven throughout this text.

Proof. Since $\sum_{k=0}^{\infty} a_k x_0^k$ converges, by the kth-term test (Proposition 4.5) we know that the terms $a_k x_0^k \to 0$, which in turn implies that these terms are bounded by some $M > 0$ (Proposition 3.20); that is, $|a_k x_0^k| \leq M$ for all $k \in \mathbb{N}$. And so, for any $x \in [-c, c]$ (implying $|x| \leq c < |x_0|$), we have

$$|a_k x^k| \leq |a_k c^k| = |a_k x_0^k| \left| \frac{c}{x_0} \right|^k \leq M \left| \frac{c}{x_0} \right|^k.$$

And we know that $\sum_{k=0}^{\infty} M \left| \frac{c}{x_0} \right|^k$ converges by the geometric series test (Proposition 4.9, with $r = \left| \frac{c}{x_0} \right| < 1$). So by the Weierstrass M-test (Corollary 9.22, with $M_k = M \cdot \left| \frac{c}{x_0} \right|^k$), we have that $\sum_{k=0}^{\infty} |a_k x^k|$ converges uniformly on $[-c, c]$, and so we are done. $\qquad \square$

Note that this lemma implies that if a power series converges at some $x_0 > 0$, then it converges on all of $(-x_0, x_0]$. And if it converges at some $x_0 < 0$, then it converges on all of $[x_0, -x_0)$. The lemma has nothing to say about $-x_0$, though. It may or it may not converge at this point. "Most" examples will have it converge at $-x_0$, but one power series where it does not converge is

$$\sum_{k=1}^{\infty} \frac{(-1)^k}{k} x^k.$$

At $x_0 = 1$ this series converges by the alternating series test (and hence by Lemma 9.26 it converges on all of $(-1, 1]$), but at $-x_0 = -1$ this is just the harmonic series, which diverges. Indeed, this power series converges only on $(-1, 1]$.

> **Theorem.**
>
> **Theorem 9.27** (*P.S. converge on an interval centered at 0*). The set of points that the power series $\sum_{k=0}^{\infty} a_k x^k$ converges at is
>
> - $\{0\}$,
>
> - $\mathbb{R}$, or
>
> - One of $(-R, R), (-R, R], [-R, R)$ or $[-R, R]$, for some $R \in \mathbb{R}$.

Proof. Let C be the collection of points for which the power series converges. We pointed out earlier that $0 \in C$. If this is all there is, then we are done. If not, then C is some nonempty set with at least one non-zero element and hence $\{|x| : x \in C\}$ either has a positive supremum or is unbounded above.

If $\{|x| : x \in C\})$ is unbounded above, then for any $x \in \mathbb{R}$ there exists some $x_0 \in C$ such that $|x| < |x_0|$. But then by Lemma 9.26 we have that $x \in C$, giving $C = \mathbb{R}$, and we are done.

If, however, $\sup(\{|x| : x \in C\}) = R$ for some $R \in \mathbb{R}$ (implying that if $|x| > R$ then $x \notin C$), then we know by the lemma that every $x \in (-R, R)$ is also in C. Moreover, if $x \notin [-R, R]$ then $x \notin C$, otherwise such an x would have the property that $|x| > R$, contradicting our assumption that $R = \sup(\{|x| : x \in C\})$. We have shown that $(-R, R) \subseteq C \subseteq [-R, R]$; that is, the only possibilities for C are $(-R, R), (-R, R], [-R, R)$, and $[-R, R]$, as desired. $\qquad \square$

> **Definition.**
>
> **Definition 9.28.** The R in the above theorem is called the *radius of (pointwise) convergence*. If the set of points of convergence is $\{0\}$, then we say that the radius of convergence is 0, and if the set of convergence is $\mathbb{R}$, then we say that the radius of convergence is ∞.

Also, notice that in Theorem 9.27 the points on which a power series converges form an interval. We call this its *interval of convergence*.

> **Definition.**
>
> **Definition 9.29.** The interval of points for which a power series converges is called its *interval of convergence*.

The above is only for pointwise convergence. What about uniform convergence?

When is that guaranteed? Turns out the answer has to do with *compact* sets. For simplicity, though, in the below we will state it just in terms of closed intervals.

> **Theorem.**
>
> **Theorem 9.30** (*P.S. converges absolutely at c $\Rightarrow$ uniformly on* $[-c, c]$). If a power series $\sum\limits_{k=0}^{\infty} a_k x^k$ converges absolutely at some $x_0 \in \mathbb{R}$, then it converges uniformly on the closed interval $\big[- |x_0|, |x_0| \big]$.

Proof. This is Exercise 9.30. $\qquad\qquad\qquad\qquad\qquad\qquad\qquad\qquad\qquad\qquad$ $\square$

It is worthwhile to note that even if you know that a power series converges uniformly on *any* closed interval within the interval of convergence, that does *not* imply that it converges uniformly on the entire interval of convergence. For example, the following geometric series converges (but not uniformly) on its interval of convergence, $(-1, 1)$:

$$1 + x + x^2 + x^3 + \cdots = \frac{1}{1 - x}.$$

9.7 Properties of Power Series

In the following table, we are considering functions $f_k(x) = a_k x^k$ on some closed interval $[-c, c]$. This allows us to apply Theorem 9.30 and the results from Table 9.2 to learn properties of $\sum\limits_{k=0}^{\infty} f_k$.

$f_k(x) = a_k x^k$ on some $[-c, c]$ have the below property	Each s_n has property?	If $\sum\limits_{k=1}^{\infty} a_k x^k$ converges uniformly, to f, must f satisfy property?
Continuous	Yes	Yes, by Proposition 9.8
Bounded	Yes	Yes, by Proposition 9.10
Uniformly Continuous	Yes	Yes, by Prop 9.12 and Ex 9.6
Integrable	Yes	Yes, by Proposition 9.16

Table 9.3: Properties of $\sum\limits_{k=1}^{\infty} a_k x^k$

Note that the above table is somewhat redundant. Since the power series was only defined on a finite, closed interval $[-c, c]$, once we deduced that it was also continuous we immediately got that it was also bounded (Corollary 6.31), uniformly continuous (Proposition 6.40) and integrable (Theorem 8.16).

Derivatives and Integrals of Power Series

Are power series differentiable and integrable? The answer is nice — it says you can differentiate or integrate term-by-term, just like with (finite) polynomials — but the proofs (given what we have developed thus far) would be long and tedious. Therefore I will tell you the answer and give a quick sketch of one of the proofs, but that's it.

> ### Theorem.
>
> **Theorem 9.31** (*Derivatives and integrals of P.S.*). Let $\sum_{k=0}^{\infty} a_k x^k$ be a power series with radius of convergence R.
>
> (i) If $R > 0$, then the power series is a differentiable function of the variable x on the domain $(-R, R)$, and
> $$\frac{d}{dx}\left(\sum_{k=0}^{\infty} a_k x^k\right) = \sum_{k=0}^{\infty}\left(\frac{d}{dx} a_k x^k\right) = \sum_{k=0}^{\infty} a_k \cdot k x^{k-1}.$$
>
> (ii) If $[a, b] \subseteq (-R, R)$, then the power series is an integrable function of the variable x on $[a, b]$, and
> $$\int_a^b \left(\sum_{k=0}^{\infty} a_k x^k\right) dx = \sum_{k=0}^{\infty}\left(\int_a^b a_k x^k \, dx\right) = \sum_{k=0}^{\infty}\left(\frac{a_k}{k+1}\left[b^{k+1} - a^{k+1}\right]\right)$$

Proof Sketch. Here is a quick sketch of how (ii) is proved.

$$\sum_{k=0}^{\infty}\int_a^b a_k x^k \, dx = \lim_{n\to\infty}\sum_{k=0}^{n}\int_a^b a_k x^k \, dx = \lim_{n\to\infty}\int_a^b \sum_{k=0}^{n} a_k x^k \, dx$$

$$= \lim_{n\to\infty}\int_a^b s_n \, dx = \int_a^b \lim_{n\to\infty} s_n \, dx = \int_a^b \sum_{k=0}^{\infty} a_k x^k \, dx.$$

Needs
justification

$\square$

9.8 New Power Series from Old

Theorem 9.31 affords us a valuable tool to turn old power series into new power series. Indeed, some of the most important power series can be obtained by manipulating geometric series using this theorem and some algebra (technically justified by the limit laws). The next example is a growing collection of such series.

Example 9.32. By the geometric series test (Proposition 4.9, with $r = x$, $a = 1$),

$$\frac{1}{1-x} = 1 + x + x^2 + x^3 + \ldots$$

with interval of convergence $(-1, 1)$. By replacing "x" with "x^3" we get

$$\frac{1}{1-x^3} = 1 + x^3 + \left(x^3\right)^2 + \left(x^3\right)^3 + \ldots \qquad (\text{🐱})$$
$$= 1 + x^3 + x^6 + x^9 + \ldots,$$

which[4] is a new power series with interval of convergence $(-1, 1)$.

We can now find a power series for $\dfrac{x^2}{1-x^3}$ (with the same interval of convergence), simply by multiplying both sides of (🐱) by x^2:

$$x^2 \left(\frac{1}{1-x^3}\right) = x^2 \left(1 + x^3 + x^6 + x^9 + \ldots\right)$$
$$\frac{x^2}{1-x^3} = x^2 + x^5 + x^8 + x^{11} + \ldots.$$

We also find a power series for $\dfrac{1}{1-x^3} - 1$ by subtracting 1 from both sides of (🐱):

$$\frac{1}{1-x^3} - 1 = \left(1 + x^3 + x^6 + x^9 + \ldots\right) - 1$$
$$\frac{1}{1-x^3} - 1 = x^3 + x^6 + x^9 + \ldots.$$

Finally, we can use calculus to find a new power series. A power series representation of $f(x) = \dfrac{3x^2}{(1-x^3)^2}$ can be obtained in the following way. First note that

$$\frac{d}{dx}\left(\frac{1}{1-x^3}\right) = \frac{d}{dx}\left(1 - x^3\right)^{-1}$$
$$= -\left(1 - x^3\right)^{-2} \cdot \left(-3x^2\right)$$
$$= \frac{3x^2}{(1-x^3)^2}.$$

---

[4]That cat symbol above was designed by my former student, April Loyd, who used it as her end-of-proof symbol. She graciously let me use it in this book.

This is helpful since we already showed that

$$\frac{1}{1-x^3} = 1 + x^3 + x^6 + x^9 + \dots.$$

Differentiating both sides (term-by-term, by Theorem 9.31),

$$\frac{d}{dx}\left(\frac{1}{1-x^3}\right) = \frac{d}{dx}\left(1 + x^3 + x^6 + x^9 + \dots\right)$$

$$\frac{3x^2}{(1-x^3)^2} = \left(0 + 3x^2 + 6x^5 + 9x^8 + \dots\right).$$

Thus we have a power series representation of $\dfrac{3x^2}{(1-x^3)^2}$ (which still has interval of convergence $(-1, 1)$). $\qquad\square$

Here's one more that uses calculus.

Example 9.33. Find a power series representation of $f(x) = \ln\left|1 - x^3\right|$.

Solution. We will make use of Example 9.32. The key is to first note is that (via w-substitution — Corollary 8.35 — with $w = 1 - x^3$)

$$\int \frac{-3x^2}{1-x^3}\, dx = \ln\left|1 - x^3\right|.$$

Then by Example 9.32,

$$\frac{1}{1-x^3} = 1 + x^3 + x^6 + x^9 + \dots$$

$$-3x^2\left(\frac{1}{1-x^3}\right) = -3x^2\left(1 + x^3 + x^6 + x^9 + \dots\right)$$

$$\frac{-3x^2}{1-x^3} = -3x^2 - 3x^5 - 3x^8 - 3x^{11} - \dots$$

$$\int \frac{-3x^2}{1-x^3}\, dx = \int \left(-3x^2 - 3x^5 - 3x^8 - 3x^{11} - \dots\right)\, dx$$

$$\ln\left|1 - x^3\right| = -x^3 - \frac{3x^6}{6} - \frac{3x^9}{9} - \frac{3x^{12}}{12} - \dots,$$

which is a power series of $\ln\left|1 - x^3\right|$. $\qquad\square$

Power series centered at c

We defined a power series to be of the form

$$\sum_{k=0}^{\infty} a_k x^k = a_0 + a_1 x + a_2 x^2 + a_3 x^3 + \dots.$$

Note that this power series is centered along the y-axis; the polynomials a_0, $a_1 x$, $a_2 x^2$, $a_3 x^3$, and so on are all symmetric about the y-axis or rotationally symmetric about the y-axis. What if we want our power series to be centered on some other vertical line, $x = c$? This is the idea of a "power series centered at c," and as you'll see in the following section, it is an important idea.

Definition.

Definition 9.34. A *power series centered at c* is an infinite series of the form

$$f(x) = \sum_{k=0}^{\infty} a_k (x - c)^k = a_0 + a_1(x - c) + a_2(x - c)^2 + a_3(x - c)^3 + \dots$$

where each $a_k \in \mathbb{R}$.

Therefore what we earlier just called a "power series" can also be called a "power series centered at 0."

9.9 Taylor and Maclaurin series

Recall.

Recall 9.35. The kth derivative of f is denoted $f^{(k)}$.

I strongly recommend you begin by watching an excellent video motivating *Taylor series*. It is a video from the inimitable Grant Sanderson, through his 3Blue1Brown channel on YouTube. Here is a link to the video: `https://youtu.be/3d6DsjIBzJ4`.

The video explains intuitively how, for "nice" functions, the function equals its Taylor series where it converges, and intuitively and visually why it does so. That is,

$$f(x) = \sum_{k=0}^{\infty} \frac{f^{(k)}(c)}{k!}(x - c)^k$$

for every x in some interval of convergence I centered at $x = c$.

Seriously, if you haven't yet, go watch that video. I know it sounds like I'm making someone else do my job for me, but Grant simply does it better than I could hope to do it. Also, this is a topic where seeing animations of the ideas makes a huge difference, and his animations are top notch. You won't be disappointed.

Ok good. Now that you have watched his excellent motivation and developed some intuition, we will get into the formal development of this theory. First up is a definition that is truly Talyor-made.

Definition.

Definition 9.36. Assume $f^{(k)}(c)$ exists for all $k \in \mathbb{N}$. The *Taylor series* of f about $x = c$ is

$$\sum_{k=0}^{\infty} \frac{f^{(k)}(c)}{k!}(x - c)^k.$$

If $c = 0$, then the series is called a *Maclaurin series*.[5]

Here is an example of finding a function's Taylor/Maclaurin series.

Example 9.37. Here we find the Maclaurin series of $f(x) = \cos(x)$.

$$
\begin{array}{lll}
f(x) = \cos(x) & f(0) = 1 & a_0 = 1 \\
f'(x) = -\sin(x) & f'(0) = 0 & a_1 = 0 \\
f''(x) = -\cos(x) & f''(0) = -1 & a_2 = -\frac{1}{2!} \\
f'''(x) = \sin(x) & f'''(0) = 0 & a_3 = 0 \\
f^{(4)}(x) = \cos(x) & f^{(4)}(0) = 1 & a_4 = \frac{1}{4!} \\
f^{(5)}(x) = -\sin(x) & f^{(5)}(0) = 0 & a_5 = 0
\end{array}
$$

It is evident that

$$f^{(k)}(0) = \begin{cases} 0, & \text{if } k \text{ is odd,} \\ (-1)^{k/2}, & \text{if } k \text{ is even.} \end{cases}$$

So the Maclaurin series for $f(x) = \cos(x)$ is

$$\sum_{k=0}^{\infty} \frac{f^{(k)}(0)}{k!}(x)^k = 1 - \frac{x^2}{2!} + \frac{x^4}{4!} - \frac{x^6}{6!} + \cdots.$$

---

[5] Named after Scottish mathematician Colin Maclaurin, who was a pioneer in the field of analysis and, presumably, a pioneer in the art of rockin a dope afro:

(He was also considered the youngest professor in history, until 2008 when American materials scientist Alia Sabur claimed the crown 3 days before her 19th birthday.)

Likewise, one can find the Maclaurin series of $g(x) = \sin(x)$ and $h(x) = e^x$ to be

$$\sum_{k=0}^{\infty} \frac{g^{(k)}(0)}{k!}(x)^k = x - \frac{x^3}{3!} + \frac{x^5}{5!} - \frac{x^7}{7!} + \cdots$$

and

$$\sum_{k=0}^{\infty} \frac{h^{(k)}(0)}{k!}(x)^k = 1 + x + \frac{x^2}{2!} + \frac{x^3}{3!} - \frac{x^4}{4!} + \cdots.$$

$\square$

It is worth underscoring that at this point we have *not* shown that $\cos(x)$ (or $\sin(x)$ or e^x) actually equals its Maclaurin series. This will turn out to be true, but we still have work to do to show this.[6]

Indeed, only nice functions equal their Maclaurin series. So what does "nice" mean here? Well, the formula itself requires that we are able to differentiate f arbitrarily many times; that is, they are *infinitely differentiable*. So "nice" certainly must include this property. And as you saw in the video[7], if a function is infinitely differentiable, then there is good reason to suspect that it might equal its Taylor series. However, that alone is not enough. What other properties are needed? Let's investigate.

Define the *Taylor polynomial of f of degree n at c* to be

$$T_{x=c}^n(f) = \sum_{k=0}^{n} \frac{f^{(k)}(c)}{k!}(x - c)^k.$$

By construction, the first n derivatives of $T_{x=c}^n(f)$ agree with the first n derivatives of f at the point $x = c$. They share a lot in common, but this degree-n Taylor polynomial is likely not equal to f, and in fact they differ by

$$E_n(x) = f(x) - T_{x=c}^n(f),$$

called the *error function*. And, as probably makes intuitive sense, f equals its Taylor series on I if and only if this error tends to 0 as $n \to \infty$. (To emphasize, it's as $n \to \infty$, *not* as $x \to c$. This is a subtle but important distinction.)

---

[6]Among the many series that were discovered first by Isaac Newton, the series for $\sin(x)$ and $\cos(x)$ were not two of them; Newton was beat by a couple hundred years, in fact. These series were written down (in verse!) by Jyesthadeva and Niklakantha Somayaji in the early 1500s in Kerala, India. They in turn attributed them down Madhava of Sangamagrama from about a century before.

[7]I'm going to keep passive aggressively encouraging you to watch it, in case you haven't yet.

> **Lemma.**
>
> **Lemma 9.38** (*f equals its Taylor series* $\Leftrightarrow E_n \to 0$). Assume that f is infinitely differentiable in an interval I, and $c \in I$. Then for $x \in I$,
>
> $$f(x) = \sum_{k=0}^{\infty} \frac{f^{(k)}(c)}{k!}(x-c)^k \qquad \text{if and only if} \qquad E_n(x) \to 0 \quad \text{pointwise.}$$

The proof simply comes down to remembering that a (numerical) series is equal to the limit of its partial sums.

Proof. Fix any x. Note that

$$f(x) = \sum_{k=0}^{\infty} \frac{f^{(k)}(c)}{k!}(x-c)^k \iff f(x) = \lim_{n \to \infty} \sum_{k=0}^{n} \frac{f^{(k)}(c)}{k!}(x-c)^k$$

$$\iff f(x) = \lim_{n \to \infty} T_{x=c}^n(f)$$

$$\iff f(x) - \lim_{n \to \infty} T_{x=c}^n(f) = 0$$

$$\iff \lim_{n \to \infty} \left[f(x) - T_{x=c}^n(f) \right] = 0$$

$$\iff E_n(x) \to 0.$$

And since this worked for any fixed x, we are done. $\qquad\square$

Two forms of the error function

The principle way to determine whether a function f equals its Taylor series is by checking its error function and applying Lemma 9.38, and thus it is important to get as good a handle on this function as possible. There are several ways to express this error, but here I will mention two. We begin with an integral expression for it.

> **Theorem.**
>
> **Theorem 9.39** (*Integral form of the error function*). Assume that f is infinitely differentiable in an interval I, and $c \in I$. Then for $x \in I$ we have
>
> $$E_n(x) = \frac{1}{n!} \int_c^x (x-t)^n f^{(n+1)}(t) \, dt.$$

Proof Sketch. The proof is by induction on n, and uses FTC and IBP. It's fairly long and technical, but the ($n = 1$) base case highlights the important ideas, so here we just sketch that. That is, we will show that $E_1(x) = \int_c^x (x-t) f''(t) \, dt$.

To this end, first note that by the definition of the error function,

$$E_1(x) = f(x) - T^1_{x=c}(x) = f(x) - \sum_{k=0}^{1} \frac{f^{(k)}(c)}{k!}(x-c)^k$$

$$= f(x) - f(c) - f'(c)(x-c)$$

By the fundamental theorem of calculus,

$$= \int_c^x f'(t)\, dt - \int_c^x f'(c)\, dt$$

$$= \int_c^x \left(f'(t) - f'(c)\right) dt$$

Now we use integration by parts $\left(\int u\, dv = uv - \int v\, du\right)$. In it, recall that when finding v from dv, we can pick *any* antiderivative.

$$u = f'(t) - f'(c) \qquad\qquad v = -(x-t)$$
$$du = f''(t)\, dt \qquad\qquad dv = dt$$

This allows us to continue our above string of equations for $E_1(x)$.

$$= \left(f'(t) - f'(c)\right)(t-x)\Big|_{t=c}^{t=x} - \int_c^x -(x-t)f''(t)\, dt$$

$$= 0 + \int_c^x (t-x)f''(t)\, dt,$$

as desired. This completes the base case, $n = 1$. The rest of the proof is by induction. By a little algebra one can show that

$$E_{k+1}(x) = E_k(x) - \frac{f^{(k+1)}(c)}{(k+1)!}(x-c)^{k+1}.$$

Using this, the inductive hypothesis, and one more application of integration by parts, one can finish the proof. $\qquad\square$

There is one more representation for the error function that we will discuss, and it is known as the *Lagrange form*, after its founder. The theorem says that at any particular point x_0, f is equal to its $(n-1)^{\text{st}}$ Taylor polynomial evaluated at x_0, plus the nth term of the Taylor series evaluated at some other nearby point.

> **Theorem.**
>
> **Theorem 9.40** (*Lagrange form of the error function*). Assume that f is infinitely differentiable in an interval I, and $c \in I$. Then for any other $x_0 \in I$ there exists some α_n between x_0 and c (which depends on x_0 and c and n) such that
>
> $$f(x_0) = \left(\sum_{k=0}^{n-1} \frac{f^{(k)}(c)}{k!} (x_0 - c)^k \right) + \frac{f^{(n)}(\alpha_n)}{n!}(x_0 - c)^n.$$
>
> That is, $E_{n-1}(x_0) = \dfrac{f^{(n)}(\alpha_n)}{n!}(x_0 - c)^n$, for some α_n between x_0 and c.

The proof follows from the integral form of the error function (Theorem 9.39), but also make use of two other important theorems from earlier in the text.

Proof. Assume, without loss of generality, that $c < x_0$. Since $f^{(n)}$ is differentiable (and hence continuous) on $[c, x_0]$, by the extreme value theorem (Theorem 6.32) it attains its supremum and infimum. Set

$$m = \inf_{t \in [c, x_0]} f^{(n)}(t) \qquad \text{and} \qquad M = \inf_{t \in [c, x_0]} f^{(n)}(t).$$

Since $m \leq f^{(n)}(t) \leq M$, by the integral form of the error function (Theorem 9.39),

$$\frac{m}{(n-1)!} \int_c^{x_0} (x_0 - t)^{n-1}\, dt \leq E_{n-1}(x_0) \leq \frac{M}{(n-1)!} \int_c^{x_0} (x_0 - t)^{n-1}\, dt.$$

And by integrating the left and right sides of the inequality,

$$\frac{m(x_0 - c)^n}{n!} \leq E_{n-1}(x_0) \leq \frac{M(x_0 - c)^n}{n!}.$$

We know that $f^{(n)}(t)$ is a continuous function on $[c, x_0]$, it is bounded by $m \leq f^{(n)}(t) \leq M$ on this interval, and it achieves these bounds. By simply multiplying by the positive constant $\frac{(x_0 - c)^n}{n!}$, we see that $\frac{f^{(n)}(t) \cdot (x_0 - c)^n}{n!}$ is also a continuous function on $[c, x_0]$, it is bounded

$$\frac{m(x_0 - c)^n}{n!} \leq \frac{f^{(n)}(t) \cdot (x_0 - c)^n}{n!} \leq \frac{m(x_0 - c)^n}{n!}$$

on this interval, and it achieves these bounds. Therefore by the intermediate value theorem (Theorem 6.38), there must be some α_n between x_0 and c for which $\frac{f^{(n)}(\alpha_n) \cdot (x_0 - c)^n}{n!}$ is equal to the number $E_{n-1}(x_0)$, completing the proof. $\square$

Note that, as $n \to \infty$, the $(x_0 - c)^n$ term is growing at the rate of an exponential function, while the $n!$ in the denominator is of course growing at the rate of a factorial.

And since factorials grow faster than exponentials, the only way that this error does not converge to zero is if the $f^{(n)}(\alpha_n)$ term is growing at really fast.

One particular conclusion that is useful in practice: If, for some M, we have $|f^{(n)}(\alpha_n)| \leq M$ for all $x \in I$, then $E_n(x) \to 0$ pointwise on I, and hence by Lemma 9.38, f equals is Taylor series on I.

9.10   A beautiful application

I wanted to wrap up this chapter, as well as the main content of this book, by sharing a rightfully-famous application of this work (with some justification omitted). Recall that in Example 9.37 we noted that $\cos(x)$, $\sin(x)$ and e^x have the following Maclaurin series.

$$\cos x = 1 - \frac{x}{2!} + \frac{x^4}{4!} - \frac{x^6}{6!} + \frac{x^8}{8!} - \frac{x^{10}}{10!} + \cdots$$

$$\sin x = x - \frac{x^3}{3!} + \frac{x^5}{5!} - \frac{x^7}{7!} + \frac{x^9}{9!} - \cdots \tag{$\circledast$}$$

$$e^x = 1 + x + \frac{x^2}{2!} + \frac{x^3}{3!} + \frac{x^4}{4!} + \frac{x^5}{5!} + \frac{x^6}{6!} + \frac{x^7}{7!} + \cdots$$

One can also prove the above really are equalities — that is, these three functions do indeed equal their Maclaurin series. The fact that $\cos(x)$ does was argued in the video that I trust by now I have pestered you all to watch, and the fact that $\sin(x)$ does is very similar.

To see that $h(x) = e^x$ does, pick any interval $[-d, d]$; we will prove that the error function converges to 0 on this interval. For any $x_0 \in [-d, d]$, by the Lagrange form of the error function (Theorem 9.40),

$$E_n(x_0) = \frac{h^{(n+1)}(\alpha_{n+1})}{(n+1)!}(x_0 - c)^{n+1}$$

for some α_{n+1} between x_n and 0. This in particular means that $|\alpha_{n+1}| \leq d$, and since also $|h^{(n+1)}(\alpha_{n+1})| = e^{\alpha_{n+1}} \leq e^d$,

$$|E_n(x_0)| \leq \left| \frac{e^d}{(n+1)!}(x_0)^{n+1} \right| \leq e^d \cdot \frac{d^{n+1}}{(n+1)!} \to 0 \text{ as } n \to \infty,$$

since factorials grow faster than exponentials (or by looking at the ratio of consecutive terms). And so the error function does indeed tend to 0 on $[-d, d]$. Moreover, since d was arbitrary, the error function tends to 0 on all of $\mathbb{R}$. And hence, by Lemma 9.38, the e^x function does indeed equal its Maclaurin series, as asserted:

$$e^x = 1 + x + \frac{x^2}{2!} + \frac{x^3}{3!} + \frac{x^4}{4!} + \frac{x^5}{5!} + \frac{x^6}{6!} + \frac{x^7}{7!} + \cdots.$$

This already gives an amazing result. And when you plug $x = 1$ into the above, you learn that this crazy number e — which pops up all over the place, from probability to statistics to calculus to asymptotics to economics — has a remarkable expression:

$$e = 1 + 1 + \frac{1}{2!} + \frac{1}{3!} + \frac{1}{4!} + \frac{1}{5!} + \frac{1}{6!} + \frac{1}{7!} + \dots.$$

But don't pack up just yet; there's still more gold out there. We now have three new expressions for our old friends, $\cos(x)$, $\sin(x)$ and e^x. In fact, when you take *complex analysis* you will learn that this expression for e^x even holds for complex values! The imaginary number i has probably been defined to you as $\sqrt{-1}$; that is, it has the special property that $i^2 = -1$, and hence $i^3 = -i$, $i^4 = 1$, $i^5 = i$, and so on. Plugging in ix to the Maclaurin series of e^x gives this:

$$
\begin{aligned}
e^{ix} &= 1 + ix + \frac{(ix)^2}{2!} + \frac{(ix)^3}{3!} + \frac{(ix)^4}{4!} + \frac{(ix)^5}{5!} + \frac{(ix)^6}{6!} + \frac{(ix)^7}{7!} + \cdots \\
&= 1 + ix + \frac{i^2 x^2}{2!} + \frac{i^3 x^3}{3!} + \frac{i^4 x^4}{4!} + \frac{i^5 x^5}{5!} + \frac{i^6 x^6}{6!} + \frac{i^7 x^7}{7!} + \cdots \\
&= 1 + ix - \frac{x^2}{2!} - \frac{ix^3}{3!} + \frac{x^4}{4!} + \frac{ix^5}{5!} - \frac{x^6}{6!} - \frac{ix^7}{7!} + \cdots.
\end{aligned}
$$

We now regroup based on which terms contain an i and which do not. That is,

$$e^{ix} = \left(1 - \frac{x^2}{2!} + \frac{x^4}{4!} - \frac{x^6}{6!} + \dots\right) + \left(ix - \frac{ix^3}{3!} + \frac{ix^5}{5!} - \frac{ix^7}{7!} + \dots\right).$$

Pulling out an i,

$$e^{ix} = \left(1 - \frac{x^2}{2!} + \frac{x^4}{4!} - \frac{x^6}{6!} \dots\right) + i\left(x - \frac{x^3}{3!} + \frac{x^5}{5!} - \frac{x^7}{7!} \dots\right),$$

and suddenly the Maclaurin series for $\sin(x)$ and $\cos(x)$ from (⊗) appear! The above is equivalent to

$$e^{ix} = \cos x + i \sin x.$$

Once again, you might be packing up your bags, thinking things can't get any better than this. But hang tight, because there's one more value to plug in. Watch what happens when we set $x = \pi$:

$$
\begin{aligned}
e^{i\pi} &= \cos \pi + i \sin \pi \\
e^{i\pi} &= (-1) + i(0) \\
e^{i\pi} &= -1
\end{aligned}
$$

Moving the -1 to the left gives one of the most remarkable equations in history, known as *Euler's identity*; a single equation that simply and beautifully combines perhaps the five most important constants in all of mathematics: 0, 1, e, π and i:

$$e^{i\pi} + 1 = 0.$$

— Notable Exercises —

- Exercise 9.11 asks you to prove the *Cauchy Criterion for Uniform Convergence*:

    Let $f_n : A \to \mathbb{R}$. The sequence (f_n) converges uniformly if and only if for every $\varepsilon > 0$ there exists some $N \in \mathbb{N}$ such that $|f_m(x) - f_n(x)| < \varepsilon$ for all $m, n \geq N$ and all $x \in A$.

    As with numerical sequences from Chapter 3, the advantage to a Cauchy-criterion is that you can show something is converging (uniformly) without knowing what it converges to.

- Exercise 9.30 essentially says that power series always converge in a region centered at 0. It does this by asking you to prove that if a power series converges at a point x_0, then it in fact converges on the closed interval $\big[-|x_0|, |x_0|\big]$.

— Exercises —

Exercise 9.1. Let $f_k(x) = \dfrac{x \sin(x)}{k}$ on $[0, 10]$. Prove that (f_k) converges uniformly.

Exercise 9.2. Let $f_k(x) = x^k$ on $[0, 1]$, and let

$$f(x) = \begin{cases} 0 & \text{if } x \in [0, 1), \\ 1 & \text{if } x = 1. \end{cases}$$

In Example 9.6 we argued that $f_k \to f$ pointwise. Prove that $f_k \not\to f$ uniformly.

Exercise 9.3. Let $f_k(x) = x/k$ on $\mathbb{R}$, and let $f(x) = 0$ on $\mathbb{R}$. Prove that (f_k) converges pointwise to f, but does not converge uniformly to f.

Exercise 9.4. For each of the following, determine the pointwise limit of (f_k) on the indicated interval, and state whether the convergence is uniform. You do not need thoroughly to prove your answers, but give a brief justification for your answer.

(a) $f_k(x) = \sqrt[k]{x}$, on $[0, 1]$.

(b) $f_k(x) = \dfrac{e^x}{k^2}$, on $(1, \infty)$.

(c) $f_k(x) = x^k - x^{2k}$ on $[0, 1]$.

(d) $f_k(x) = \dfrac{2kx}{1 + k + 3x}$ on $[0, \infty)$.

Exercise 9.5. Assume that, for each $k \in \mathbb{N}$, the function $f_k : A \to \mathbb{R}$ is unbounded.

(a) If $f_k \to f$ pointwise, must f be unbounded? Either prove that f must be unbounded or give a counterexample.

(b) If $f_k \to f$ uniformly, must f be unbounded? Either prove that f must be unbounded or give a counterexample.

Exercise 9.6. Assume that, for each $k \in \mathbb{N}$, the function $f_k : A \to \mathbb{R}$ is uniformly continuous. Also assume that $f_k \to f$ uniformly. Prove that f is uniformly continuous.

Exercise 9.7. Give an example of a sequence (f_k) of discontinuous functions, each defined on a domain I, for which $f_k \to f$ uniformly, for some continuous f.

Exercise 9.8. Complete Example 9.13 by showing that the f_k and f given in that example have the property that $f_k \to f$ uniformly.

Exercise 9.9. Give an example of a sequence (f_k) of integrable functions on $[a, b]$ for which $f_k \to f$ pointwise, but yet f is not integrable.

Exercise 9.10. Give an example of a sequence (f_k) of integrable functions on $[a, b]$ for which $f_k \to f$ pointwise, and f is integrable, but yet

$$\lim_{k\to\infty} \int_a^b f_k(x)\, dx \neq \int_a^b f(x)\, dx.$$

Exercise 9.11. Prove the *Cauchy Criterion for Uniform Convergence*: Let $f_k : A \to \mathbb{R}$. The sequence (f_k) converges uniformly if and only if for every $\varepsilon > 0$ there exists some $N \in \mathbb{N}$ such that $|f_m(x) - f_n(x)| < \varepsilon$ for all $m, n \geq N$ and all $x \in A$.

Exercise 9.12.

(a) Prove that if $f_k \to f$ uniformly on each of the sets $A_1, A_2, \ldots, A_n$, then f_k also converges uniformly to f on the set $\bigcup_{i=1}^{n} A_i$.

(b) Give an example to show that the above does not necessarily hold for countably many sets A_i.

Exercise 9.13. Suppose $f_k \to f$ pointwise and $g_k \to g$ pointwise, both on a domain $A \subseteq \mathbb{R}$.

(a) Prove that $(f_k + g_k) \to (f + g)$ pointwise on A.

(b) Does $(f_k \cdot g_k) \to (f \cdot g)$ pointwise? Prove or give a counterexample.

Exercise 9.14. Suppose $f_k \to f$ uniformly and $g_k \to g$ uniformly, both on a domain $A \subseteq \mathbb{R}$.

(a) Prove that $(f_k + g_k) \to (f + g)$ uniformly on A.

(b) Does $(f_k \cdot g_k) \to (f \cdot g)$ uniformly? Prove or give a counterexample.

Exercise 9.15. Give an example of a sequence (f_k) of functions where

(i) (f_k) converges uniformly on $[-10, 10]$, and

(ii) (f_k) converges pointwise on $\mathbb{R}$, but does not converge uniformly on $\mathbb{R}$.

Exercise 9.16. By Proposition 9.16, if $f_k \to f$ uniformly on $[a, b]$ and each f_k is integrable on $[a, b]$, then $\int_a^b f_k(x)\, dx \to \int_a^b f(x)\, dx$. We also showed in Example 9.6 and thereafter than $f_k(x) = x^k$ does *not* converge uniformly on $[0, 1]$. In this exercise, prove that uniform convergence is not a *necessary* condition for the convergence of a sequence of integrals by showing that if $f_k(x) = x^k$ on $[0, 1]$ and f is its pointwise limit, then

$$\int_0^1 f_k(x)\, dx \to \int f(x)\, dx$$

Exercise 9.17. Given an example of (1) a sequence of functions f_k on $[0,1]$, (2) a function f on $[0,1]$, and (3) a sequence (a_k) where each $a_k \in [0,1]$ and $a_k \to a$ for some a, such that the following conditions hold.

(i) Each f_k is continuous;

(ii) $f_k \to f$ pointwise; and

(iii) $f(a_k) \not\to f(a)$.

Exercise 9.18. Suppose (f_k) is a sequence of functions on $[a,b]$, each of which is monotone increasing. Suppose further that $f_k \to f$, for some f on $[a,b]$. Prove that f is monotone increasing on $[a,b]$.

Exercise 9.19. Give an example of a sequence (f_k) of functions on $[0,1]$, each of which is discontinuous at every point in $[0,1]$, but which converges uniformly to a continuous function.

Exercise 9.20. Let $f_k : A \to \mathbb{R}$. A sequence of functions (f_k) is said to be *uniformly bounded* if there exists some $C \in \mathbb{R}$ such that $|f_k(x)| \le C$ for all $k \in \mathbb{N}$ and all $x \in A$.

(a) Explain why the property "each f_k is bounded" is different than the property "(f_k) is uniformly bounded."

(b) Give an example of a set A and a sequence (f_k) for which each f_k is bounded on A but the sequence (f_k) is not uniformly bounded.

Exercise 9.21. Prove the *Cauchy criterion for series of functions*. That is, let $f_k : A \to \mathbb{R}$, and prove that the series $\sum\limits_{k=1}^{\infty} f_k$ converges uniformly on A if and only if for every $\varepsilon > 0$ there exists some $N \in \mathbb{N}$ such that

$$\left| \sum_{k=m}^{n} f_k(x) \right| < \varepsilon$$

for all $n \ge m \ge N$ and all $x \in A$.

Exercise 9.22.

(a) Suppose that $f : \mathbb{R} \to \mathbb{R}$ is uniformly continuous, and define a sequence of functions by $f_n(x) = f(x + \frac{1}{n})$. Show that (f_n) converges uniformly to f.

(b) Give an example to show that this fails if we assume only that f is continuous and not uniformly continuous.

Exercise 9.23. Recall that $\int \dfrac{1}{1+x^2}\, dx = \tan^{-1}(x)$. Use this in the following.

(a) Show that, for $x \in (-1, 1)$,

$$\tan^{-1}(x) = x - \frac{x^3}{3} + \frac{x^5}{5} - \frac{x^7}{7} + \dots.$$

(b) Use part (a) to find a series expression for π.

Exercise 9.24. Show that the following series converge uniformly on the given interval.

(a) $\displaystyle\sum_{k=1}^{\infty} \frac{\sin(k^2 x)}{k^2}$ on $(-\infty, \infty)$.

(b) $\displaystyle\sum_{k=1}^{\infty} \frac{x}{k^2}$ on $[-10, 10]$.

Exercise 9.25. Prove that if the series of functions $\displaystyle\sum_{k=1}^{\infty} f_k$ converges uniformly on a set A, then the sequence of functions (f_k) converges uniformly to 0 on A.

Exercise 9.26. Prove that the series $\displaystyle\sum_{k=1}^{\infty} \frac{k^2 + x^4}{k^4 + x^2}$ converges to a continuous function $f : \mathbb{R} \to \mathbb{R}$.

Exercise 9.27. Prove that the series

$$\sum_{k=1}^{\infty} 4^k \sin\left(\frac{1}{5^k x}\right)$$

converges uniformly on $[2, \infty)$. You may use the fact that $\displaystyle\lim_{x \to 0} \frac{\sin(x)}{x} = 1$.

Exercise 9.28. Give an example of a series of functions $\displaystyle\sum_{k=1}^{\infty} f_k$ on some set A which converges uniformly, but for which the Weierstrass M-test fails.

Exercise 9.29. Determine whether the following converse to the Weierstrass M-test holds. Suppose that (f_k) is a sequence of nonnegative, bounded function on a set A, and let $M_k = \sup\left(\{f_k(x) : x \in A\}\right)$. If $\displaystyle\sum_{k=1}^{\infty} f_k$ converges uniformly on A, does it follow that $\displaystyle\sum_{k=1}^{\infty} M_k$ converges?

Exercise 9.30. Prove that if a power series $\displaystyle\sum_{k=0}^{\infty} a_k x^k$ converges absolutely at some $x_0 \in \mathbb{R}$, then it converges uniformly on the closed interval $\big[-|x_0|, |x_0| \big]$.

Exercise 9.31. Prove that if a power series $\displaystyle\sum_{k=0}^{\infty} a_k x^k$ has the property that $0 < m \le |a_k| \le M < \infty$ for some m and M, and for all k, then the power series has a radius of convergence of 1.

Exercise 9.32. Recall that the Fibonacci sequence $1, 1, 2, 3, 5, 8, \dots$ is defined by $a_1 = a_2 = 1$ and $a_k = a_{k-1} + a_{k-2}$.

(a) Show that for every $k \in \mathbb{N}$ the inequality $\frac{a_{k+1}}{a_k} \le 2$ holds.

(b) Show that the power series

$$\sum_{k=1}^{\infty} a_k x^{k-1} = 1 + 1x + 2x^2 + 3x^3 + 5x^4 + 8x^5 + \cdots$$

converges for $|x| < 1/2$.

(c) Prove that for $|x| < 1/2$ the power series in the previous part converges to

$$f(x) = \frac{-1}{x^2 + x - 1}.$$

(d) Use partial fractions to obtain another power series for $f(x)$.

(e) The two power series must be the same. Conclude that

$$a_k = \frac{1}{\sqrt{5}} \left(\frac{1 + \sqrt{5}}{2} \right)^n - \frac{1}{\sqrt{5}} \left(\frac{1 - \sqrt{5}}{2} \right)^k.$$

(f) Reflect on the strange beauty of the formula from the previous part.

Exercise 9.33. Find the Taylor series for the following functions about the given point.

(a) $f(x) = \sin(x)$ about $x = 0$.

(b) $g(x) = \cos(2x)$ about $x = \pi/6$.

(c) $h(x) = x^3$ about $x = 2$.

(d) $s(x) = \frac{1+x}{1-x}$ about $x = 0$.

(e) $t(x) = \sqrt{x}$ about $x = 9$.

(f) $u(x) = \frac{1}{x}$ about $x = 4$.

Exercise 9.34. Use Theorem 9.31 and the Maclaurin series expansions for $\sin(x)$, $\cos(x)$ and e^x to show that

$$\frac{d}{dx}e^x = e^x \qquad \text{and} \qquad \frac{d}{dx}\sin(x) = \cos(x) \qquad \text{and} \qquad \frac{d}{dx}\cos(x) = -\sin(x).$$

Exercise 9.35. Use the Maclaurin series expansion for e^x to show $e^x e^y = e^{x+y}$.

Exercise 9.36. Let $f(x) = \ln|x+1|$.

(a) Find the Maclaurin series for f.

(b) State where this series converges.

(c) Use this to find a series representation for $\ln(2)$.

Exercise 9.37. In this exercise you will prove that e is irrational.

(a) Show that for any n,

$$e = e^1 = 1 + \frac{1}{1!} + \frac{1}{2!} + \frac{1}{3!} + \frac{1}{4!} + \cdots + \frac{1}{n!} + E_n(1),$$

where $0 < E_n(1) < \frac{3}{(n+1)!}$.

(b) Assume for a contradiction that $e = a/b$, for some $a, b \in \mathbb{N}$. That is,

$$\frac{a}{b} = 1 + \frac{1}{1!} + \frac{1}{2!} + \frac{1}{3!} + \frac{1}{4!} + \cdots + \frac{1}{n!} + E_n(1).$$

Since this holds for all n, pick an n so that the rest of the proof works out.

(c) Multiply both sides of the above by $n!$. For your chosen n, argue that $n!E_n(1)$ must be an integer. Then explain why this is this a contradiction.

— Open Questions —

Question 1. Assume that $f(x) := \sum_{k=1}^{\infty} a_k x^k$ is convergent for all x. What is a necessary and sufficient condition on the sequence (a_k) for f to be a bounded function?[8]

Question 2. Let $R(x) = \dfrac{P(x)}{Q(x)}$ where $P(x)$ and $Q(x)$ are polynomials with integer coefficients and $Q(0) \neq 0$. Is there an algorithm that, given $P(x)$ and $Q(x)$ as input, always halts and correctly decides if the Taylor series of $R(x)$ at $x = 0$ has a constant coefficient of 0?

---

[8] As two examples, note that if all but finitely many of the a_k are zero, then f is a non-constant polynomial which is certainly unbounded. However, if $a_k = \frac{(-1)^k}{k!}$ for odd k and $a_k = 0$ for even k, then $f(x) = \sin(x)$, which is bounded. As you can see, there has to be some intricate cancellation occurring to achieve boundedness.

Appendices

Appendix A: Construction of $\mathbb{R}$

"Far out in the uncharted backwaters of the unfashionable end of the western spiral arm of the Galaxy lies a small unregarded yellow sun. Orbiting this at a distance of roughly ninety-two million miles is an utterly insignificant little blue green planet whose ape-descended life forms are so amazingly primitive that they still think digital watches are a pretty neat idea." This is how Douglas Adams began his classic book *The Hitchhiker's Guide to the Galaxy*. He began its sequel, "The story so far: In the beginning the Universe was created. This has made a lot of people very angry and been widely regarded as a bad move." Douglas Adams can certainly set a tone.

"It was a pleasure to burn" is the first sentence of Ray Bradbury's *Fahrenheit 451*. "I'm pretty much fucked" is the start to Andy Weir's *The Martian*. And, "All happy families are alike; each unhappy family is unhappy in its own way" is how Leo Tolstoy began *Anna Karenina*.

If the first chapter of the *Harry Potter* series were a dry introduction to wand movements and cauldron thickness, few would read on. Yes, the spells and potions rely on those details, but it's not where the magic really is. Therefore J.K. Rowling's first chapter contained exhibitions of magic, mention of a dark wizard so terrifying that most fear to speak his name, allusions to this wizard's sadistic reign which mysteriously and abruptly just ended, and a significant baby boy, dropped on the doorstep of the least magical people possible. Lots of questions, few answers.

Where am I going with this? The material in Appendix A is the 'wand movements' and 'cauldron thickness' part of real analysis. The spells and potions we developed in this text rely on the foundational aspects we are about to discuss, but if I started the book with them you might have fallen asleep before Chapter 2. And just like how, after reading (and rereading[1]) the books, hardcore Harry Potter fans crave to know the small — but fundamental — details about the world they entered, perhaps now you will be motivated to learn the underpinnings of the world of real analysis.

Everything we proved relied on the existence of the real numbers. They needed to be there and to have the properties we discussed, and the rigor of mathematics demands we don't simply assume their existence. Which axioms are necessary to give us the world we want, and why? Doing this thoroughly (in long-form style) might take 50 pages, and even hardcore analysis fans would likely fall asleep. This appendix is just a 7 page sketch of it. Its goal is simply to illustrate for you the meticulous process which needed to be done to formally begin the theory of real analysis.[2]

---

[1] And rereading

[2] And to make you appreciate the brave soul who took one for the team, and painstakingly did it.

A.1   Axioms of Set Theory

It is hard to get more simple than the notion of a set, but the mathematical world did a double take in 1901 when Bertrand Russell discovered a paradoxical set. Recalling that the elements of a set can themselves be sets, he considered the set

$$R = \{x : x \notin x\}.$$

Symbolically, there was no reason to disallow such a set definition, but yet you can work out the strange contradiction that $R \in R$ if and only if $R \notin R$; indeed, if R is not a member of itself, then according to its definition it must be a member of itself, and if it does contain itself, then it contradicts its own definition.[3]

When mathematicians reach a logical impasse, they often trace their reasoning backwards; the longer they trace, the more nervous they get, as they fear for the foundations of their field. These nerves caused some great mathematicians of the early twentieth century to rework set theory from the ground up. In the end, *Zermelo-Fraenkel+Choice set theory* (ZFC) was born—nine axioms for the field of set theory, and in turn for much of mathematics.

1. Axiom of extensionality: Two sets are equal if they have the same elements.

2. Axiom of regularity: Every non-empty set A contains an element B such that A and B are disjoint sets.

3. Axiom schema of specification: You can create subsets using set builder notation.

4. Axiom of pairing: If A and B are sets, there exists a set which contains A and B as elements.

5. Axiom of union: You can take union over the elements of a set. (E.g., the union of the elements of the set $\{\{1,3\}, \{3, \odot\}\}$ is $\{1, 3, \odot\}$.)

6. Axiom schema of replacement: Given a function f whose domain is a set A, $f(A)$ is also a set.

7. Axiom of infinity: There exists a set having infinitely many members.

8. Axiom of power set: For any set A, there is a set B that contains every subset of A.

9. Axiom of Choice:[4] First, a *choice function* is a function f, defined on a collection X of nonempty sets, such that for every set A in X, $f(A)$ is an element of A. Now, the axiom is: For any set X of nonempty sets, there exists a choice function f defined on X.

These are the axioms of set theory, and on their shoulders the reals are built.

---

[3] A classic riddle works along the same lines: "A barber cuts the hair of everyone in his town who does not cut their own hair. Does the barber cut his own hair?"

[4] Axioms 1-8 you probably think are completely obvious and fair to assume, while Axiom 9 is a bit confusing. Indeed, due to its less-than-painstakingly-obvious stature, it has been blamed for all the paradoxical-sounding theorems of set theory—the final example in Appendix B, most notably.

A.2 Constructing $\mathbb{N}$

Everything we build is going to rely on nothing more than the axioms of set theory. Indeed, we begin by constructing the natural numbers — which are themselves *sets*.

We represent the number 0 as the empty set, $\emptyset$; this may seem weird, but bear with me here. We then define a *successor function* $S(x)$ of a set x to be the set

$$S(x) = x \cup \{x\}.$$

In the naturals, the successor to 0 is 1, and indeed we are going to represent 1 as the set $S(0) = S(\emptyset) = \emptyset \cup \{\emptyset\} = \{\emptyset\}$; note that this equals $\{0\}$. Then, 2 is going to be represented as the set $S(1) = S(\{\emptyset\}) = \{\emptyset\} \cup \{\{\emptyset\}\} = \{\emptyset, \{\emptyset\}\}$; note that this equals $\{0, 1\}$.

Continuing in this way, we get a collection of sets which are going to represent the natural numbers for us.

$$0 = \emptyset$$
$$1 = \{\emptyset\} = \{0\}$$
$$2 = \{\emptyset, \{\emptyset\}\} = \{0, 1\}$$
$$3 = \{\emptyset, \{\emptyset\}, \{\emptyset, \{\emptyset\}\}\} = \{0, 1, 2\}$$
$$4 = \{\emptyset, \{\emptyset\}, \{\emptyset, \{\emptyset\}\}, \{\emptyset, \{\emptyset\}, \{\emptyset, \{\emptyset\}\}\}\} = \{0, 1, 2, 3\}$$
$$\vdots$$

We define $\mathbb{N}_0$ to be the set of all these numbers (represented as sets).

Of course, the natural numbers are not just isolated objects that we need to make up representations for — they also have an order to them and an algebraic structure. Beginning with the former, if we are going to describe the natural numbers as sets, we must also describe what "$n < m$" means in terms of these sets. One reason why the naturals were constructed like the above is that it is now easy to describe their order. Notice how, as sets, each number (represented as a set) is *an element of* each later number (represented as a set). Indeed, we say

$$n < m \qquad \text{if} \qquad n \in m.$$

For example, $1 < 3$ since it is true that $\{\emptyset\} \in \{\emptyset, \{\emptyset\}, \{\emptyset, \{\emptyset\}\}\}$. We now have an ordering on $\mathbb{N}_0$!

Another important property of the numbers in $\mathbb{N}_0$ that we need to capture is that we can add them together. How should we define the addition $2 + 5$ by using only the tools of set theory applied to the sets representing 2 and 5? This one is a little more complicated to write down, but the basic idea is to think about "$m + n$" as $m + 1 + 1 + \cdots + 1$, where you have n copies of 1. With this frame of mind, "$m + n$" is defined to be $S(S(S(\ldots S(m) \ldots)))$, where you apply the successor function n times. The only problem being that, at this point 'n' is a set, so "n times" doesn't make sense. Instead, we must create a recursive function S_n that equals a composition of

'n' successor functions.

$$S_1(x) := S(x)$$
$$S_2(x) := S(S(x))$$
$$S_3(x) := S(S(S(x)))$$
$$S_4(x) := S(S(S(S(x))))$$
$$S_5(x) := S(S(S(S(S(x)))))$$
$$\vdots$$

Then, "$m + n$" can be defined to be $S_n(m)$.

As for multiplication, we define $n \cdot 0 = 0$, and for other multiplication problems we iteratively turn them into an addition problem by defining $n \cdot (m+1) = n \cdot m + n$.

One can then spend 15 pages proving the following, where $m, n, k \in \mathbb{N}_0$.

- $(m + n) + k = m + (n + k)$

- $m + n = n + m$

- $n + 0 = n$

- $(m \cdot n) \cdot k = m \cdot (n \cdot k)$

- $m \cdot n = n \cdot m$

- $n \cdot 1 = n$

- $m \cdot (n + k) = (m \cdot n) + (m \cdot k)$

- If $m < n$, then $m + k < n + k$

- If $m < n$ and $0 < k$, then $k \cdot m < k \cdot n$

- If $m < n$ and $n < k$, then $m < k$

- Exactly one of the following holds: $m < n$, $m = n$, or $n < m$.

As you'll see, I am beginning to really lean into the fact that this is a "sketch" of how to construct $\mathbb{R}$. Properties like the above are a pain and a half to prove, and even harder to type up in LaTeX without making a typo. But, it can be done.

Once completed, you can proudly claim to have shown that $(\mathbb{N}_0, +, \cdot, <)$ is what is called an *ordered semi-ring*. This is a completed construction of the natural numbers (and 0), but we still have a ways to go. An ordered semi-ring is good, but to remove the "semi," we construct negative numbers.

A.3   Constructing $\mathbb{Z}$

Axiomatically, $\mathbb{Z}$ is the smallest ordered ring (containing a copy of $\mathbb{N}_0$), but to construct it explicitly, it suffices to introduce subtraction. The difference between two numbers is going to have to be something new that breaks the mold established by the successor function. Indeed, we define the difference "$m - n$" simply to be the ordered pair $(m, n) \in \mathbb{N}_0 \times \mathbb{N}_0$.

If we stopped here, we would have the problem where we would consider "$6 - 2$" to be different from "$7 - 3$"; after all, as ordered pairs $(6, 2)$ and $(7, 3)$ are certainly different. To get around this, we define an *equivalence relation* on ordered pairs, where we want $(m, n) \sim (k, \ell)$ provided "$m - n = k - \ell$." But since we haven't defined subtraction yet, we move the 'n' and 'ℓ' over, giving the condition we want: $m + \ell = k + n$. That is, we say

$$(m, n) \sim (k, \ell) \qquad \text{if and only if} \qquad m + \ell = k + n.$$

One would then prove that this is indeed an equivalence relation. To make it easier to read along, though, let's write the ordered pair (m, n) as $[m - n]$. This will vastly help one follow along — just don't forget that these are technically ordered pairs, and that minus sign is just notation.

Of course, one would have to formally prove that $\sim$ does indeed form an equivalence relation... but it does. You can also do a mental check that the following make sense as definitions of addition, multiplication and ordering.

Define the following,[5]

$$[m - n] + [k - \ell] := [(m + k) - (n + \ell)]$$
$$[m - n] \cdot [k - \ell] := [(m \cdot k + n \cdot \ell) - (m \cdot \ell + n \cdot k)]$$
$$[m - n] < [k - \ell] \quad \text{if and only if} \quad m + \ell < k + n$$

But here, the plot annoyingly thickens again: Given that $[m - n]$ represents an entire equivalence class, for our definitions of '$+$', '$\cdot$' and '$<$' to be well-defined they have to be playing nicely with this equivalence relation. The proofs of these are actually reasonably short (under a page for all three), but are enormously boring and so we will again skip them.

Negatives also distribute, so we define $-[m - n]$ to be $[n - m]$. Moreover, we can simplify our notation of these elements, by defining $\boldsymbol{n} := [n - 0]$. The set of all such $\boldsymbol{n}$ we define to be the set of *integers*, $\mathbb{Z}$.

Then, for instance, we can say $\boldsymbol{n}$ is *positive* if $\boldsymbol{n} > 0$. And is *negative* if $\boldsymbol{n} < 0$. Moreover, one can prove that every non-zero integer is either of the form $\boldsymbol{m}$ or $-\boldsymbol{m}$, and that multiplication handles the signs as expected. And, finally, at this point one could struggle through a 3-page proof that $(\mathbb{Z}, +, \cdot, <)$ satisfies the additive, multiplicative and order properties to be an ordered ring.

And with that, it has been fully sketched how to construct the integers.

---

[5]Technically it's an abuse of notation to cavalierly reuse '$+$', '$\cdot$' and '$<$' without ensuring that doing so is consistent with our previous uses of it... but with some elbow grease it does work out.

A.4 Constructing $\mathbb{Q}$

Axiomatically, $\mathbb{Q}$ is the smallest ordered field (containing a copy of $\mathbb{Z}$), but we again describe its construction, which has many similarities to the construction of $\mathbb{Z}$.

The construction of $\mathbb{Z}$ was focused on incorporating subtraction; here we focus on incorporating division. To describe this division, we again use ordered pairs. The fraction $\frac{m}{n}$ that we are after will be given by the ordered pair (m, n). Since we do not want the denominator to be zero, and we only have to allow either the numerator or denominator to be negative, we begin by considering the ordered pairs $\mathbb{Z} \times \mathbb{N}$, where $\mathbb{N} = \mathbb{N}_0 \setminus \{0\}$.

We now need another equivalence relation. Intuitively, $(m, n) = (k, \ell)$ when $\frac{m}{n} = \frac{k}{\ell}$, which we can write without division as $m \cdot \ell = k \cdot n$. So we define the equivalence relation $\sim$ on $\mathbb{Z} \times \mathbb{N}$ by

$$(m, n) \sim (k, \ell) \qquad \text{if and only if} \qquad m \cdot \ell = k \cdot n.$$

One can again prove that this is an equivalence relation, and then can define $\mathbb{Q}$ to be the set of all such equivalence classes. Again, to make it easier to read, we will write the ordered pair (m, n) as $[m/n]$, understanding that this is just notation. You can check by rewriting them as fractions that the below definitions make sense.[6]

- $[m/n] + [k/\ell] := [(m \cdot \ell) + (k \cdot n) \, / \, n \cdot \ell]$

- $[m/n] \cdot [k/\ell] := [m \cdot k \, / \, n \cdot \ell]$

- $[m/n] < [k/\ell]$ if and only if $m \cdot \ell < k \cdot n$

Once again, we are applying a definition to an equivalence class but only stating the definition via a representative of that class — therefore one must prove that the above operations are well-defined. (Spoiler: They are.)

This set contains a copy of $\mathbb{Z}$, where you identify $[n/1]$ with $\boldsymbol{n}$.

One can then define positive and negative numbers similar to before, and the additive inverse of $[m/n]$ is $[-m/n]$.

The multiplicative inverse (for "non-zero" classes) must be defined a little more carefully, since we demand that any negative sign be in the numerator (which is the first coordinate of the ordered pair and in the bracket notation).

$$[m/n]^{-1} := \begin{cases} [n/m] & \text{if } m > 0; \\ [-n/-m] & \text{if } m < 0. \end{cases}$$

And, finally, one can work out a long proof that $(\mathbb{Q}, +, \cdot, <)$ satisfies the algebraic properties to form an ordered field.

And with that, $\mathbb{Q}$ is constructed.

---

[6] Again, with an abuse of notation.

A.5   Constructing $\mathbb{R}$

Here we are, the moment you've been waiting for. We have the ordered field of rational numbers, but they aren't complete—there are holes everywhere, and to get to $\mathbb{R}$ we must fill in these gaps. There are several ways to do this. Cantor used Cauchy sequences of rationals: If you demand that each Cauchy sequence converges to something, and pairs of sequences whose difference is Cauchy must converge to the same thing, then by adding these limits into your set $\mathbb{Q}$ (by identifying them with an equivalence class of Cauchy sequences converging to them), you in effect complete $\mathbb{Q}$, giving $\mathbb{R}$. Lots of details missing, but that's the big picture idea.

There is also a non-standard approach using so-called *ultrafilters*, another using *hyperrationals*, and several more beyond that. But the most common method, which we discuss now, uses *Dedekind cuts*.

Just like with all our previous constructions, each real number is going to be a set; at this point we have the rationals constructed, so each real number is going to be represented by a set of rationals. The way you want to think about it is this: the real number x is going to be represented by the set of all rational numbers strictly less than x. These sets are going to be called *cuts*, and while we discuss them you can start convincing yourself that each real number will indeed correspond to a unique cut, and each cut corresponds to a unique real number. But first, let's formally discuss cuts.

A *cut* should be thought of as the set $(-\infty, b) \cap \mathbb{Q}$. That is, all rational numbers up to a certain point. Formally, it is defined as any set C_b satisfying the following three conditions.

1. $C_b \subseteq \mathbb{Q}$, but $C_b \neq \emptyset$ and $C_b \neq \mathbb{Q}$;

2. If $p \in C_b$ and $q \notin C_b$, then $p < q$;

3. If $p \in C_b$, then there exists some $q \in C_b$ where $p < q$.

And then $\mathbb{R}$ is defined as the set of all cuts.[7]

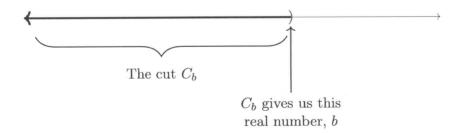

The cut C_b

C_b gives us this
real number, b

Although we have an intuitive picture of a cut, at the moment it is just a set satisfying the above properties. And we again are interested in putting an algebraic

---

[7]So real numbers are infinite sets of rational numbers, which themselves are pairs of integers, which themselves are pairs of natural numbers, which themselves are unions of empty sets and sets containing empty sets and so on. That's what a single real number is constructed to be.

structure on this collection of cuts. Addition and order work quite smoothly. For a pair of cuts C_a and C_b, define the following.[8]

- $C_a + C_b := \{p + q : p \in C_a, q \in C_b\}$

- $C_a < C_b$ if and only if $C_a \subset C_b$

One can verify that the addition of two cuts is still a cut, and that addition is commutative and associative, that the 0 cut behaves as it should (in turn giving *positive* and *negative* cuts) and that additive inverses exist. The inequality has the property that, given any two cuts C_a and C_b, exactly one of the following holds: $C_a < C_b$, $C_a = C_b$ or $C_b < C_a$.

Defining multiplication is trickier, because if you simply multiply the two sets together you'll have massive negative numbers multiplying against each other, creating massive positive numbers. Intuitively, you want $C_a \cdot C_b = C_{ab}$. That is, the cut $(-\infty, a) \cap \mathbb{Q}$ times the cut $(-\infty, b) \cap \mathbb{Q}$ should equal the cut $(-\infty, a \cdot b) \cap \mathbb{Q}$; but cuts aren't defined with such a and b—they produce a and b. One way around this is to first define multiplication for positive cuts. That is, if C_a and C_b are both positive (larger than the cut $\{q \in \mathbb{Q} : q < 0\}$), then define

$$C_a \cdot C_b := \{p \cdot q : p \in C_a, q \in C_b \text{ with } p, q \geq 0\} \cup \{q \in Q : q < 0\}.$$

With this you can then define a product of two negative cuts by setting the product equal to the product of the two corresponding positive cuts. To define multiplication between a positive cut and a negative cut, you know the product should be negative so one approach is to consider the multiplication when both are positive, and then translate the result to the corresponding negative cut. It's a hassle to write out, but that's the idea.

One can then check that the product of two cuts is still a cut, that multiplicative inverses exist, and that multiplication is commutative and associative, as well as the remaining multiplicative/additive distributive and order properties. These would be quite annoying to work out in detail, but you can smile knowing someone carefully checked them.

With our set built, and with the algebraic properties defined and their properties verified, we now know that $\mathbb{R}$ is an ordered field. All that is left is to show that it is complete. This is done by showing that cuts satisfy the *least upper bound property*. That is, if $\mathcal{C}$ is a collection of cuts which is bounded above (meaning there exists some cut D such that $C \leq D$ for all $C \in \mathcal{C}$; i.e., $C \subseteq D$ for all $C \in \mathcal{C}$), then there exists a least upper bound (meaning there is a cut M such that $M \leq D$ for all upper bounds D). The proof is this: Let $M = \bigcup_{C \in \mathcal{C}} C$, and show that

1. M is a cut and therefore $M \in \mathbb{R}$;

2. M is an upper bound of $\mathcal{C}$, but is smaller than all other upper bounds.

With those, $\mathbb{R}$ is a complete ordered field, and hence the real numbers are constructed.[9]

---

[8]One last time with a slight abuse of notation.

[9]*drops mic*

Appendix B: Peculiar and Pathological Examples

Jerry Seinfeld: Oh you're crazy.
Cosmo Kramer: Am I? Or am I so sane that you just blew your mind?
Jerry Seinfeld: It's impossible.
Cosmo Kramer: Is it? Or is it so possible that your head is spinning like a top?
Jerry Seinfeld: It can't be.
Cosmo Kramer: Can't it? Or is your entire world just crashing down all around you?
Jerry Seinfeld: Alright that's enough.
Cosmo Kramer: Yeaaaaaaaah!

The world of mathematics can be a strange place, and when your mind is exploring this world and finds itself in some dark, foreign region, you risk falling into pits, being eaten by monsters, and missing a beautiful forest for the boring trees.

It has been said that one of the most important goals of learning real analysis is to collect as many bizarre examples as you can, and to keep them in your back pocket. From a practical standpoint they will inform your conjectures and guide your proofs, but they will also help to demonstrate why real analysis is such a great subject.

When little kids learn about the world, many are fascinated by dinosaurs because they show how strange and large and monstrous life on Earth can be. Flowers are also nice. I like flowers. Your course abstract algebra, for instance, is filled with theorems which are small and pretty and I am happy to see them and learn about them. But I prefer the monsters.[1] We have seen many awesome monsters already in this book, but if your back pocket has some room left, this appendix has a few more.

---

[1]Don't @ me, researchers of the Monster group.

For instructors/students looking to insert these these examples into the earlier chapters, below is where each example can be inserted, consistent with the material.

Note: There were some examples which in some small way referenced later material but could easily be discussed earlier. For these I noted a couple chapters which would work well.

The last example is based on the famous Banach-Tarski paradox, and I think that would make for an especially nice example to end a course on. No material after Chapter 2 is needed, so it can be discussed toward the end of just about any course using this text.

P&P Example	Suggested Chapter
The Curious Case of the Cantor Set	2 or 3
Doubled Digits of Diametrical Degrees	6
Structuring Stuff from its Shadows' Shapes	1 or 8
Menger's Matterless Material	1 or 3
Obtainable Outrageousness in an Orderly Overhang	4
A Composition Conundrum	8
Turning the Tables on your Teetering Troubles	6
A Devilish De-Descent	7
A Pack of Pretty Proofs by Picture	4
An Abundant Addition Aboundingly Ascends	4
A Smooth and Spiky Solution	7 or 9
Finding ϕ for First-place Finishes	3
Peculiar and Pathological Perimeters	1 or 8
Fractal Functions Filling Foursquare Frames	6
A Prestigious Proof of a Primal Puzzle	4
A Topological Treatment with a Tremendous Twist	5
Modern Measuring's Misfit Member	8
Tarski's Teriffic Talents Times Two	Last

B.1 The Curious Case of the Cantor Set

The *Cantor set* has some fascinating properties. To describe this set, the construction is typically done in stages.

Stage 1 is the interval $[0, 1]$, represented by this bar "of length 1."

Stage 2 is found by taking the above and removing the middle third. This gives $[0, 1/3] \cup [2/3, 1]$.

Next you take each of the intervals $[0, 1/3]$ and $[2/3, 1]$ and remove the middle third of each of those. This gives $[0, 1/9] \cup [2/9, 1/3] \cup [2/3, 7/9] \cup [8/9, 1]$.

Continuing in this way gives a sequence of sets.

Stage 1:

Stage 2:

Stage 3:

Stage 4:

Stage 5:

As you can see, we are removing more and more points. The *Cantor set* is defined to be the limit of this procedure. That is, the Cantor set is the set of points in $[0, 1]$ which are never going to be removed by any of the above stages, even if you continue this cutting-out procedure forever.

A reasonable question would be: Are there *any* points in the Cantor set? At first glance it seems possible that *all* the points will eventually be removed! This is not the case, though. For instance, you can quickly verify that 0 and 1 will always be there. Also 1/3 and 2/3 will always be there. Indeed, if a point is ever the end point of one of the intervals, then it will never be removed.

Here is another way to think about it. Given a number $x \in [0, 1]$, that number is typically written in base 10. For instance 0.2748 means this:

$$0.2748 = 2 \cdot 10^{-1} + 7 \cdot 10^{-2} + 4 \cdot 10^{-3} + 8 \cdot 10^{-4}.$$

But you could easily write numbers in other bases too. For example, you can write 7/9 in base 3:

$$7/9 = 2 \cdot 3^{-1} + 1 \cdot 3^{-2}.$$

An alternative perspective: Since this number is in $[0, 1]$, the first digit after the decimal tells you if your number is between 0/3 and 1/3, or 1/3 and 2/3, or 2/3 and 3/3. Once you know which of these intervals of length $\frac{1}{3}$ you are in, then divide *that* interval into 3 subintervals, each of length 1/9. Which of those your number falls in determines the next digit, and so on.

Fact B.1. The elements in the Cantor set are precisely the numbers in $[0, 1]$ which can be written[2] in base 3, using only 0s and 2s.

So the number whose base-3 representation is 0.02202022 is in the Cantor, but the number whose base-3 representation is 0.21012021 is not in the Cantor set.

Now, we mentioned that the endpoints of each interval are in there. So is that it? If so, then among other things we could prove that the Cantor set is countable: Go through each stage, enumerating the new endpoints; in this way you list out all the endpoints one after another, proving that the collection of endpoints is countable.

But, no, there are numbers in the Cantor set which are *not* endpoints! For example you can show that 1/4 is in the Cantor set; 3/10 is also in there. (For instance, 1/4 has base-3 representation 0.02020202....) And it's not just a few points either — there are *uncountably many* points in the Cantor set. So looking just at the endpoints, we missed essentially all of them!

Fact B.2. There are uncountably many points in the Cantor set.

One way to prove this is to find a bijection between the Cantor set and all the numbers in $[0, 1]$, since we know the latter set is uncountably infinite. The bijection is pretty straightforward too: Take any base-3 representation of a number which consists of only 0s and 2s, and change all the 2s to 1s; note that the result is a string of 0s and 1s. If you interpret this string as a number in base-2, then this procedure turns a base-3 number into a unique base-2 representation of a number in $[0, 1]$. Conversely, take any base-2 representation of a number and by switching the 2s to 3s you get a base-3 representation of a number with only 0s and 2s, and hence a number in the Cantor set. In this way, you can show that the Cantor set is uncountable.

---

[2]Sometimes there is more than one way to write a number. You have maybe seen the fact (in base 10) that 0.9999... = 1. Likewise, in base 3, 0.02222... = 0.1

At the start of this discussion we wondered whether there were any points at all in the Cantor set, and now we are saying that there are as many in the Cantor set as there are in all of $[0, 1]$?? Indeed, one of the amazing things about the set is that it is both "big" and "small" at the same time. It's big in that there are uncountably many points in it, but yet it is small in that there are not any "clumps" of points anywhere![3] The way this is usually stated is in terms of "length."

The length of $[a, b]$ is $b - a$. It's just the length of the interval. Likewise, the set $[0, 1] \cup [6, 9]$ has length $1 + 3 = 4$. So Stage 1 of the Cantor set is $[0, 1]$ and hence has length 1. Stage 2 of the Cantor set is $[0, 1/3] \cup [2/3, 1]$ and hence has length $2/3$; another way to think about it is that we removed 1/3rd of the set, leaving 2/3rds left. Stage 3 has length $(2/3)^2$, since again we found each interval and cut out 1/3 of it. Stage 4 has length $(2/3)^3$ because again we cut 1/3 out of each interval we had.

Continuing like this we see that Stage n of the Cantor set has length $(2/3)^n$, and hence the Cantor set has length

$$\lim_{n \to \infty} \left(\frac{2}{3}\right)^n = 0.$$

So in this "length" sense, the Cantor set is tiny!

Fact B.3. The Cantor set is both big and small at the same time. It is big because it has uncountable many points, but tiny in that it has zero length.

And even though it's tiny in length, if we denote the Cantor set by $\mathcal{C}$, then what is $\mathcal{C} + \mathcal{C}$? (That is, which numbers can be represented as $x + y$, where $x, y \in \mathcal{C}$? Amazingly... $\mathcal{C} + \mathcal{C} = [0, 2]$! And entire interval! And here's even more:

Fact B.4. The Cantor set is *totally disconnected*, but yet is *perfect*.

Totally disconnected means that given any two points in the set, there are numbers between these which are not in the Cantor set. The reason is that at some point a "middle third" got removed between these two. And perfect means that each point in the set is a limit of a sequence of other points in the set. So even though *every* point is in some sense "separated" from any other point, with an interval of missing points in-between them, still every point is the limit point of points in the set!

Finally, it is of course fractal-like, which is usually called being *self-similar*. So if you zoom in on any interval from any stage of the Cantor set, what you are looking at looks exactly like the original Cantor set. What a wacky set.

A set is never big. Nor is it small. It is precisely the size it means to be.

3

B.2   Doubled Digits of Diametrical Degrees

One of my favorite aspects of mathematics is when a simple idea has a surprising consequence. The intermediate value theorem (Theorem 6.38) is such an idea, and one such consequence is the temperature problem, which says this: There are two *antipodal* points on the Earth (meaning they are diametrically opposite to each other — a line connecting them would go through the center of the Earth) which have *exactly* the same temperature! Here, temperature can be viewed as having infinite precision or rounded to any level of precision you desire.

Here's why it works. First, consider any great circle of the Earth.[4] Pick any point on your great circle to correspond to 0 radians, and identify all other points using radians in the usual way: π radians corresponds to its antipodal point (which is also on this great circle), $\pi/2$ and $3\pi/2$ radians correspond to the points a quarter turn away, and so on. Let $\text{temp}(\theta)$ be the temperature of each point to infinite precision, which is continuous function. Also, let

$$T(\theta) = \text{temp}(\theta) - \text{temp}(\theta + \pi),$$

which is a difference of continuous functions and hence is continuous by the continuity limit laws (Proposition 6.20). Notice that since $\text{temp}(0) = \text{temp}(2\pi)$,

$$T(0) = \text{temp}(0) - \text{temp}(\pi) = -[\text{temp}(\pi) - \text{temp}(2\pi)] = -T(\pi).$$

And $T(0) = -T(\pi)$ means that either $T(0) = T(\pi) = 0$ or these two have opposite signs, which by Proposition 6.37 (or the intermediate value theorem, with $\alpha = 0$), implies that there must be a point $c \in (0, \pi)$ for which

$$T(c) = 0$$
$$\text{temp}(c) - \text{temp}(c + \pi) = 0$$
$$\text{temp}(c) = \text{temp}(c + \pi).$$

And thus the points corresponding to c and $c + \pi$ are antipodal and have the exact same temperature.[5]   Cool!

It's worth noting that every great circle would have a pair of points with this property, and so these temperature-equivalent antipodal points are all over the place! Indeed, you can even prove that there exists an entire curve around the globe where every point on that curve has its antipodal point also on the curve, and which has the same temperature as its antipodal point.

---

[4]A *great circle* is like the equator, which is a largest-possible circle around the Earth — one that would cut the Earth in two. There is another great circle going through the north and south poles, and, of course, there are infinitely others.

[5]And if something about the discrete orbital energy levels of atoms means that temperature precision must be rounded to, say, the nearest 0.0000001 degree celsius, then these two points would round to that same point, and so they still agree.[6]

[6]But if you start rounding antipodal locations because of something about Planck lengths, then you're just trying to spoil our fun.

Next, observe this:

- The above didn't need to be for temperature. For example, we could have done the above argument for the humidity or barometric pressure at every point on the globe.

- From a topological perspective, this curve around the globe — where every point on the curve has the same temperature as its antipodal point, which is also on the curve — looks just like another great circle!

Combining the above two observations, we can take this curve and apply the above argument — except for, say, humidity — to find a pair of antipodal points *on this curve* which have exactly the same humidity. Being on this curve means they have the same temperature, and they were just found to have the same humidity, and so we even found a pair of points on the globe which have *exactly the same temperature <u>and</u> exactly the same humidity!*

All this is a special case of a much more general theorem known as the Borsuk-Ulam theorem, which says that if you have n continuous functions on an n-dimensional sphere,[7] then there exist a pair of antipodal points which take the same value on all of these functions.

The Borsuk-Ulam theorem also has some great applications, the first of which is called *the ham sandwich theorem.* In three dimensions, the theorem says that if you have any three compact objects with positive volume, there is some (two dimensional) plane which simultaneously slices all three objects in half.

The theorem is called the ham sandwich theorem because it implies that if you have a (super bland) sandwich consisting of a single slice of ham between of two slices of bread, then even if the ham is thicker in some parts than others, and even if it was sloppily put together so that the slices of bread aren't perfectly aligned and some of the ham is spilling out one side, it is always possible to cut it into two pieces so that, with just that single cut, each piece received exactly half of the first slice of bread, exactly half of the ham, and exactly half the second slice of bread.

The ham sandwich theorem in n dimensions says that given any collection of n compact objects in $\mathbb{R}^n$ which have positive volume, there exists a hyperplane which simultaneously bisects all of these objects. It can be proven by constructing a clever function from the n-dimensional sphere $\mathbb{S}^n$ (representing the possible angles of your hyperplane) to $\mathbb{R}^n$ (which are n-tuples whose i^{th} coordinate representing the area on the positive side of one of these angled hyperplanes).

A fun, discrete application of this theorem is called the *necklace splitting problem.* Suppose a pair of thieves have stolen a necklace containing n types of gems, and each gem appears an even number of times. The gems are in some order around the necklace, and the necklace can be cut between the gems. How many cuts are needed so that both people get the same number of gems of each type? By another clever application of the Borsuk-Ulam theorem (or more directly, by the ham sandwich theorem), the answer is n, which to me seems surprisingly small!

---

[7]Note that a *sphere* is just the surface of a ball. So a 1-dimensional sphere is a circle, a 2-dimensional sphere is the surface of a 3-dimensional ball (like the Earth), and so on.

B.3   Structuring Stuff from its Shadows' Shapes

There is software now which, if you can take a bunch of pictures of some object (from all sides), and it can take those pictures and create a 3D rendering of that object — turning 2D perspectives into a 3D object.

CAT scans also work on this basic principle. They take many different x-ray "pictures" at a given level to generate a 2D cross-section at that level. A computer then pieces these all together to form a 3D rendering of a patient's organ.

But think now about shadows. Given the shadows of something, can you construct the object? Or, more to the point, given a collection of *desired* shadows, does there exist an object that has those shadows? If you want the shadows to be "G", "E" and "B" then the answer is yes (as the book cover to *Godel, Escher, Bach* shows):

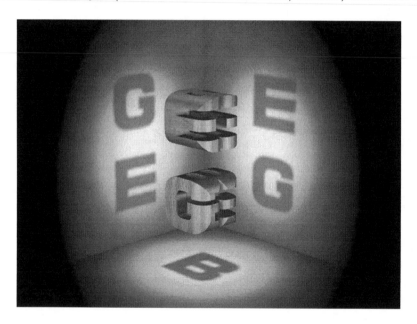

In fact, there is a theorem that says that this is *almost*-always *nearly* possible. If you don't understand all the jargon, don't sweat it.

Theorem (2D version).   Let L_θ denote the line in the xy-plane which passes through the origin at an angle of θ, and for a set A let $\text{proj}_\theta(A)$ denote the projection of A onto the line L_θ. If for each $\theta \in [0, \pi)$ we have a set $G_\theta \subseteq L_\theta$ for which $\bigcup_\theta G_\theta$ is a measurable set in the plane, then there exists a set $F \subseteq \mathbb{R}^2$ such that

- $G_\theta \subseteq \text{proj}_\theta F$;

- $\text{proj}_\theta F \setminus G_\theta$ has measure 0 for almost all $\theta \in [0, \pi)$.

Essentially, you can pick any sets G_θ that you wish, and there exists some 2D set F whose projections at nearly every angle θ is the set G_θ (with only imperceptible errors)! This theorem can also be generalized to three dimensions: Given a collection of 2D sets (which will be our desired shadows), there is a 3D set for which essentially all of its shadows are what you want — and in order too! This generalization has a fantastic consequence; suppose the shadows that you want look like the digits on a clock:

$$1{:}00$$

$$1{:}01$$

$$1{:}02$$

$$1{:}03$$

$$1{:}04$$

$$\vdots$$

Then there exists a 3D object whose shadows are these times — and you can even pick each angle you want for each time. Which means... that you could make a sundial whose shadows — as the sun moves across the sky — are precisely the time! In fact, you could even have it show the seconds too, and each second, as the sun moves, a new shadow is produced representing the new time at that precise second. It's a digital sundial!!

The technology to put this theory into practice is impressive, but admittedly is nowhere near refined enough produce readable per-second changes to the shadows. That said, one can now 3D print a digital sundial which shows the time rounded to the nearest 10-minute mark.[8]

As in the picture above, the time will switch from 11:40 to 11:50 to 12:00 to 12:10 and so on, and will change every 10 minutes. Pretty amazing stuff.

---

[8]Thank you to my former student Trenton Deacon for 3D printing me one of these, which is pictured above in my driveway, showing 11:40.

B.4   Menger's Matterless Material

The *Sierpinski carpet* is a fractal-like set constructed as follows. Begin with a square.

Cut this square into 9 equally-sized smaller squares, and remove the middle piece:

Now consider the remaining 8 squares, cut each of these into 9 equally-sized pieces, and remove each middle piece:

And continue.

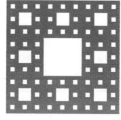

Do this forever. The *Sierpinski carpet* is the set of points that will never be removed. Here's what one more stage looks like:

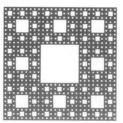

The Sierpinski carpet has some wonderful properties. As one example, you might wonder whether forever poking holes in the carpet will, in the limit, leave nothing left. This doesn't happen, though—like with the Cantor set, lots and lots of points remain. So what's the area of what's left? Notice that at each step, each square chunk of area is reduced by 8/9. And if each piece of area shrinks by 8/9 at each stage, then the whole area also shrinks by 8/9 at each stage. If the area starts out as 1, then the area after stage 1 is 8/9. After stage 2 it is $(8/9) \cdot (8/9) = (8/9)^2$. After stage 3 it is $(8/9)^2 \cdot (8/9) = (8/9)^3$. And so on. Indeed, after stage n, the area is $(8/9)^n$. So what's the area[9] after all the stages? It's

$$\lim_{n \to \infty} (8/9)^n = 0.$$

So as we have pointed out, the carpet does not completely disappear—uncountably many points will never be removed. And yet, as the above calculation shows, no area remains![10]

We could discuss the Sierpinski carpet much more, but for now let's turn to the 3-dimensional version of this, called the *Menger sponge*. The idea is similar, except instead of beginning with a square we begin with a cube.

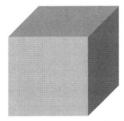

Now divide each face into 9 equally-sized squares and punch-out each middle one.

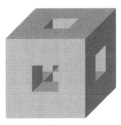

And, like before, continue this procedure.

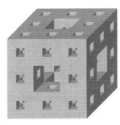

---

[9]"Area" in standard Lebesgue measure.

[10]This is particularly convenient if you plan to sew a Sierpinski carpet for yourself, since you won't need any thread.

Here are levels three and four of the sponge:

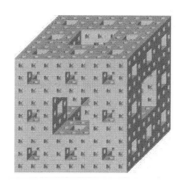 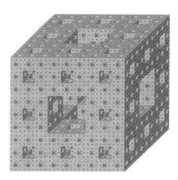

The *Menger sponge* is the limiting shape of this process. There are many fascinating facts about the Menger sponge, but one of my favorites is what happens when you cut the sponge diagonally. Here is the video showing the awesome answer: `https://youtu.be/fWsmq9E4YC0`, courtesy of the Simons Foundation. (Make sure to pause the video to guess what shapes you're going to see.)

There are many self-similar, fractal-like geometric constructions (See: *Iterated function systems* and the amazing consequences of the *chaos game*). One particularly neat example is called *Barnsley's fern*.

Such shapes are easily and quickly created graphically using just a few basic rules. Such shapes are used in computer graphics to create natural, beautiful shapes with little processing power, and I love how such things mimic nature. Barnsley's fern resembles a common real-world fern called *black spleenwort*, which is native to all sorts of places, from Norway to Egypt to Hawaii to Turkey.

B.5   Obtainable Outrageousness in an Orderly Overhang

Assume you have a bunch of ordinary blocks ('ordinary' means that each is a rectangular cuboid, has uniform density, and the blocks are identical). If you hang one block off a table, up to half of it can be hanging off the edge without it falling:

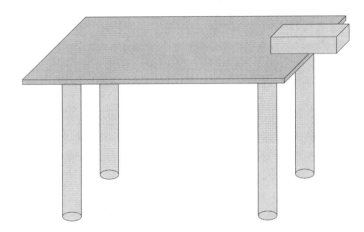

For convenience, let's say that each block is 1 foot long, meaning that we achieved $\frac{1}{2}$ foot of overhang. Here's the big question we are interested in asking: Using more blocks, can we achieve more overhang? If so, how much overhang can we achieve?

If you have blocks or books nearby, even if they aren't completely uniform go ahead and try this out. Come back and continue reading once you've experimented.

If you don't have any blocks or books, call up everyone you know and invite them to a pizza party at your house in 45 minutes. Then order 31 pizzas. Once all the pizza has been eaten you can use the pizza boxes for the experiment.

$$\vdots$$

What did you find out? First, you probably quickly noticed that with two blocks you can achieve more than $\frac{1}{2}$ foot of overhang. Indeed, if a second block is placed on the edge of the table and with just a couple inches of overhang, you can now place the first block on top of this block so that half of it is overhanging the block below it.

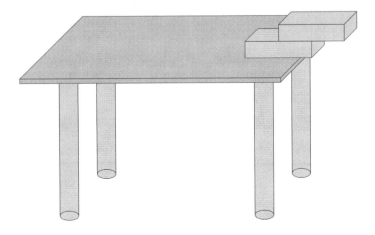

The bottom block "extends the table" in some sense. Here is a simple zoomed-in profile picture of this successful situation:

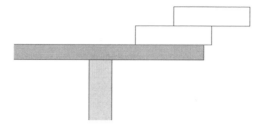

Of course, there is a limit to how much overhang the bottom block can have. For example, if the bottom block had $\frac{1}{2}$ foot of overhang (from the table), and if the top block had an additional $\frac{1}{2}$ foot of overhang (from the bottom block), then there's no way it would stay up. Here is a profile picture of this bad situation:

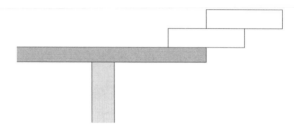

So if the bottom can have some overhang, but not as much as in the above, where is the tipping point? It all comes down to the *center of mass*. With a single block, you one can have up to half of it hanging off the edge because the center of mass of the block is exactly in the middle, and provided the center of mass is above the table, it will not topple. If you stack two blocks and neither is sticking out more than halfway from what is below it, then the only way it will topple is if their *collective* center of mass is not above the table.

To proceed, it'll be useful to recall two basic facts about centers of mass. First, to find the center of mass of a collection of objects one may assume that each object's entire mass occurs at its center of mass — this is called a *point mass*. Second, to find the (x-coordinate of the) center of mass of two objects, one can use a *weighted average*, which in the figure below is $\frac{m_1 d_1 + m_2 d_2}{m_1 + m_2}$, where the two point masses (represented with circles) have masses m_1 and m_2.

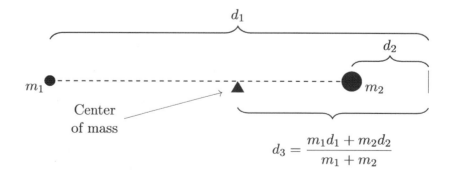

In our problem, all the blocks are identical and therefore have the same weight. For simplicity, let's assume each weighs 1 pound. Then we can compute the location of the center of mass of two blocks; the optimal placement of the blocks will be such that their collective center of mass is right at the table's edge.

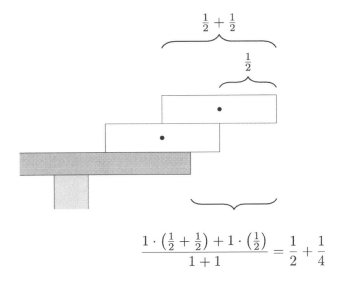

$$\frac{1 \cdot \left(\frac{1}{2} + \frac{1}{2}\right) + 1 \cdot \left(\frac{1}{2}\right)}{1 + 1} = \frac{1}{2} + \frac{1}{4}$$

So if the top block hangs out $\frac{1}{2}$ feet from the bottom block, up to $\frac{1}{4}$ feet of the bottom block can hang off the table, giving a combined $\frac{1}{2} + \frac{1}{4}$ feet of overhang. (You'll soon see why I'm writing it as a summation, rather than just $\frac{3}{4}$ feet.)

With three blocks, where the top two use the above (optimal) arrangement, you can view the top two blocks as being a single point mass, with a mass of 2 pounds and a center of mass located $\frac{1}{2} + \frac{1}{4}$ feet from the right side of this two-block, which is optimally placed directly above the edge of the bottom-most block.

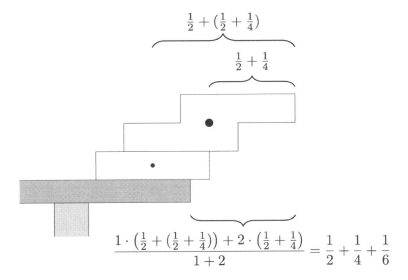

$$\frac{1 \cdot \left(\frac{1}{2} + \left(\frac{1}{2} + \frac{1}{4}\right)\right) + 2 \cdot \left(\frac{1}{2} + \frac{1}{4}\right)}{1 + 2} = \frac{1}{2} + \frac{1}{4} + \frac{1}{6}$$

So a third block can safely overhang the table by $\frac{1}{6}$ feet, if one stacks the 2-block solution from before on top of it. In general, the nth block can overhand the table by

$\frac{1}{2n}$ feet. Indeed, one can prove this by induction. If n blocks can overhang

$$\frac{1}{2} + \frac{1}{4} + \frac{1}{6} + \cdots + \frac{1}{2n}$$

feet, then using the above center-of-mass argument one can show that $n + 1$ blocks, in feet, can overhang

$$\frac{1 \cdot \left(\frac{1}{2} + \left(\frac{1}{2} + \cdots + \frac{1}{2n}\right)\right) + n \cdot \left(\frac{1}{2} + \cdots + \frac{1}{2n}\right)}{1 + n} = \frac{1}{2} + \frac{1}{4} + \cdots \frac{1}{2n} + \frac{1}{2(n+1)}.$$

And now, if you've read this far, it's time for a huge payoff. We asked how far this stack can hang out from the table. For example, can you stack enough blocks so that no part of the top block is above the table? Amazingly, yes! In fact, just four blocks is enough, because with four blocks the total overhang, in feet, is

$$\frac{1}{2} + \frac{1}{4} + \frac{1}{6} + \frac{1}{8} \approx 1.04 > 1.$$

Since there is more than a 1 foot overhang, the top box is entirely to the right of the table!

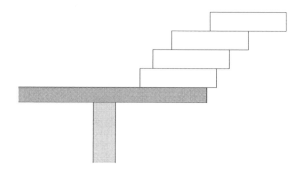

Moreover, because

$$\frac{1}{2} + \frac{1}{4} + \frac{1}{6} + \cdots + \frac{1}{2(31)} \approx 2.01 > 2,$$

with 31 blocks there can be an entire block-length separating the top block and the table! In fact, the only limit to how big the overhang can be is the limit to how big

$$\frac{1}{2} + \frac{1}{4} + \frac{1}{6} + \frac{1}{8} + \cdots = \frac{1}{2} \cdot \sum_{k=1}^{\infty} \frac{1}{k}$$

can be. And in Chapter 4 we showed that $\sum_{k=1}^{\infty} \frac{1}{k} = \infty$, and called this series the *harmonic series*. So, since this series diverges to ∞, there is no limit to the size of the overhang! If you stack enough blocks, you can have a mile of overhang, or a million miles, or anything else. It'll take a lot of blocks, since the harmonic series diverges suuuuuppper slowly, but because it diverges, any overhang distance is possible.

I will leave you with the image of 31 blocks stacked on top of each other, which I mentioned is the first case in which the top block is separated from the table by an entire block-length.

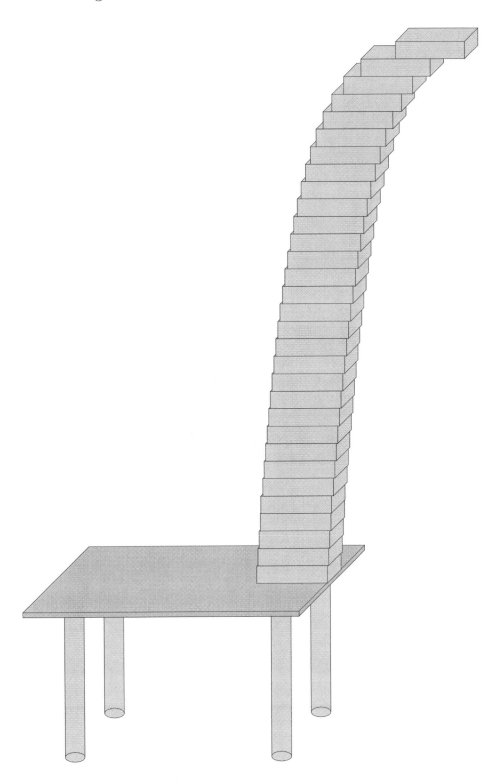

B.6   A Composition Conundrum

Here is an example of integrable functions f and g, where their composition $f \circ g$ is *not* integrable.

At least some of the functions involved are going to have to be pretty weird. Since $f \circ g$ is not integrable, already we know that $f \circ g$ has to be weird, and therefore f and g can't both be super nice.

We do have a few weird functions at our disposal that we know how to handle. The Chapter 8 exercises (twice) ask you to prove that Thomae's function

$$t(x) = \begin{cases} 1/n & \text{if } x \neq 0 \text{ and } x = m/n \in \mathbb{Q} \text{ in lowest terms with } n > 0 \\ 0 & \text{if } x \notin \mathbb{Q} \\ 1 & \text{if } x = 0 \end{cases}$$

is integrable. And in case you can't be bothered to flip to this book's cover, below is the function's graph on $[0, 1]$.

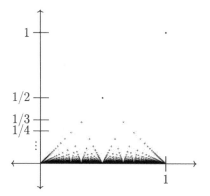

And there's another function that isn't so different than this which we know is *not* integrable: Dirichlet's function. Indeed, in Example 8.12 we showed that

$$d(x) = \begin{cases} 1 & \text{if } x \in \mathbb{Q} \\ 0 & \text{if } x \notin \mathbb{Q} \end{cases}$$

is *not* integrable. Here's its graph:

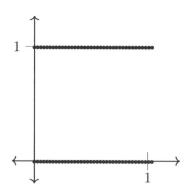

So can we choose t to be either f or g, and pick the other one cleverly so that their composition equals the Dirichlet function? There is a way!

Let $g(x)$ be Thomae's function on $[0, 1]$. How do we choose f to turn g into the Dirichlet function? The irrationals are already good-to-go. So when we take $f(g(x))$, if $g(x) = 0$ we don't want f to change this fact. That is, we want $f(0) = 0$.

What else? Well, all the rationals are getting mapped to values $1/n$, for some n. And Dirichlet wants those value to all equal 1. That is, if $g(x) > 0$, we want $f(g(x))$ to equal 1. So for any $x > 0$, we want $f(x) = 1$. Make sense? This suggests that we define $f : [0, 1] \to \mathbb{R}$ to be

$$f(x) = \begin{cases} 1 & \text{if } x > 0 \\ 0 & \text{if } x = 0. \end{cases}$$

This f has a very simple form:

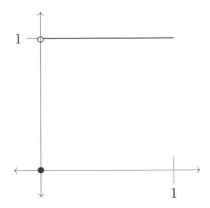

And by Porism 8.18, f is integrable. So we have that f and g are both integrable. We now simply note that $f \circ g$ is the Dirichlet function, and hence is not integrable.

$$f(g(x)) = \begin{cases} f(1/n) & \text{if } x \in \mathbb{Q} \\ f(0) & \text{if } x \notin \mathbb{Q} \end{cases}$$

$$= \begin{cases} 1 & \text{if } x \in \mathbb{Q} \\ 0 & \text{if } x \notin \mathbb{Q} \end{cases}$$

So indeed, despite f and g both being integrable, $f \circ g$ wound up being the Dirichlet function which is *not* integrable.

B.7   Turning the Tables on your Teetering Troubles

You're at a café, you've just ordered a delicious drink and you head to the patio to enjoy the beautiful day and learn some mathematics. You sit down at the last open table and quickly discover why no one had taken it — the ground there is a little uneven, and so the table wobbles. For a non-mathematician, their entire day would have been utterly ruined; even if they try the classic strategy of stuffing a napkin under one leg, before long it will compress a bit and the wobble will return. Lucky for you, though, there is an easy, mathematically-verified solution — just rotate the table slightly, and like magic you will immediately find a spot where all four legs are completely stable on the ground. Allow me to explain.

Suppose first that we have a square table whose legs are all the same length but due to some uneven ground, the table wobbles. Let's say that legs 1 and 3 wobble, while legs 2 and 4 are on the ground.

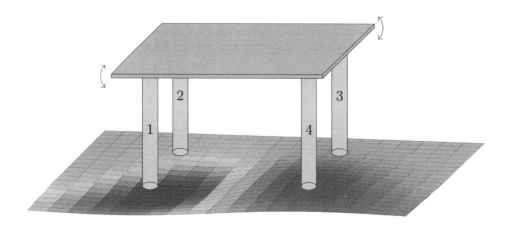

Push down near the wobbly leg 3 so that legs 2, 3, and 4 are on the ground, while leg 1 is suspended above the ground. Let's say leg 1 is d inches above the ground.

Now rotate the table like a disc, moving leg 1 towards where leg 2 is, leg 2 towards where leg 3 is, and so on. We claim that before this 90° turn is complete, the table will reach a point where it is completely stable.

When you rotate, keep legs 2, 3 and 4 flush to the ground, and let $d(t)$ be the distance above the ground of leg 1 at each moment of time — therefore we want to prove that at some point $d(t) = 0$. When discussing the proof, it's helpful to imagine that leg 1 is a ghost leg, which is able to pass into the ground. With this perspective, every turn is possible, and $d(t)$ will be negative if leg 1 moves into the ground.

Suppose the 90° turn takes 1 second. At $t = 0$ (the moment before the rotation begins) we have said that $d(t) = d$. After the 90° rotation, can you see what $d(t)$ will be? Legs 1 and 3 (previously wobbly) have switched spots with legs 2 and 4 (previously sturdy). So legs 2 and 4 are now on wobbly ground, but remember that we are insisting on keeping legs 2, 3 and 4 on the ground — and the only way to take two wobbly legs and make them stable on the ground, is to force one of the other legs *into the ground*. Leg 1 — the ghost leg — must therefore be inside the ground after the turn. That is, $d(1) < 0$. We have shown this:

$$d(0) = d > 0 \qquad \text{while} \qquad d(1) < 0.$$

And since this is a café patio ground in the real world (not egregiously bumpy, no steps, etc.), $d(t)$ will be a continuous function. So we have a continuous function which is positive at $t = 0$ and negative at $t = 1$, and hence by the intermediate value theorem (in particular by Proposition 6.37) there must exist some $c \in (0, 1)$ such that $d(c) = 0$. And therefore, at $t = c$, your table will be perfectly stable. You're welcome.

If a table has a circular top but its legs form a square, then of course the above works the same way. What about other table shapes? If the legs form a rectangle,[11] it works almost just the same. Unlike with square tables where after a 90° turn the legs will have collectively returned to their original location, with rectangular legs you have to turn the table 180° to get that property — and it takes up to a 180° turn to stabilize the table. The proof is a little more complicated as well, since after that 180° turn, legs 1 and 3 occupy the same locations (as a pair), rather than the locations of legs 2 and 4.

The proof in this rectangular case is by contradiction,[12] and goes something like this: If legs 1 and 3 were always hovering and legs 2 and 4 were always on the ground, then at some point the table top is closest to the ground. This is a contradiction, though, since we could then rotate the table until legs 2 and 4 are where legs 1 and 3 were — but at this point we have replaced hovering legs with legs touch the ground, which lowers the table top further, and shows that our purported "closest to the ground" location was not actually minimal. And that's a contradiction.

I have used this table-turning technique about a dozen times over the years and it has always worked. That said, it is *possible* that the ground is just too uneven. Indeed, if the ground is super wavy, and at points the ground's steepness is larger than $\arctan\left(\frac{1}{\sqrt{2}}\right) \approx 35°$, then it could fail. But that is *reeeally* steep. Even the steepest streets in the world — including Baldwin Street in Dunedin, New Zealand and Filbert Street in San Francisco, United States — only slope at about 20°. In practice you will rarely, if ever, have a problem.

---

[11] Which includes many non-tables too. For instance, a wobbly step ladder should never be climbed without rotating to find a stable footing.

[12] For details, check out the paper *Mathematical Table Turning Revisited* by Baritompa, Löwen, Polster, and Ross. See here: `https://arxiv.org/pdf/math/0511490.pdf`

B.8   A Devilish De-Descent

Here's one more example that in some ways relates to the weirdness of the Cantor set. It's a function called *The Devil's Staircase*. And like the Cantor set, it's best to describe it in stages.

Consider the graph of the straight-line function from (0,0) to (1,1); this is Stage 1 of the creation of the Devil's Staircase.

Stage 1

In Stage 2, you take the middle third of the function and replace it with a horizontal line (halfway up; so at $y = 1/2$), and then redraw the diagonal lines in the first and last third to meet this horizontal line.

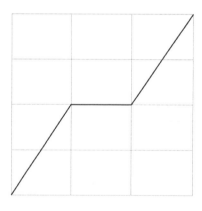

Stage 2

In Stage 3 you do the same thing to each diagonal line. That is, you take the middle third of the line in $[0, 1/3]$ and the line in $[2/3, 1]$, and replace each with a horizontal line (halfway up; so at $y = 1/4$ and $y = 3/4$, respectively), and then redraw the diagonal lines in the first and last third to meet this horizontal line.

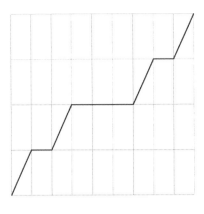

Stage 3

Now continue this. For each diagonal line, replace it with a horizontal line halfway up, and redraw the diagonal lines to make it continuous. Below are the next 4 iterations.

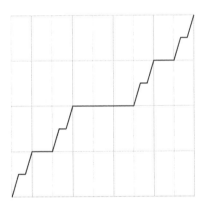

Stage 4

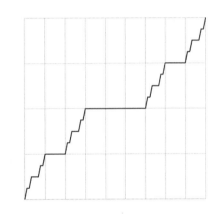

Stage 5

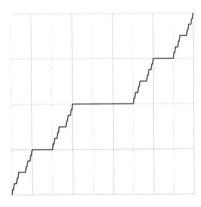

Stage 6

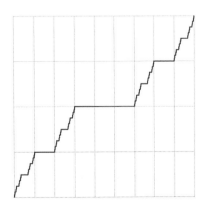

Stage 7

Continuing this process forever, the resulting sequence of functions converges to the function which is called *The Devil's Staircase*. And this function has some fascinating properties.

- Note first that if you get a flat spot at any stage, it remains a flat spot forever. And the flat spot after the first stage is $[1/3, 2/3]$. The flat spots in Stage $n+1$ are those in Stage n, as well as this: Given any diagonal region in Stage n, the middle third becomes flat in Stage $n+1$. That is, the flat spots are exactly the regions removed in the construction of the Cantor set! And therefore, the only points which will never be part of (the interior of) a flat spot are precisely the points in the Cantor set.

- And since the Cantor set has length 0, the Devil's staircase is flat almost everywhere. Said differently, the Devil's Staircase has a derivative of 0 at almost every point. But this is odd, because despite being horizontal nearly everywhere... somehow it still manages to climb from $(0,0)$ to $(1,1)$!

Now, you might say that you can think of a function that is flat nearly everywhere but still manages to climb from $(0,0)$ to $(1,1)$. A (stair-)step function, for instance:

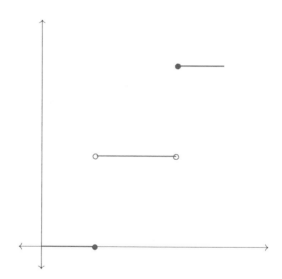

But this function is clearly not continuous, and to make it continuous we would need to connect the steps together, producing intervals on which the function is increasing (with a positive derivative).

And yet the Devil's Staircase somehow manages to never have an interval where it is strictly increasing, and at no point does its derivative take a positive value (in fact, its derivative is zero almost everywhere)... and yet it still manages to somehow climb from the origin to $(1,1)$. Amazing.

- And it's worth reiterating that amazing property: For a continuous function to get bigger, you would think there would have to be some interval in which it is getting bigger. But alas, that's not the case. You can get bigger, be continuous, and yet have no interval in which you are strictly rising. WHAAAAA?!?!?!?

- Another interesting fact about this function is its length. The straight-line function from $(0,0)$ to $(1,1)$ has a length of $\sqrt{2}$. Meanwhile, the line segment from $(0,0)$ to $(1,0)$, plus the segment from $(1,0)$ to $(1,1)$, also makes it from $(0,0)$ to $(1,1)$, but has length 2. So what's the length of the Devil's staircase? You might imagine that since it seems to be a mix of these two that its length is somewhere in between, but in fact its length is also 2. Indeed, it starts at Stage 1 with a length of $\sqrt{2}$, but with each successive iteration its length gets closer and closer to 2, where it ends up.

- Final thoughts: This function is also called the Cantor function. Note that since the function is continuous, it has the intermediate value property. Hence not only does it climb from $(0,0)$ to $(1,1)$, but it hits every y-value in between. It is continuous and also uniformly continuous, but it is not what is called *absolutely continuous*.

- P.S. While researching about the Devil's Staircase I learned about a series of waterfalls in the coastal rainforest of Oregon that are called The Devil's Staircase. It is a gorgeous area that has been labeled an "area of critical environmental concern" and as former congressman Jim Weaver says, "[the area] deserves no less than the highest protection that can be afforded under law," and I gotta say I completely agree. Here are two angles of the Devil's Staircase in the wild:

B.9   A Pack of Pretty Proofs by Picture

Theorem. FOILing: $(a+b)^2 = a^2 + 2ab + b^2$.

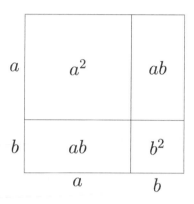

Theorem.   The Pythagorean theorem.

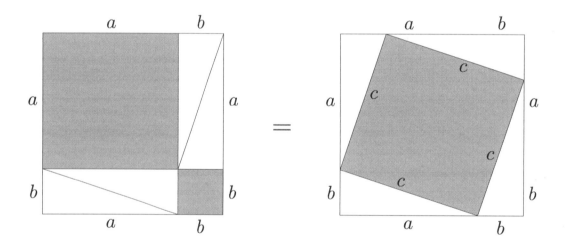

Theorem.   There is a bijection f between the interval $(0,1)$ and the interval $(0, \infty)$. That is, $|(0,1)| = |(0, \infty)|$.

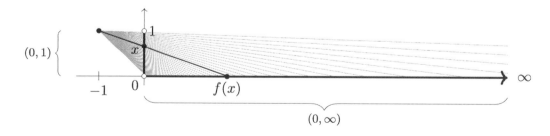

<u>Theorem.</u> Geometric series test: If $0 < r < 1$, then $\displaystyle\sum_{k=0}^{\infty} ar^k = \frac{a}{1-r}$.

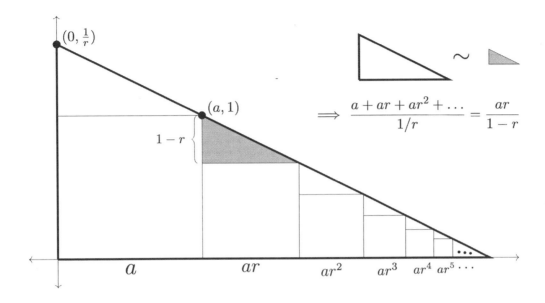

<u>Theorem.</u> $\dfrac{3}{4} + \dfrac{3}{16} + \dfrac{3}{64} + \dfrac{3}{256} + \cdots = 1$.

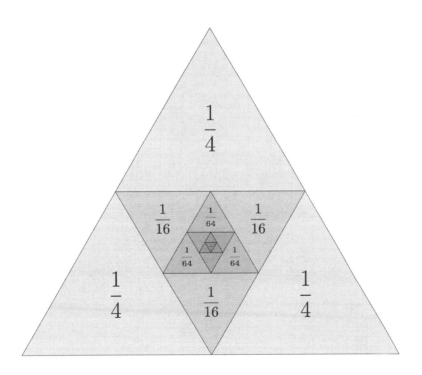

<u>Theorem.</u> $1 - \dfrac{1}{3} + \dfrac{1}{9} - \dfrac{1}{27} + \dfrac{1}{81} - \dfrac{1}{243} - \cdots = \dfrac{3}{4}.$

Area $= 1$

Area $= 1 - \dfrac{1}{3}$

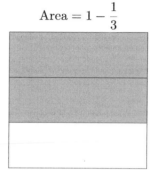

Area $= 1 - \dfrac{1}{3} + \dfrac{1}{9}$

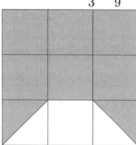

Area $= 1 - \dfrac{1}{3} + \dfrac{1}{9} - \dfrac{1}{27}$

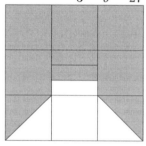

Area $= 1 - \dfrac{1}{3} + \dfrac{1}{9} - \dfrac{1}{27} + \dfrac{1}{81}$

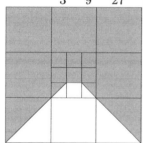

Area $= 1 - \dfrac{1}{3} + \dfrac{1}{9} - \dfrac{1}{27} + \dfrac{1}{81} - \dfrac{1}{243}$

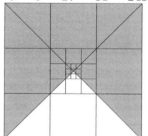

Area $= 1 - \dfrac{1}{3} + \dfrac{1}{9} - \dfrac{1}{27} + \dfrac{1}{81} - \dfrac{1}{243} - \cdots = \dfrac{3}{4}$

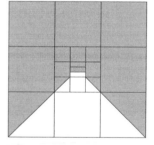

B.10 An Abundant Addition Aboundingly Ascends

In Chapter 4 we saw many examples in which we summed up *infinitely* many positive numbers and yet got a *finite* number out. For example,

$$\frac{1}{2} + \frac{1}{4} + \frac{1}{8} + \frac{1}{16} + \cdots = 1.$$

Notice, though, that each of our examples only involved a *countably* infinite collection of numbers, and we have yet to talk about what happens when you sum up an *uncountably* infinite collection of numbers. Can such a sum also add up to a finite number?

First, you might ask what we even mean by this. How can you sum up uncountably many things when sums like

$$\frac{1}{2} + \frac{1}{4} + \frac{1}{8} + \frac{1}{16} + \cdots$$

are inherently countable—we are adding up one thing, then the next, then the next, and so on, and we are asking what that sequence of partial sums converges to. And doing one thing after another guarantees that the procedure has only countably many steps.

As it turns out, we don't even have to get into the details about how to generalize our notion of countable sums to uncountable sums. The answer to our original question is 'no'—every sum of uncountably many positive numbers must equal ∞, and in fact something even stronger always holds: The uncountable set is even guaranteed to contain a countable subset which itself sums to ∞!

To see this, let S be an arbitrary set of uncountably many positive numbers, which we want to add up. Since everything in S is positive, we can group these elements into the following sets. Let S_1 be the numbers in S which are at least 1. Let S_2 be the elements of S which are between 1/2 and 1. Let S_3 be the elements of S which are between 1/3 and 1/2. Let S_4 be the elements of S which are between 1/4 and 1/3. And so on. In general, for $n \geq 2$ let

$$S_n = \left\{ x \in S : \frac{1}{n} \leq x < \frac{1}{n-1} \right\}.$$

Now simply note that we just put uncountably many numbers into countably many sets. Clearly at least one of these sets must have infinitely many numbers in it (in fact, at least one will have uncountably many numbers in it, but this is stronger than we need). Suppose S_N has infinitely many numbers in it. Adding up any countable collection from this set will be at least

$$\frac{1}{N} + \frac{1}{N} + \frac{1}{N} + \frac{1}{N} + \frac{1}{N} + \cdots,$$

which equals ∞, completing the proof.[13] □

---

[13] *Shhh... Don't tell anyone, but this is also essentially a solution to Exercise 6.59.*

B.11   A Smooth and Spiky Solution

Recall that Theorem 7.6 says that if f is differentiable, then f is also continuous. So differentiability is a *stronger* condition; differentiability implies continuity. And in fact, it is *strictly* stronger as well, since there are functions which are continuous at every point, but are not differentiable at every point. The absolute value function $A_0(x) = |x|$ on $[-1, 1]$ is a classic example.

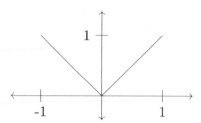

This function is continuous everywhere, but there is one point where it is not differentiable: at the origin. Indeed, applying the definition of the derivative of A_0 at the point $x = 0$, the limit

$$A_0'(0) = \lim_{h \to 0} \frac{A_0(0 + h) - A_0(0)}{h}$$

equals two different values depending on whether h approaches zero from the left or from the right. From the left, $A_0(x) = -x$ and so

$$\lim_{h \to 0^-} \frac{A_0(0 + h) - A_0(0)}{h} = \lim_{h \to 0} \frac{-h - 0}{h} = \lim_{h \to 0} -1 = -1.$$

While from the right, $A_0(x) = x$ and so

$$\lim_{h \to 0^+} \frac{A_0(0 + h) - A_0(0)}{h} = \lim_{h \to 0} \frac{h - 0}{h} = \lim_{h \to 0} 1 = 1.$$

Both of these make sense, as the line approaching zero from the left is $y = -x$ which has slope -1, while the line approaching zero from the right is $y = x$ which has slope 1. And because those are different, the two-sided limit

$$A_0'(0) = \lim_{h \to 0} \frac{A_0(0 + h) - A_0(0)}{h}$$

does not exist.

So we can indeed find a function A_0 where there is a point which is continuous but not differentiable. Moreover, if we extend A_0 to be periodic on all of $\mathbb{R}$, then we get a function with "spikes" at every integer, and hence a function which is non-differentiable at every integer:

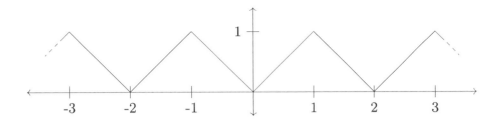

But how non-differentiable can a continuous function be? We could of course make the non-differentiable points occur more often. Consider the function which we'll call A_1:

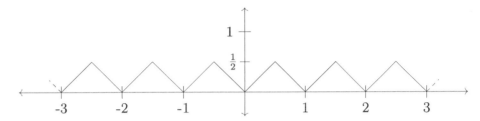

Note that A_1 is discontinuous at every point of the form $\frac{m}{2}$, where m is an integer. Better yet, here's A_2:

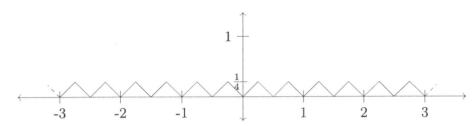

Note that A_2 is discontinuous at every point of the form $\frac{m}{2^2}$, where m is an integer. Indeed, you could keep doing this and A_k would have the property that it is discontinuous at every point of the form $\frac{m}{2^k}$.

In some sense this is making things weirder, but in another it's not doing much: Is it so different to have every 8th term be discontinuous compared to every 64th term? It's still a discrete set where you have fixed gap-sizes between the points.

The holy grail example would be to find a continuous function that is differentiable *nowhere*. Now take a minute to appreciate how peculiar such a function would be. In all our examples thus far, being non-differentiable at a point was obtained by having two lines (which are not only continuous, but also differentiable) meet at a single point, forming a peak. The other "elementary" technique to get a non-differentiable point is to have a jump discontinuity there, but this can't be used since we want our function to be continuous. In Example 7.15 we saw another approach to maintain continuity while losing differentiability, but this approach seems to have no advantages over a simple peak, while at the same time being much harder to grasp.

Peaks seem the best approach, and it is reasonable to think that in order to have a peak, you need to have a region around that peak where you have two nice, differentiable lines meeting each other. And of course, if it were true that you need to have a region of differentiability around each non-differentiable point, then we would be able to conclude that it is not possible to be both continuous at every point and differentiable at no point.

And yet! And yet! It is possible! There do exist functions which are continuous everywhere and differentiable nowhere. Moreover, we are already close to writing down an example...

Consider the function

$$f(x) = \sum_{k=0}^{\infty} A_k(x).$$

This is a good example to introduce series of functions, so I won't assume that you're an expert in the material from Chapter 9. Moreover, that material is quite deep and nuanced, so we will only scratch the surface here. Briefly, though: For any number x_0, and each fixed k, you can clearly plug this number into A_k and get out some other number $A_k(x_0)$; furthermore, note that they are all non-negative outputs. What is $f(x_0)$? It's what you get when you add up all these numbers: $A_1(x_0) + A_2(x_0) + \dots$. That is, recall from Chapter 4 that a series of non-negative numbers can either add up to some number or to ∞. In this case, each $f(x_0)$ is finite as it is less than $\sum_{k=0}^{\infty} \frac{1}{2^k}$. (Formally, use the Weierstrass M-test (Corollary 9.22) with $M_k = \frac{1}{2^k}$.)

Moreover, one can show that f would also be continuous. And yet — and this is more challenging — one can also prove that f is differentiable nowhere.

In short: If x_0 is of the form $\frac{m}{2^n}$ for integers m and n, then f's graph is intuitively peak-like at $(x_0, f(x_0))$, and is therefore non-differentiable at that point. If x_0 is not of this form, then you should instead consider a sequence of points x_n converging to x_0 from the left, each of which *is* of the form $\frac{m}{2^n}$, and a second sequence y_n converging to x_0 from the right, each of which is *also* of that form. If you do so such that

$$x_n = \frac{p_n}{2^n} < x_0 < \frac{p_n + 1}{2^n} = y_n$$

for each n, then the derivative limit along these two sequences are not equal, implying that $f'(x_0)$ does not exist.

Finally, in closing, here is a look at the graph of f, which is a more discrete version of an example published by Weierstrass in 1872. We begin by showing partial sums of its definition on $[0, 4]$. First, here is $A_0 + A_1$:

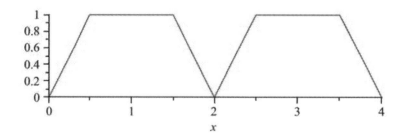

And now here is $A_0 + A_1 + A_2$:

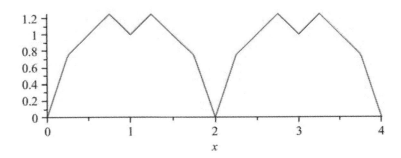

Here is $A_0 + A_1 + A_2 + A_3$:

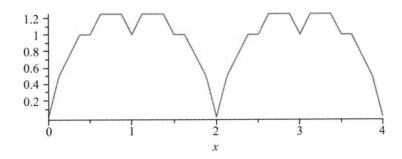

Skipping ahead a bit, here is $A_0 + A_1 + A_2 + A_3 + A_4 + A_5 + A_6$:

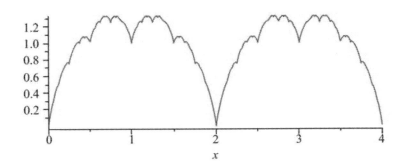

Note that A_7 is a function with amplitude $\frac{1}{2^7} \approx .0078$, and it only gets smaller from there. So the above is nearly what the graph of f looks like. The later sums play an extremely important role, but you'd have to zoom in quite a bit to see how they change the graph.

B.12   Finding ϕ for First-Place Finishes

The recurrence $a_{n+1} = a_n + 2$, where $a_1 = 1$, grows linearly. The sequence begins $1, 3, 5, 7, \ldots$. But if the sequence is a recurrence involving a sum of the last *two* terms, it grows *much* faster. If $a_{n+2} = a_{n+1} + a_n + 2$, where $a_1 = a_2 = 1$, then the sequence begins $1, 1, 4, 7, 13, 22, 37, \ldots$. It is hard to spot at this point, but this sequence is growing faster than any polynomial grows—it is growing *exponentially*. The classic exponential function is 2^n, and if you think this function looks innocent, think again. Here's my favorite illustration of exponential growth:

- A standard piece of paper is about 0.05 millimeters thick. If you fold the paper in half, the thickness is now 0.1 millimeters thick, being two pieces thick. A third fold and it is 0.2 millimeters thick.

- With 13 folds you would set a new world record for number of folds.[14] If you accomplished this, the thickness would (theoretically) be about 16 inches thick (although about twice this in practice).

- If you fold it 22 times it'll be about a mile thick.

- If you fold it 43 times it would be as thick as the distance to the moon.

- If you fold it 54 times it would be as thick as the distance to the sun.

- If you fold it 103 times it would be as thick as *the width of the observable universe*.

The moral: Repeated doubling amounts to exponential growth, and exponential growth is super, super fast growth. And indeed, one can show that an innocent looking recurrence like $a_{n+2} = a_{n+1} + a_n + 2$, where $a_1 = a_2 = 1$, is growing exponentially.

Slightly more simply, the Fibonacci sequence obeys the recurrence $F_{n+2} = F_{n+1} + F_n$, where $a_1 = a_2 = 1$. It begins $1, 1, 2, 3, 5, 8, 13, 21, 34, \ldots$. This sequence presents itself in the real world. First introduced to enumerate forms of poetry, it has since found numerous applications in music, plant biology and elsewhere.

As claimed, it is also true that this sequence is approximated by an exponential function ϕ^n, for some number ϕ. Let's find ϕ. Notice that for exponential functions like 2^n, the base number is the ratio of consecutive terms: $\frac{2^{n+1}}{2^n} = 2$. For functions which grow like an exponential but aren't precisely of this form, the ratio of consecutive terms *approaches* some fixed number. That is,

$$\phi = \lim_{n \to \infty} \frac{F_{n+1}}{F_n}.$$

But how do we use this to find ϕ? It's really clever. First note that since clearly

$$\lim_{n \to \infty} \frac{F_{n+1}}{F_n} = \lim_{n \to \infty} \frac{F_n}{F_{n-1}}, \qquad \text{we also have} \qquad \phi = \lim_{n \to \infty} \frac{F_n}{F_{n-1}},$$

---

[14]You may use a very large piece of paper, but it must still be 0.05 millimeters thick

which by the limit laws implies that

$$\frac{1}{\phi} = \lim_{n \to \infty} \frac{F_{n-1}}{F_n}.$$

Applying this, as well as the Fibonacci recurrence $F_{n+1} = F_n + F_{n-1}$,

$$\phi = \lim_{n \to \infty} \frac{F_{n+1}}{F_n} = \lim_{n \to \infty} \frac{F_n + F_{n-1}}{F_n} = \lim_{n \to \infty} \left(1 + \frac{F_{n-1}}{F_n} \right) = 1 + \frac{1}{\phi}.$$

That is, by using a little bit of trickery with limits we were able to show that whatever ϕ is, it must satisfy the equation

$$\phi = 1 + \frac{1}{\phi}.$$

And here is one of those wonderful moments when the key to finishing off a problem relies only on the math you learned in high school. If ϕ must satisfy the above equation, then we can solve that equation to determine what ϕ must be! Rewriting the above,

$$\phi^2 = \phi + 1.$$

Moving everything to the left,

$$\phi^2 - \phi - 1 = 0.$$

And how do you solve a equation like this that doesn't easily factor? The quadratic formula! The solutions are

$$\frac{-(-1) \pm \sqrt{(-1)^2 - 4(1)(-1)}}{2(1)} = \frac{1 \pm \sqrt{5}}{2}.$$

Clearly the Fibonacci sequence is increasing which means ϕ has to be at least 1, and so we must have

$$\phi = \frac{1 + \sqrt{5}}{2}.$$

This number might look familiar. It's better known as *the golden ratio*. It too has popped up in unsuspecting places, including prominently in architecture. However, my favorite application of ϕ is to distance. First, note that the ratio of consecutive Fibonacci terms approach ϕ pretty quickly:

$$\frac{3}{2} = 1.5 \quad , \quad \frac{5}{3} = 1.66 \quad , \quad \frac{8}{5} = 1.6 \quad , \quad \frac{13}{8} = 1.625 \quad , \quad \frac{21}{13} = 1.615 \quad , \quad \ldots$$

while $\phi = 1.618\ldots$. And here's just a silly coincidence: 1 mile is about 1.61 kilometers. Said differently, 1 mile is roughly ϕ kilometers. So if you want to convert from miles to kilometers, you can use the Fibonacci sequence! For example, 3 miles is about 5 kilometers (as runners know), and 5 miles is about 8 kilometers, and 8 miles is about 13 kilometers, and 13 miles is about 21 kilometers, and so on. Golden.

B.13   Peculiar and Pathological Perimeters

Chapter 8 began with a procedure in which the area of polygons converged to the area of a circle, which we used to find the formula for the area of a circle. First, even though most of our attention was on the inscribed n-gons, it was quite important that we noted then that the areas of the inscribed n-gons *and* the areas of the circumscribed n-gons were *both* converging to the same thing.

Interestingly, perimeter and area do not behave well together. If I have a sequence of shapes containing a circle whose areas are getting closer and closer to the area of a circle, it need not be the case that the *perimeters* of those shapes are getting closer and closer to the *perimeter* of the circle. Indeed, consider the following procedure to converge to a circle. For example, consider the following sequence of shapes.

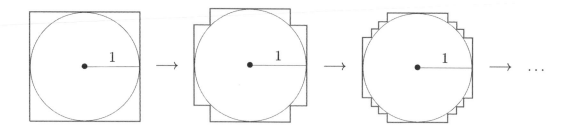

Our first shape is a square that is circumscribed about the unit circle. For our next shape we cut off each of the corners of the square, and replaced them by straight line segments to the circle. This gave a new shape containing the circle, and this new shape now has 8 corners sticking out. We now repeat, cutting off all 8 of these corners in the same way. Repeating this indefinitely, one can see that the zig-zag shape is, in some way, *converging* to the circle. If you calculated the areas of the these zig-zag shapes, for instance, you would find that they are indeed converging to the area of the circle.

But notice now that the perimeter of the first square is 8. The perimeter of the next zig-zag shape is... also 8! The straight line on the top of the square, for instance, is still there, it's just now in three pieces:

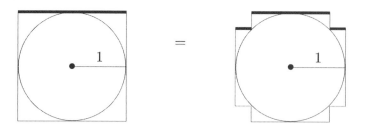

So you have a sequence of shapes, each with perimeter 8, that are "converging" to a shape with perimeter 2π... so $8 = 2\pi$? And so $\pi = 4$? No! Because perimeters simply do not converge in the same way that areas do.

Here's another example. Note that each drawing below has the same perimeter. Let's say the first drawing is a 1×1 square. Then every drawing has perimeter 4. In particular, the stair-step portion of each has length 2.

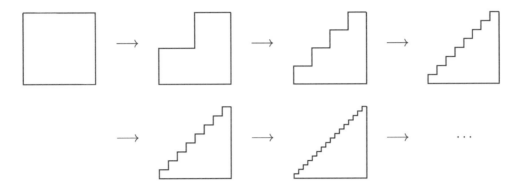

However, as the steps get increasingly tiny, it sure looks like it is approaching the hypotenuse of a right triangle...

...but in the limit, the length-2 stair steps turn into a straight line hypotenuse of length $\sqrt{2}$. This time, if perimeters and limits behaved nicely, this would imply that $\sqrt{2} = 2$!

Similar phenomena exist in functional analysis. For instance, it is possible for a pair of functions f and g to be very close to each other, while f' and g' are very far from each other. Likewise, it's possible for two functions to be very close while their arc lengths are far apart.

Dog owners know this from personal experience. If you take an energetic dog on a walk, even if the leash is pretty short (ensuring you two are "close" together at every moment), the dog still manages to hop from side to side, smell this, pee on that... the dog's path length is much longer than yours! I bet my mom's dog walks twice the distance she does when they go on a walk together, despite her short leash.

There is one final interesting manifestation of this that I'd like to mention, called *the coastline paradox*. If you fly above a coastline and take pictures of it, and then later use those pictures to measure the coastline's length, the answer you'll get *depends on how good of a camera you have*. If you have a better camera your picture will include more small zigs and zags and consequently you will measure a longer distance than if your resolution is low, and the details run together. Its length is like some Schrödinger's *stat*, whose value doesn't exist until observed.[15]

---

[15] ...Now, if you want to know how the coastline paradox affects the evil art of Gerrymandering, take my future class on the Math of Politics.

B.14   Fractal Functions Filling Foursquare Frames

A classic magical feat involves the magician seemingly creating something out of nothing. Whether it is a rabbit out of an empty hat, a person out of empty space, or lighter fluid out of a sleeve,[16] such tricks always rely on a small lie — perhaps a hidden compartment or a sleight of hand.

In math, though, lies are forbidden. Can one still create something from nothing? Let's talk about area. A line segment is one-dimensional and so has no area.

But a filled-in square does.

All lines have no area, as well as all parabolas and (non-filled in) circles. In fact, the graph of any continuous function you can think of probably has no area. Suppose, for instance, we wanted a continuous *surjective* (or *onto*) function that maps from the line segment $[0, 1]$ to the (filled-in) square $[0, 1] \times [0, 1]$ — thus, by just moving the points around, creating area out of no area. We have a domain of $[0, 1]$ to work with — what if we, for example, began by mapping $[0, 0.1]$ to the middle portion of the square, it would look like this:[17]

We have used up 10% of our domain and, in an area sense, we made zero progress toward our goal of magically creating area. The area of the "remaining" square is still 1, as we have only covered a single line, which had zero area.

---

[16] *"But where did it come from?"*

[17] An explicit function that can do this is $f : [0, 0.1] \to [0, 1] \times [0, 1]$, where $f(x) = (10x, 0.5)$.

Indeed, it seems quite reasonable that there is no way around this, and you would be forgiven for believing that there can't possibly be a continuous function which surjectively maps $[0,1]$ to $[0,1] \times [0,1]$. But, as happens so often in real analysis, the magic exists — provided you look to the infinite.

Consider the following infinite procedure. Here is one continuous function (call it f_1) from $[0,1]$ to $[0,1] \times [0,1]$:

It only covers three of the four borders, but that's where we start.[18] And even though our function's graph has no area, we actually have made some progress. To see why, what we do next is create a new function (call it f_2) where we change three of the flat-spots into some orientation of the ⊔ shape that we started with, like so:

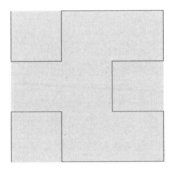

We continue. For every ⊔ shape that you see (in any orientation), replace it with the above. The below is f_3.

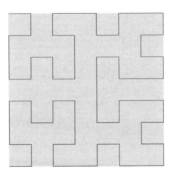

---

[18] An explicit $f_1 : [0,1] \to [0,1] \times [0,1]$ giving this: $f_1(x) = \begin{cases} (3x, 1) & \text{if } x \in [0, 1/3]; \\ (1, 2-3x) & \text{if } x \in (1/3, 2/3]; \\ (3-3x, 0) & \text{if } x \in (2/3, 1]. \end{cases}$

Continuing this procedure, here are the next three stages

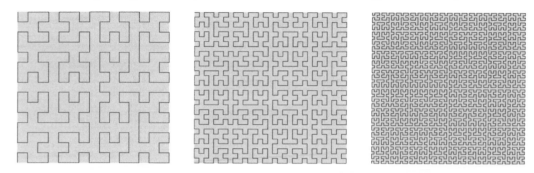

Each of these functions still has zero area, but this sequence of functions is famous because it *converges*, the limit is continuous, and every point (a, b) in the $[0, 1] \times [0, 1]$ square is a limit point of some sequence $f_1(x_0), f_2(x_0), f_3(x_0), \ldots$! Indeed, this is the true brilliance of this sequence. Suppose you had the idea to have a sequence of functions that you hoped would fill all the area of a square. It might be natural to consider a sequence like this:

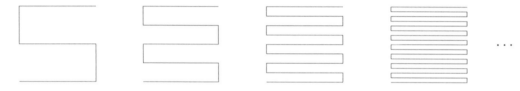

It's certainly easier to describe and seems much easier to study. The problem is, this sequence does not converge. If g_k is the kth function in this procedure, then, for example, the sequence $(g_k(0.1))$ does not converge. These points are jumping all over the place.

The dots' vertical locations are stabilizing nicely (to about 90% of the vertical way up), but its horizontal location is surely not. It will be bouncing around from left to right forever, and so this sequence won't converge.

For the sequence $(f_1(0.1))$, though, things are better behaved.

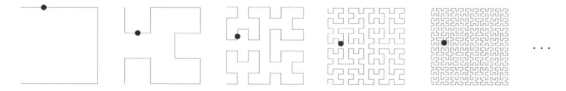

Notice how the dots' are converging to a point somewhere around $(0.17, 0.54)$. And so even though every f_k has 0 area, their limit magically has area 1.

B.15   A Prestigious Proof of a Primal Puzzle

Back in the third century BC, Euclid proved that there are infinitely many primes, and today his argument is perhaps the world's most widely known proof. If we let $\mathcal{P}$ be the collection of all prime numbers, then his proof began by assuming $|\mathcal{P}|$ is finite. If it is, and we let

$$Q = \prod_{p \in \mathcal{P}} p,$$

then we have obtained an integer which contains every prime as a factor. The contradiction then comes by considering $Q + 1$, which according to the *fundamental theorem of arithmetic* must also be prime or product of primes. But since $Q + 1 \equiv 1 \pmod{p}$ for every $p \in \mathcal{P}$, our collection of primes must be incomplete. Ta-da!

One consequence of the above reasoning is the following fact.

Fact B.5. If Q is any finite product of primes and $n \in \mathbb{N}$, then $nQ + 1$ is not divisible by any of those primes.

Before preceding, recall the following two results from Chapter 4.

- **Proposition 4.9** (*Geometric series test*). Assume that a and r are non-zero real numbers. Then $\displaystyle\sum_{k=0}^{\infty} a \cdot r^k = \begin{cases} \dfrac{a}{1-r} & \text{if } \ |r| < 1; \\ \text{diverges} & \text{if } \ |r| \geq 1. \end{cases}$

- **Proposition 4.15.** The harmonic series $\displaystyle\sum_{n=1}^{\infty} \frac{1}{n}$ diverges.

There are infinitely many primes, but they become sparse as you move to larger and larger integers. Indeed, the celebrated *prime number theorem* says that the number of primes in $\{1, 2, 3, \ldots, N\}$ is about $N/\log(N)$. In particular, if you pick a large N, the probability that it is prime is about $1/\log(N)$. Since $1/\log(N) \to 0$ as $N \to \infty$, this means that fewer and fewer numbers are prime.

Now, we saw in Chapter 4 that not only does the harmonic series $\sum_{n=1}^{\infty} \frac{1}{n}$ diverge, but in some sense it just barely diverges — according to the series p-test (Proposition 4.16), $\sum_{n=1}^{\infty} \frac{1}{n^r}$ will converge for *any* $r > 1$.

Here is an interesting question: What happens if instead of considering

$$\sum_{n=1}^{\infty} \frac{1}{n} = 1 + \frac{1}{2} + \frac{1}{3} + \frac{1}{4} + \frac{1}{5} + \frac{1}{6} + \frac{1}{7} + \frac{1}{8} + \frac{1}{9} + \frac{1}{10} + \frac{1}{11} + \frac{1}{12} + \frac{1}{13} + \cdots$$

we only considered reciprocals of primes?

$$\sum_{p \in \mathcal{P}} \frac{1}{p} = \frac{1}{2} + \frac{1}{3} + \frac{1}{5} + \frac{1}{7} + \frac{1}{11} + \frac{1}{13} + \frac{1}{17} + \frac{1}{19} + \frac{1}{23} + \frac{1}{31} + \frac{1}{37} + \frac{1}{41} + \cdots$$

Notice that even at the start a lot of terms are missing, and once you go way down the sum there are a *lot* fewer numbers. Since the harmonic series just barely diverged, and the sum of reciprocals of primes seems to be growing much slower than the harmonic series, it would be reasonable to guess that this second sum might converge. Interestingly, as the great Leonard Euler proved, it does not.

Theorem B.6. $\displaystyle\sum_{p\in\mathcal{P}}\frac{1}{p}=\infty.$

There are many known proofs of this. The one I will show you is quite nice and also combines some important tools we developed in Chapter 4.

Proof. Assume for a contradiction that $\sum_{p\in\mathcal{P}}\frac{1}{p}$ converges to some real number L. Since each $1/p > 0$, the sequence of partial sums is monotonically increasing to L, and by the definition of sequence convergence, at some point the partial sums get above $L-\frac{1}{2}$; said differently, at some point the "tail" of this sum is less than $1/2$. That is, if we enumerate the primes in order, $\mathcal{P}=\{p_1,p_2,p_3,\dots\}$, then there must exist some k for which

$$\sum_{i>k}\frac{1}{p_i}<\frac{1}{2}. \qquad (🦆)$$

Let

$$Q=\prod_{i\le k}p_i.$$

Being the product of all primes at most p_k, by Fact B.5 we see that $nQ+1$ is not divisible by any of $\{p_1,p_2,\dots,p_k\}$, and so must either be a prime in $\{p_{k+1},p_{k+2},p_{k+3},\dots\}$, or a product of primes from this set.

Neat Observation B.7. Recall that $(x+y)^2=(x+y)(x+y)=(xx+xy+yx+yy)$. Likewise, $(x+y+z)^2=(x+y+z)(x+y+z)=(xx+xy+xz+yx+yy+yz+zx+zy+zz)$; in particular, the answer is a sum of all ways to pick one thing from $(x+y+z)$ multiplied with one thing from $(x+y+z)$. More generally, $(p_{k+1}+p_{k+2}+p_{k+3}+p_{k+4}+\dots)^2$ is equal to a sum of all numbers which are a product of two (not necessarily distinct) primes from $\{p_{k+1},p_{k+2},p_{k+3},\dots\}$ (with some repetition). Even more generally, $(p_{k+1}+p_{k+2}+p_{k+3}+p_{k+4}+\dots)^j$ is equal to a sum of all numbers which are a product of j (not necessarily distinct) primes from $\{p_{k+1},p_{k+2},p_{k+3},\dots\}$ (with some repetition).

We are almost done. Next note that $(\frac{1}{p_{k+1}}+\frac{1}{p_{k+2}}+\frac{1}{p_{k+3}}+\frac{1}{p_{k+4}}+\dots)^j$ is equal to a sum of all numbers of the form $\frac{1}{m}$ where m is product of j primes from $\{p_{k+1},p_{k+2},p_{k+3},\dots\}$ (with some repetition). And, lastly, that

$$\sum_{j=1}^{\infty}\left(\sum_{i>k}\frac{1}{p_i}\right)^j$$

is the sum of all numbers of the form $\frac{1}{m}$ where m is product of *any number* of (not necessarily distinct) primes from $\{p_{k+1},p_{k+2},p_{k+3},\dots\}$ (with some repetition).

That neat observation is useful because we also know that

$$\sum_{j=1}^{\infty} \left(\sum_{i>k} \frac{1}{p_i} \right)^j$$

converges by the geometric series test. Indeed, if we let $\tilde{r} = \sum_{i>k} \frac{1}{p_i}$, then combining

its positivity with (🦆) we have that

$$0 < \tilde{r} < \frac{1}{2},$$

and by the geometric series test $\sum_{j=1}^{\infty} r^j$ converges whenever $|r| < 1$. Since it converges, let's say that it converges to $L \in \mathbb{R}$:

$$\sum_{j=1}^{\infty} \left(\sum_{i>k} \frac{1}{p_i} \right)^j = L.$$

To review:

- Since $Q = \prod_{i \leq k} p_i$, by Fact B.5 we know that for any $n \in \mathbb{N}$, $nQ+1$ is a product only of primes in the set $\{p_{k+1}, p_{k+2}, p_{k+3}, \dots\}$. Therefore $\sum_{n=1}^{\infty} \frac{1}{nQ+1}$ is the sum of *some* of the reciprocals of products of primes from $\{p_{k+1}, p_{k+2}, p_{k+3}, \dots\}$.

- By Neat Observation B.7, $\sum_{j=1}^{\infty} \left(\sum_{i>k} \frac{1}{p_i} \right)^j$ is the sum of *all* reciprocals of products of primes from $\{p_{k+1}, p_{k+2}, p_{k+3}, \dots\}$. Moreover, this sum converges to L.

Putting this all together,

$$\sum_{n=1}^{\infty} \frac{1}{nQ+1} \leq \sum_{j=1}^{\infty} \left(\sum_{i>k} \frac{1}{p_i} \right)^j = L. \qquad (🦆_\pi)$$

But since $0 < nQ + 1 \leq n(Q+1)$ for all $n \in \mathbb{N}$, we see that $\frac{1}{n(Q+1)} < \frac{1}{nQ+1}$. Furthermore, since the harmonic series diverges, we have that

$$\sum_{n=1}^{\infty} \frac{1}{n(Q+1)} = \frac{1}{Q+1} \sum_{n=1}^{\infty} \frac{1}{n} = \infty. \qquad (🦆)$$

Combining (🦆_\pi) and (🦆) gives the contradiction that $\infty \leq L$:

$$\infty = \frac{1}{Q+1} \sum_{n=1}^{\infty} \frac{1}{n} = \sum_{n=1}^{\infty} \frac{1}{n(Q+1)} \leq \sum_{j=1}^{\infty} \left(\sum_{i>k} \frac{1}{p_i} \right)^j = L.$$

$\square$

B.16   A Topological Treatment with a Tremendous Twist

A sequence of numbers like $2, 5, 8, 11, 14$ is an *arithmetic progression* because consecutive numbers differ by the same constant — in this case, 3. Another arithmetic progression is $-10, -5, 0, 5, 10, 15, 20$. Likewise, an *infinite arithmetic progression* is an *infinite* list of numbers with this common difference property. For example,

$$\ldots, -21, -17, -13, -9, -5, -1, 3, 7, 11, 15, 19, 23, 27, \ldots.$$

Notice that if you have an infinite arithmetic progression like the one above, and a second infinite arithmetic progression like

$$\ldots, -23, -20, -17, -14, -11, -8, -5, -2, 1, 4, 7, 10, 13, 16, 19, 22, 25, \ldots,$$

then the numbers

$$\ldots, -29, -17, -5, 7, 19, 31, \ldots,$$

that are in *both* infinite arithmetic progressions also forms an infinite arithmetic progression. Indeed, since the first had a common difference of 3 and the second had a common difference of 4, the common difference of their "intersection" will be 12, which is the least common multiple of 3 and 4. As you can imagine, if you intersected any *finite collection* of infinite arithmetic progressions, you will either get yet another infinite arithmetic progression or will get the empty set.[19]

In Chapter 5 we learned about the topology of $\mathbb{R}$, but there are many other topologies in the world, which are *defined* to satisfy the properties we *proved* in that chapter. Indeed, a topology of $\mathbb{Z}$ is any collection sets (called *open sets*) where

(i)   The union of any collection these open sets is another open set, and

(ii) The intersection of any *finite collection* of these open sets is another open set.

For technical reasons we also insist that $\mathbb{Z}$ and $\emptyset$ are considered open. And if a set A is open, then its complement $\mathbb{Z} \setminus A$ is called *closed*.

What does this have to do with infinite arithmetic progressions? Well, we said at the top that infinite arithmetic progressions also have the property that if you intersect any finite collection of them, you get another infinite arithmetic progression. Are infinite arithmetic progressions the open sets of a topology on $\mathbb{Z}$?? Sadly, they don't satisfy the union condition; if you pick almost any two infinite arithmetic progressions you will find that their union is not an infinite arithmetic progression.

However, if we declare a set U to be open if it's either the empty set *or* is an infinite arithmetic progression *or* is a *union* of infinite arithmetic progressions, then

---

[19]The empty set is obtained when, for example, you intersect the infinite arithmetic progression

$$\cdots - 40, -30, -20, -10, 0, 10, 20, 30, 40 \ldots$$

with the infinite arithmetic progression

$$\ldots, -45, -35, -25, -15, -5, 5, 15, 25, 35, 45, \ldots.$$

you *can* show that this collection forms the open sets of a topology on $\mathbb{Z}$, called the *evenly spaced integer topology*

For $m \neq 0$ and any a, we will use $a + m\mathbb{Z}$ to denote the infinite arithmetic progression

$$\ldots, a - 3m, a - 2m, a - m, a, a + m, a + 2m, a + 3m, \ldots.$$

So not only are $3\mathbb{Z}$ and $5 + 4\mathbb{Z}$ open sets in the evenly spaced integer topology, but $3\mathbb{Z} \cup (5 + 4\mathbb{Z})$ is also considered to be open in this topology, even though it is not an infinite arithmetic progression.

So we know that $m\mathbb{Z}$ is open in this topology, since it is an infinite arithmetic progression, but we can also show that it must be closed![20] Being closed means your compliment is open, and the set $m\mathbb{Z}$ is closed since it is the compliment of

$$(1 + m\mathbb{Z}) \cup (2 + m\mathbb{Z}) \cup \cdots \cup ((m - 1) + m\mathbb{Z}),$$

which is the union of open sets and hence is open.

Ok, fine, we have constructed a strange topology and noted that it has the property that any $m\mathbb{Z}$ is a closed set. Where is this headed? You're not going to believe this, but *we are just two short paragraphs from proving that there are infinitely many prime numbers!* What?? Check this out.

Suppose there were finitely many primes, $p_1, p_2, \ldots, p_k$. Since we noted above that $p_i\mathbb{Z}$ is closed in the evenly spaced integer topology, this means that

$$p_1\mathbb{Z} \cup p_2\mathbb{Z} \cup \cdots \cup p_k\mathbb{Z} \qquad \qquad (\text{🏛})$$

is a finite union of closed sets and hence is also closed. Moreover, since every number except for 1 and -1 is a product of prime numbers and hence is a multiple of a prime, this would imply that (🏛) includes every integer except for 1 and -1. Said differently,

$$\left(p_1\mathbb{Z} \cup p_2 Z \cup \cdots \cup p_k\mathbb{Z}\right)^c = \{-1, 1\},$$

and so $\{-1, 1\}$ is the complement of a closed set and hence is open.

But wait! Every nonempty open set is supposed to contain an infinite arithmetic progression, which $\{-1, 1\}$ clearly does not, and so we have a contradiction. Therefore there must be infinitely many primes. $\qquad \square$

This proof was discovered by Hillel Furstenberg in 1955, when he was an undergraduate student. It's pretty neat that 2000 years after Euclid first proved the infinitude of primes, more proofs are still being discovered.

---

[20] A set in a topology is called *clopen* if it is both open and closed. In the topology of $\mathbb{R}$ that we studied in Chapter 5, the only clopen sets were $\emptyset$ and $\mathbb{R}$, but oftentimes other topologies have many such sets.

B.17   Modern Measuring's Misfit Member

Measurability was introduced in Chapter 8 as a generalization of the *length* of an interval. In Chapter 8 we focused only one what it means for a set to have measure zero. Here's a reminder of how we defined those.

- The *length* of the interval $[a, b]$ is equal to $b - a$, and is denoted $\mathscr{L}([a, b])$. This is also the length of $(a, b), (a, b]$ and $[a, b)$. E.g., $\mathscr{L}([-2, 4]) = 6$, $\mathscr{L}([3, \infty)) = \infty$.

- A set $A \subset \mathbb{R}$ has *measure zero* if for all $\varepsilon > 0$ there exists a countable collection $I_1, I_2, I_3, \ldots$ of intervals such that

$$A \subseteq \bigcup_{k=1}^{\infty} I_k \qquad \text{and} \qquad \sum_{k=1}^{\infty} \mathscr{L}(I_k) < \varepsilon.$$

Intuitively, having measure zero means that you can cover the set with a countable collection of intervals whose collective length is as close to 0 as you want. Sets can also have measures larger than zero. For instance, the measure of $[1, 4]$ (which we denote $\mu([1, 4])$) is equal to 3, which is its length. Also, $\mu([2, 3] \cup [7, 9]) = 3$. More complicated sets can also have a measure equal to 3. Intuitively, this occurs when you can cover your set with a countable collection of intervals whose collective length is as close to 3 as you want, *and also* you can find a countable collection of intervals which are covered *by* the set and whose collective length is as close to 3 as you want.[21]

In the end, one obtains a non-negative function μ on subsets of $\mathbb{R}$ which does indeed generalize the notion of length, and has the following four important properties.

1. It gives the correct answer for intervals: If A is an interval, then $\mu(A) = \mathscr{L}(A)$.

2. Bigger sets get bigger measures: $A \subseteq B$, then $\mu(A) \leq \mu(B)$.

3. Shifting a set does not change its measure: If $x_0 \in \mathbb{R}$, then $\mu(A + x_0) = \mu(A)$. ($A + x_0$ means that we add x_0 to everything in A; $A + x_0 = \{a + x_0 : a \in A\}$.)

4. If a set can be broken into pieces, the measure of the set equals the sum of the measures of its parts: If A and B are disjoint, then $\mu(A \cup B) = \mu(A) + \mu(B)$. And more generally, if $A_1, A_2, A_3, \ldots$ are pairwise disjoint, then

$$\mu\left(\bigcup_{i=1}^{\infty} A_i\right) = \sum_{i=1}^{\infty} \mu(A_i).$$

These are all properties that you would demand if μ is sensibly measuring the sizes of sets. One might hope that every set under the sun can be measured. Sadly, this is not possible, but the good news is that only really weird sets are non-measurable. Below we construct such a set. The argument works best if we assume for a contradiction that every set is measurable and then construct a set that can not possibly satisfy the above four properties.

---

[21]For details, go to grad school.

The idea is to create a (countably) infinite collection of disjoint sets $E_1, E_2, E_3, \ldots$, each set having the same measure (by using the third property), and whose union has the property $[0,1] \subseteq \bigcup_{i=1}^{\infty} E_i \subseteq [-1,2]$ (which by properties 1 and 2 then imply $1 \leq \mu\left(\bigcup_{i=1}^{\infty} E_i\right) \leq 3$). If each $\mu(E_i) = \alpha$ (some $\alpha \geq 0$), then this will imply

$$1 \leq m\left(\bigcup_{i=1}^{\infty} E_i\right) = \sum_{i=1}^{\infty} \mu(E_i) = \sum_{i=1}^{\infty} \alpha \leq 3,$$

which you might already be able to see is impossible, giving the contradiction.

That's where we are headed. The main challenge is to construct these disjoint sets E_i which all have the same measure and whose union is between $[0,1]$ and $[-1,2]$.

Assume for a contradiction that every subset of $\mathbb{R}$ is measurable. Define an equivalence relation $\sim$ on $\mathbb{R}$ by $x \sim y$ if $x - y \in \mathbb{Q}$. Each equivalence class is then of the form $\{x + r : r \in \mathbb{Q}\}$, for some x. For example, the equivalence class containing π also contains $\pi - 3$ and $\pi + 24.85$, but doesn't contain $\pi - \sqrt{2}$ or $e + 7$.

Note that each equivalence class contains many numbers in $[0,1]$. For each equivalence class, pick one such number, and let E be the set of these numbers. There are of course lots of options for E. For instance, $\pi - 3$ might be in there (but if so, then $\pi - 2.874$ and $\pi - 3.1$ are not), and $\sqrt{2} - 0.8$ and $e - 2.4$ might be in there too. This all implies a few things:

Suppose $\{r_1, r_2, r_3, \ldots\}$ is an enumeration of the rationals in $[-1,1]$, and $E_i = E + r_i$. Then,

- $E_i \subseteq [-1,2]$ (smallest possible number is 0+(-1), biggest is 1+1);

- $\mu(E) = \mu(E_i)$ (by Property 3);

- $E_i \cap E_j = \emptyset$ for all $i \neq j$ (by definition of E and E_i);

- $[0,1] \subseteq \bigcup_{i=1}^{\infty} E_i$ (any $x \in [0,1]$ is in some equivalence class, and therefore x is some $r_i \in [-1,1]$ away from an element of E).

Summarizing,

$$[0,1] \subseteq \bigcup_{i=1}^{\infty} E_i \subseteq [-1,2],$$

which by Properties 2 and 4 mean that

$$\mu([0,1]) \leq \sum_{i=1}^{\infty} \mu(E_i) \leq \mu([-1,2]).$$

So by Property 1 and the second bullet point above,

$$1 \leq \sum_{i=1}^{\infty} \mu(E) \leq 3.$$

That is, we have a sum of infinitely many copies of the same non-negative number, which can only possibly equal 0 or ∞, giving the contradiction.

B.18 Tarski's Terrific Talents Times Two

The great Ian Stewart imagined a *hyperdictionary*, which contains not only every word, but also contains every *possible* word. Sure, it contains "math" and "is" and "fun", but it also contains "ydac" and "faqir" and "galbhepvnx".[22] These are called *hyperwords*. The hyperdictionary contains infinitely many words, since any combination of finitely many letters counts as a hyperword. The hyperdictionary begins:

a, aa, aaa, aaaa, aaaaa, ... <INFINITELY MANY HYPERWORDS> ..., aaab, aaaba, aaabaa, ... <INFINITELY MANY HYPERWORDS> ..., aab, aaba, ... <INFINITELY MANY HYPERWORDS> ..., ab, aba, abaa, ... <INFINITELY MANY WORDS> ..., b, ba, baa, baaa, ...

The words form what is called a total ordering — meaning that any two of them can be compared to determine which comes first in the hyperdictionary — but things are still a little weird in that most pairs have infinitely many words between them.

The hyperdictionary has another amazing property — in some sense it contains itself. This is what I mean: Suppose the hyperdictionary were divided into 26 volumes, based on each word's first letter. If you want a copy of the hyperdictionary but don't have enough money to buy all 26 volumes, do this: Buy Volume A and a lot of white-out. Every word in Volume A starts with an 'a' — now go through this volume and white out that first 'a' from each word. After doing so what you're left with is the entire 26 volume hyperdictionary! For example, we don't need Volume K to get the word "kdalghsdh", because the word "akdalghsdh" was in Volume A, and after removing the first 'a' with white-out, you're left with "kdalghsdh"! This is an interesting phenomenon, and it's only possible with an infinite dictionary. It might remind you of Chapter 2 where, for instance, we showed that a set (like $\mathbb{N}$) can be in bijection to a proper subset of it (like $2\mathbb{N}$).

Now we move on to an amazing theorem, named after Stefan Banach and Alfred Tarski who in 1924 gave a construction that shocked the mathematical word.[23]

We begin by noting that a square can be divided up into five pieces, and those five pieces can then be rearranged to form two other squares:

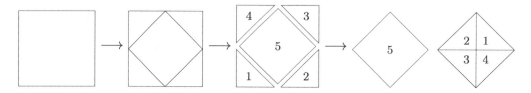

These new squares are, of course, smaller — each has half the area of the original square. However, of course, you should always be careful when saying "of course" in

---

[22]Actually, one of those is a real word!

[23]If I could have either a construction, lemma or theorem named after me, I think my preferences would be in that order. To me, proving something is cool, but proving something that is fundamental enough and used often enough to be called a lemma is even cooler, but the best would be to construct a concrete mathematical object that shocks and awes.

mathematics. The Banach-Tarski construction is amazing. It is incredible. It says that you can take a solid sphere, break it up into 5 pieces, rearrange those pieces, put those pieces together in a different way, and the result is two separate spheres *both of which are the same size as the original sphere*. Seriously. Throughout this book I've tried to point out moments where you should not just breeze by some fact, but appreciate it. Here, if you're not *marveling*, you're doing it wrong.

This theorem is often called the Banach-Tarski *paradox*, not because there isn't a proof, but because it's just so freaking unbelievable. It's the kind of result where, if it was announced tomorrow that the proof had a mistake that for some reason had been missed for a century, we'd all nod along and wonder how anyone could have ever believed something so obviously false. There are many amazing results when studying the infinite, but Banach-Tarski reigns king over them all.

The first thing to note is that when you cut the sphere into 5 pieces, they're not simple cuts like with the square example above. The cuts are *weird*. Some of the resulting pieces are non-measurable (see PP Example B.17), which helps get around the apparent preservation-of-volume contradiction. In fact, the cuts essentially take the sphere and turn it into a *hyperdictionary*!

Doing it in detail goes beyond the scope of this appendix, but the below should at least give you a feel for the proof.

First, when dividing the sphere into 5 pieces, our pieces will have this property: If a point on the surface is in one of the pieces, then all the points between that surface point and the sphere's center are also in the same piece. Basically, this allows us to only focus on the surface of the sphere, because if we know how to divide up those points, then all the internal points are in turn divided up too.

Pick a point s on the surface of the unit sphere, which we call the *starting point*. We consider a hyperdictionary using just the letters F, B, L and R, which stand for 'forward', 'backward', 'left' and 'right'. We call, for example, a point on the sphere $FLBLB$ if we can reach that point by beginning at the starting point s (facing in some direction that we call the "forward direction"), moving one step forward, one step left, one step backward, one step left, and then one step backward (here, each step is of distance $\arccos(1/3) \approx 1.23$). In a sense, each hyperword acts as a set of directions. Consider the hyperdictionary of all such directions, which in turn correspond to a bunch of points on the surface of the sphere. (But throw out directions which include, say, FB or LR, since those cancel each other out.) Notice that unlike when one moves on a plane, when on the surface of a sphere the point FRB and the point R are not the same. Indeed, every[24] such $F/B/L/R$ sequence in this hyperdictionary corresponds to a distinct point on the sphere.

Not every point can be reached in this fashion from our starting point, though. So now pick a point that wasn't hit, use that as a new starting point, and go again. To get *all* the points, we must do this with uncountably many starting points, each of which uses the hyperdictionary to produce countably many more points.

Let Collection R be the set of all points which were reached from their starting point by a sequence of steps that ends with an R (a right step). Likewise, let

---

[24]Well, nearly every. See below.

Collection L be the set of all points that are reached by a sequence ending in an L. And likewise for Collection F and Collection B. Finally, let Collection S be the set of all starting points. These five sets are nearly the five pieces that we cut the sphere in to. (There is one final issue, which is that there is a small collection of points which actually can be reached in multiple ways, but they are inconsequential in the end so we ignore them here.[25])

So where does the magic happen? Here's the important question: What happens if you take Collection R, which are the points that ended with a right step, and you simply rotate this collection one step to the left? Notice how that's the same as taking the sequences in Collection R and adding an L to the end! But adding an L after an R simply cancels out that R—almost like applying white-out to the last letter!

- R
- $LFFLBRR$
- BBR
- $FRRFR$

$\longrightarrow$

- RL
- $LFFLBRRL$
- $BBRL$
- $FRRFRL$

$\longrightarrow$

- $<$A starting point$>$
- $LFFLBR$
- BB
- $FRRF$

And what's left afterwards? Simply by rotating Collection R to the left one notch we get all of Collection R back, as well as Collections S, B and F! It's the same set of points, just rotated, but somehow this rotation turned one collection into four! This is the magic that allows us to take certain sets, move them around in space, and have them form not only a copy of themselves, but also copies of some of the other sets. If you do this carefully, you can form two copies of everything, which when pushed together just right give two copies of the original sphere. Amazing.

It's kind of like how the right branch of the fractal below, if rotated 90° and scaled properly, gives back itself, as well as the top and left branches.

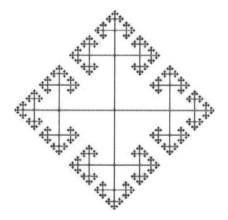

Following this amazing theorem, I'll end this book with my favorite math joke. Q: What's an anagram of Banach-Tarski? A: Banach-Tarski Banach-Tarski.

---

[25]You'll know what points I'm referring to if you've ever heard the riddle "A bear walks 1 mile north, then 1 mile east, and then 1 mile south, and ends where it started. What color is the bear?"

Index

Made in the USA
Coppell, TX
10 January 2023

10768678R00245